Endogenous Plant Rhythms

Annual Plant Reviews

A series for researchers and postgraduates in the plant sciences. Each volume in this series focuses on a theme of topical importance, and emphasis is placed on rapid publication.

Editorial Board:

Titles in the series:

Contents

Contributors

Dr Oliver Bläsing Max-Planck Institut, Molekulare Pflanzen-physiologie, Golm, D-14476, Germany

Dr Isabelle Carré Department of Biological Sciences, University of Warwick, Coventry, CV4 7AL, UK

Prof. George Coupland Max Planck Institute for Plant Breeding, Carl Von Linne Weg, 10 Cologne, D-50829, Germany

Dr Michael F. Covington Section of Plant Biology Life Sciences Addition, Room 1002, One Shields Avenue, University of California, Davis, CA 95616-5270, USA

Dr. Seth J. Davis Max Planck Institute for Plant Breeding, Carl Von Linne Weg, 10 Cologne, D-50829, Germany

Dr Antony N. Dodd Department of Plant Sciences, University of Cambridge, Downing Site, Downing Street, CAMBRIDGE, CB2 3EA, UK.

Dr Keara A. Franklin Department of Biology, University of Leicester, University Road, Leicester, LE1 7RH, UK.

Mr Michael J. Gardner Department of Plant Sciences, University of Cambridge, Downing Site, Downing Street, CAMBRIDGE, CB2 3EA, UK.

Dr Anthony Hall School of Biological Sciences, University of Liverpool, Crown Street, Liverpool, L69 7ZB, UK.

Dr Shigeru Hanano Max Planck Institute for Plant Breeding, Carl Von Linne Weg, 10 Cologne, D-50829, Germany

Dr Stacey L. Harmer Section of Plant Biology Life Sciences Addition, Room 1002, One Shields Avenue, University of California, Davis, CA 95616-5270, USA

Dr Frank Harmon   Department of Cell Biology, The Scripps Research Institute, 10550 North Torrey Pines Road, La Jolla, CA 92037, USA

Dr James Hartwell   School of Biological Sciences, University of Liverpool, Crown Street, Liverpool, L69 7ZB, UK

Mr Carlos T. Hotta   Department of Plant Sciences, University of Cambridge, Downing Site, Downing Street, CAMBRIDGE, CB2 3EA, UK.

Dr Takato Imaizumi   Department of Cell Biology, The Scripps Research Institute, 10550 North Torrey Pines Road, La Jolla, CA 92037, USA

Prof Carl H. Johnson   Department of Biological Sciences, Vanderbilt University, Box 1634 Station B, Nashville, TN 37235, USA

Prof Steve A. Kay   Department of Cell Biology, The Scripps Research Institute, 10550 North Torrey Pines Road, La Jolla, CA 92037, USA

Prof. Charalambos P. Kyriacou   Department of Genetics, University of Leicester, University Road, Leicester, LE1 7RH, U.K.

Dr Victoria S. Larner   Department of Biology, University of Leicester, University Road, Leicester, LE1 7RH, UK.

Dr Harriet G. Mc Watters   Department of Plant Sciences, University of Oxford South Parks Road, Oxford, OX1 3RB, UK

Dr Joanna Putterill   University of Auckland, School of Biological Sciences, Level 3, Private Bag 92019, Auckland, New Zealand

Mr Fabian Rudolf   Institute of Biochemistry, ETH Zurich, CH-8093 Zurich, Switzerland

Prof Dale Sanders   The Plant Laboratory, Biology Department, P.O. Box No 373, York, YO10 5YW, UK

Dr David E. Somers   Department of Plant Biology/Plant Biotechnology Center, Ohio State, Columbus, OH 43210, USA

Prof. Dorothee Staiger   University of Bielefeld. Molecular Cell Physiology University Street 25, D-33615 Bielefeld, Germany

Prof. Mark Stitt   Max-Planck   Institut,Molekulare   Pflanzen-
                               physiologie, Golm, D-14476, Germany

Ms Corinna Streitner   University of Bielefeld. Molecular Cell Physi-
                               ology University Street 25, D-33615 Bielefeld,
                               Germany

Dr Alex A. R. Webb   Department of Plant Sciences, University of
                               Cambridge, Downing Site, Downing Street,
                               CAMBRIDGE, CB2 3EA, UK.

Prof Garry C. Whitelam   Department of Biology, University of Leices-
                               ter, University Road, Leicester, LE1 7RH, UK.

Preface

The purpose of this book is to review our current knowledge of endogenous rhythms in plants. In preparing this book we have focused almost exclusively on circadian rhythms, which are the most thoroughly characterised and widespread of these. Circadian rhythms are biological rhythms that repeat once a day, persist in constant conditions, are entrained to local time and maintain a constant periodicity over a broad range of physiological temperatures. The value of a correctly adjusted circadian oscillator should be easily appreciated by anyone who has travelled between several time zones and suffered from jet-lag, the result of a rapid desynchronisation between the internal clock and the surrounding environment.

Plants have played an important role historically in the study of circadian rhythms. As early as 1727, an astronomer De Mairan observed experimentally that rhythms in leaf movement continued in the absence of daily cues. In 1835 similar experiments using *Mimosa pudica* were performed by Candolle. He confirmed the results from a century earlier and went on to demonstrate that these rhythms did not re-occur with a periodicity of exactly 24 hours (and therefore were not a direct response to day and night) but instead had a periodicity that was shorter by one to one and half hours.

In the later part of the nineteenth century Charles Darwin also became interested in the phenomenon, culminating in the publishing of his book 'The Power of Movement in Plants' in 1881. In the early part of the twentieth century Erwin Bünning lead the field of circadian biology hypothesising that circadian rhythms have an adaptive value for organisms and that the rhythms are driven by an endogenous clock. In the mid-twentieth century two researchers Pittendrigh and Aschoff drove the emerging field of chronobiology. The end of the twentieth century has seen the identification of components of molecular oscillators in the model plant *Arabidiopsis* as well as *Drosophila*, Cyanobacteria, mice and *Neurospora*. Intriguingly while the molecular mechanism appears similar across a variety of groups of organisms, there is little conservation of the components used, with only the molecular components of the *Drosophila* and mice clock sharing homology.

Much of our understanding of the mechanism of the circadian clock in plants has come from the development of *Arabidopsis thaliana* as a model circadian organism. Critical to this has been the development of robust high-through put assays for measuring circadian rhythms in this plant. Among these, the development by Millar and Kay of promoter luciferase fusions, which allow measurement of rhythmic changes in promoter activity, have provided the greatest stride forward in understanding the interaction of components of the circadian system. The ability to assay the clock in *Arabidopsis* has allowed the exploitation of a huge number of international resources developed for the dissection of the biology of this model species. Over the past five years the field has made huge advances, identifying photoreceptors involved in entrainment and key components likely to play a pivotal role in the central oscillator. Furthermore, through the pioneering use of micro-array technology, which has identified the coupling of nearly 10% of the transcriptome to the circadian clock, we now have an idea about the extent of the clock's function in regulating the biology of the plant; finally, we have also identified at the molecular level the function of the clock in the control of photoperiodism. Together this new understanding very much supports the hypothesis of Bünning, that circadian rhythms, driven by an endogenous oscillator, have an adaptive value for organisms.

This book not only describes where the field currently stands, it brings together and summarises decades of research. Moreover, all the authors are key figures at the cutting edge of plant circadian research today. The book also provides clear and thought provoking insights into the direction of future plant circadian research.

At its simplest level the circadian system can be visualised as a three component pathway. In this model, the circadian system is divided into three parts: a photoreceptor or photoreceptors, the central oscillator, and the output pathways which culminate in the overt rhythm under examination, the actual 'hands' of the circadian clock. The central oscillator generates the rhythm. This has a period length of usually a little more or less than twenty-four hours. In the absence of an appropriate *zeitgeber* ('time-giver'), such as day and night, the oscillator free-runs to reveal the endogenous period length of the circadian system; however it responds to environmental information received as input from the photoreceptors. The entrained oscillator then interacts with downstream pathways to control the many overt rhythms of the organism, such as leaf movement, activation of the photosynthesis machinery or starch mobilisation.

We have taken this simplistic model as the basis for the layout of this book. Each chapter has been contributed by authors who are actively contributing to this dynamic and fast moving area of plant science. The first three chapters focus on the central oscillator. Chapter 1 outlines the current view of the molecular mechanism of the clock, focusing on a central feedback loop made up of three proteins CCA1, LHY and TOC1. Chapter 2 discusses the roles of other members of the *TOC1/PRR* family on modulating the circadian system and its outputs; this five member gene family contains the central clock component *TOC1* but also four other closely related genes, each of which affects multiple circadian outputs. Chapter 3 considers additional oscillators downstream of and controlled by the central clockwork, in particular the so-called 'slave oscillators', circadian feedback loops resembling those of the clock but whose rhythms are set by the central clockwork.

Chapters 4 and 5 consider photoperception and the mechanisms of light signal transduction (input) to the oscillator are described. In chapter 4, evidence for a role for phytochromes and crytochrome, together with a novel set of blue light photoreceptors in entrainment of the clock, is summarised. This chapter goes on to describe how the clock itself can regulate or "gate" light input blurring the boundary between input, oscillator and output. Finally, it describes our current understanding of thermal entrainment, including recent research identifying a role for *APRR7* and *APRR9*. Chapter 5 gives an in-depth overview of the huge field of plant photoperception, describing the structure and function of each class of photoreceptor and our current understanding of the mechanisms of signal transduction.

The set of chapters discuss the control of processes (outputs) by the circadian clock. In chapter 6, circadian regulation of the transcriptome is described, highlighting the broad range of biological processes regulated by the plant clock. This chapter summarises recent metabolic data and tries to reconcile discrepancies between the transcription regulation by the clock and rhythms in metabolism. Chapter 7 describes the role of the clock in the photoperiodic regulation of flowering. In chapter 8 the role of Ca^{2+} in the architecture of the mammalian clock is described, contrasting this with our current understanding of the function of Ca^{2+} in the plant circadian system. It is not yet clear where Ca^{2+} fits in the plant circadian system; however the authors consider possible functions in transducing light signals, the core oscillator function and conferring rhythmicity on physiological processes. Chapter 9 summarises our current understanding of the clock in a class of plants capable of crassulacean acid metabolism (CAM), focusing on how CAM metabolism itself is regulated by the clock. Finally, chapter 10 discusses how and why the circadian clock has evolved, considering the evidence that possession of a clock enhances fitness in an organism.

We hope this book will provide an exciting and stimulating insight into this rapidly developing field of plant circadian biology. Our book is designed to appeal to the circadian researchers, primarily those based in plant sciences, but we expect it to be of interest and value to the wider circadian field as this is a discipline where research in one model system frequently informs work on another. It is becoming apparent that the circadian clock is having a broad influence on the biology of the plant and as such we hope this book will be of interest to the wider plant science research and teaching communities. It will also be of use to undergraduates and graduate students taking courses in areas of plant signalling, development and environmental responses.

1 The plant circadian clock: review of a clockwork *Arabidopsis*

Frank G. Harmon, Takato Imaizumi and Steve A. Kay

1.1 Introduction

Organisms from all the major kingdoms of life exhibit periodic rhythms of approximately 24 hours in vital cellular and physiological processes, including animal locomotor activity and photosynthesis in cyanobacteria and plants. These rhythms sustain most of their properties under constant environmental conditions, which is a hallmark of the intracellular pacemakers known as circadian clocks. The existence of the circadian clock in plants has been known for hundreds of years; however, only recently has work been focused on understanding the molecular makeup of the plant clockworks. Using the model plant *Arabidopsis thaliana*, significant progress has been made in identifying key genes in the oscillator (Salome & McClung, 2004). As anticipated from the more mature work on the *Neurospora* and *Drosophila* clocks, the minimal *Arabidopsis* clock is composed of a negative feedback loop.

Even with the immense progress made in the past 10 years, the molecular mechanism of the plant circadian clock remains unclear, especially in comparison to the details available for fungal and animal clocks (Glossop & Hardin, 2002). Therefore, it is the task of the plant clock community to add color and depth to this relatively bare canvas by defining the function of the proteins now in hand, identifying the factors missing from the current picture, and connecting the function of each to generate a more comprehensive model of the plant oscillator. This chapter summarizes the work that has brought the plant clock field to this point, and goes on to illuminate avenues likely to be fruitful in the future. A current view of the plant oscillator is also presented based on the list of circadian genes identified so far.

1.2 How plant circadian biology got to where it is today

1.2.1 *Forward genetics*

1.2.1.1 *Direct circadian screens – TOC1, ZTL, TEJ, TIC, LUX*

Although physiological and biochemical analysis in plants played an important historical role in describing overt circadian rhythms, the successful forward genetic approaches performed in *Drosophila* and *Neurospora* illuminated the next direction for plant chronobiology (Rosbash & Hall, 1989; Dunlap, 1990). However, plant clock researchers lacked robust circadian phenotypes that would serve as facile markers for forward genetic screens, which would pinpoint the molecular components of the central oscillator. Kay and colleagues discovered a way around this roadblock during analysis of the *cis* elements responsible for light-dependent regulation of the gene for *chlorophyll a/b-binding protein* (*CAB*); namely, a 320-bp fragment of the *Arabidopsis CAB2* promoter was sufficient to mediate transcriptional regulation of CAB gene expression by both light and the circadian clock (Millar & Kay, 1991). Subsequent construction of a transcriptional fusion between this short region of the *CAB2* promoter and the firefly *luciferase* (*luc*) coding sequence created a real-time molecular reporter (*cab2::luc*) that mirrored the circadian expression of the endogenous *CAB2* gene (Millar *et al.*, 1992). Consequently, the *cab2::luc* reporter provided a means to monitor the circadian regulation of *CAB2* expression in transgenic seedlings without destruction of the experimental material.

Taking advantage of the *cab2::luc* reporter line, Kay and colleagues screened a population of mutagenized *Arabidopsis* seedlings for aberrant circadian expression of this reporter under continuous white light conditions (Millar *et al.*, 1995a). Accordingly, the first suite of circadian clock mutants isolated by this direct forward genetic screen was the *timing of cab expression* (*toc*) mutants.

Of these *toc* mutants, *toc1* was the first to be analyzed in detail (Somers *et al.*, 1998b). Both weak and strong alleles of *toc1* show a 3–4 hours shortening of circadian period as measured by several methods, including *cab2::luc* expression, leaf movement rhythms, stomatal conductance, and mRNA expression of all circadian-regulated genes (Millar *et al.*, 1995a; Somers *et al.*, 1998b; Strayer *et al.*, 2000). In addition, TOC1 loss-of-function seedlings display photomorphogenic phenotypes, including elongated hypocotyls under continuous red or far-red light conditions and limited light-mediated induction of the light-regulated *circadian clock associated* 1 (CCA1; see Section 1.2.2) gene (Mas *et al.*, 2003a). Light plays a crucial role in resetting the circadian clock and a decrease in light intensity lengthens the free-running period (Aschoff, 1979). The period for *cab2::luc* expression exhibited by the semi-dominant *toc1-1* mutant remained 2–3 hours shorter than that of wild-type plants when these lines were compared over a range of red light fluence rates (Somers *et al.*, 1998b). It was initially interpreted that TOC1 protein functions in or close to the circadian oscillator rather than in the light input pathway into the clock. In addition, severe *TOC1* RNA interference lines and the *toc1-2* mutant display arrhythmic phenotypes in the dark and under red light (Mas *et al.*, 2003a), further demonstrating that TOC1 is a crucial factor for sustaining rhythms under certain conditions.

Positional cloning of the *toc1* locus demonstrated that the *TOC1* gene encodes a nuclear protein, which possesses a unique combination of domains (Strayer *et al.*, 2000). The N-terminal portion of the TOC1 protein contains an atypical receiver domain, which is similar to those found in response regulator proteins, whereas the C-terminal half of the protein consists of an acidic domain and a basic motif. The latter domain is conserved amongst the CONSTANS family of transcription factors, suggesting that TOC1 may be involved in transcriptional regulation. Expression of the *TOC1* transcript is under the control of the circadian clock (Strayer *et al.*, 2000). Strong, constitutive overexpression of the *TOC1* transcript abolishes rhythmic expression of all genes examined, regardless of the light condition (Makino et al., 2001). Moderate overexpression, on the other hand, significantly lengthens free-running period (Mas et al., 2003a). These data support the notion that TOC1 is an essential component of the circadian clock, but its molecular activities within the oscillator remain unknown.

Other than the toc1 mutants, four other distinct mutants have been identified through their altered cab2::luc rhythms: ztl, tic, tej, and lux. The best characterized of these mutants is zeitlupe (ztl), which has a long free-running period for several different circadian outputs; however, the period length effect of the ztl mutation is strongly dependent on fluence rate (Somers et al., 2000). Map-based cloning of the ztl locus identified a protein containing single light, oxygen or voltage (LOV) and F-box motifs, as well as several kelch repeat domains. LOV domains are found in blue light photoreceptors like phototropins (Huala et al., 1998; Christie et al., 1999) and the Neurospora circadian photoreceptor White Collar-1 (Froehlich et al., 2002; He et al., 2002). Like the phototropins, the LOV domain of ZTL binds to the chromophore flavin mononucleotide in vitro (Imaizumi et al., 2003). Typically, a combination of an F-box and kelch repeats is found in Arabidopsis ubiquitin ligases of the skp1 cullin F-box (SCF)-type (Vierstra, 2003; Petroski & Deshaies, 2005). These observations, together with the light condition-sensitive phenotype of ztl, suggested that ZTL is involved in the light-dependent turnover of one or more clock component(s). Indeed, as discussed elsewhere in detail (see Section 1.3.2), ZTL was recently shown to regulate TOC1 stability in a light-dependent manner (Mas et al., 2003b).

Another long-period mutant is tej, which was identified in the original muta-genized cab2::luc population that also produced the toc1-1 allele (Panda et al., 2002c). The tej mutation has a global effect on circadian rhythms: all known clock-controlled genes are long period in this background. Positional cloning of the tej locus identified a mutation in a highly conserved region of a gene encoding for a poly(ADP-ribose) glycohydrolase (PARG). PARG typically degrades polymers of rADP as part of a system that regulates the activity of transcription factors. Thus, poly(ADP-ribosyl)ation appears to contribute to the setting of period length in Arabidopsis.

The time for coffee (tic) mutant was originally isolated for its short-period phe-notype and lower amplitude cab2::luc rhythms (Hall et al., 2003). Paradoxically, the same mutation in the tic locus causes a slightly longer period for evening phased expression of cold and circadian regulated 2 (CCR2). Loss of TIC activity appears to affect circadian gating of light responses in a fashion analogous to early flow-ering 3 (ELF3; see Section 1.2.1.2), but TIC appears to act during a different part of the day. This idea is supported by the observation that a tic elf3 double mutant has an additive phenotype: the tic elf3 background is fully arrhythmic, unlike the conditional arrhythmia of an elf3 line. Further analysis of TIC function awaits the identification of the gene that is affected in the tic background.

Finally, the arrhythmic mutant lux arrythmo (lux) was recently isolated through the use of the cab2::luc reporter (Hazen et al., 2005b). The lux mutant is long hypocotyl and early flowering in short day conditions. Furthermore, the expression of all circadian genes so far tested, including *CAB2*, *CCR2*, *CCA1*, *LHY* and *TOC1*, is arrhythmic in continuous light conditions. Mapping of the mutation responsible for the phenotype revealed a lesion in the gene for a single Myb-like transcription factor. The single DNA binding domain present in this protein falls into the same class as that found in CCA1 and *late elongated hypocotyl* (LHY; see Section 1.2.1.2). The *LUX* gene is circadian regulated and peak expression coincides with that of *TOC1* in the evening. Like *TOC1*, the *LUX* promoter contains an important *cis* circadian element (see Section 1.5.2) and this sequence is specifically bound by CCA1 and LHY *in vitro*. Taken together, these data make a tempting case for LUX acting together with TOC1 to positively influence *CCA1* and *LHY* expression.

1.2.1.2 Non-circadian forward genetic screens – LHY, ELF3, ELF4

Beyond direct forward genetic screens for circadian mutants, many genetic screens aimed at uncovering components of clock-regulated physiology have fortuitously revealed clock components. For example, the circadian clock plays an important role in day-length measurement for the photoperiodic flowering pathway (Yanovsky & Kay, 2003); therefore, mutants with defective circadian clocks often do not respond normally to changes in day length. In several cases, mutants that were originally isolated as being defective in day-length dependent regulation of flowering were subsequently found to have aberrant clock function.

A case in point is *lhy*, which was identified as a clock component after its initial discovery as a dominant day-length insensitive late-flowering mutant (Schaffer *et al.*, 1998). In the *lhy-1* background, the *LHY* gene is constitutively overexpressed due to the presence of a Ds element just upstream of the gene, which accounts for its dominant phenotype. The flowering phenotype of this mutant was suggestive of a clock dysfunction. As expected, *lhy-1* displays generalized arrhythmia in both circadian gene expression and rhythmic leaf movement. Identification of the *LHY* gene by the use of transposon tagging revealed that the LHY protein was a single Myb-like transcription factor with significant identity to CCA1 (see Section 1.2.2). *LHY* expression is light and clock regulated, with its own expression levels being very low in the *lhy-1* background. Null mutants in *LHY*, obtained in the form of either an RNAi line (Alabadi *et al.*, 2002) or an intragenic suppressor of *lhy-1* (Mizoguchi *et al.*, 2002), are short period by approximately 2 hours. Subsequent analysis demonstrated that LHY and CCA1 are partially redundant clock components (see Section 1.3.1).

The *early flowering 3* (*elf3*) mutant was isolated as a mutant that is early-flowering under both long- and short-day conditions (Zagotta *et al.*, 1996). Analysis of *elf3* for circadian phenotypes revealed that normal ELF3 function is necessary for sustained rhythms in the continuous light, but not darkness: expression of clock-regulated genes is arrhythmic only under continuous light conditions (Hicks *et al.*, 1996; Covington *et al.*, 2001). Positional cloning of the *elf3* locus identified the ELF3 protein as a nuclear protein with no distinct protein motifs. Interestingly, ELF3 physically interacts with phytochrome B (phyB), and acts downstream of phyB in red light signaling (Liu *et al.*, 2001; Hicks *et al.*, 2001). Indeed, ELF3 functions as a circadian gating factor that attenuates light signals to limit resetting of the oscillator during the subjective evening (McWatters *et al.*, 2000).

The *early flowering 4* (*elf4*) mutant is also an early-flowering mutant, and its flowering phenotype is superficially similar to that of the *elf3* mutant (Doyle *et al.*, 2002). Whereas *elf3* seedlings display conditional arrhythmia, circadian expression of clock outputs in *elf4* is profoundly compromised under both continuous light and dark conditions. In addition, *cca1::luc* expression, and presumably that of *CCA1* itself, is severely diminished in the *elf4* mutant. Thus, ELF4 is either directly or indirectly involved in the regulation of gene expression of this core clock component. Identification of the *ELF4* gene has shed little light on the function of this protein, since this small protein lacks conserved domains.

The *gigantea* (*gi*) mutant is a late-flowering mutant that is affected under both long day and short day conditions. Further investigation revealed that GI protein is required to maintain the accuracy and robustness of circadian rhythms (Park *et al.*, 1999). The *GI* gene encodes a nuclear protein with several potential membrane-spanning domains (Fowler *et al.*, 1999; Huq *et al.*, 2000). Although its molecular function remains unknown, GI may have a role in regulating expression of *CCA1* and *LHY*, since mRNA levels for both these genes are very low in a *gi* background (Park *et al.*, 1999; Mizoguchi *et al.*, 2002).

1.2.2   Reverse genetics – CCA1, PRR3-PRR9

In addition to direct forward genetic screens from circadian mutants, clock components have also been found indirectly through reverse genetics. Moreover, several loci that contribute to the oscillator have been suggested through the study of recombinant inbred lines (RIL).

CCA1, a paralogue of LHY, was initially identified as a DNA binding activity that specifically recognized a CA-rich motif in the *Lhcb1*3* (originally known as *cab140*) promoter that is responsible for phytochrome-regulated expression of this gene (Ha & An, 1988; Sun *et al.*, 1993; Kenigsbuch & Tobin, 1995). To identify this protein, Wang and Tobin screened a cDNA library for proteins binding to this specific element, and found that this protein was a single Myb-like transcription factor, dubbed CCA1 (Wang *et al.*, 1997). The first suggestion that CCA1 was involved in the clock came with the discovery that the gene itself is both light and clock regulated. Overexpression of CCA1 (*cca1-ox*) abolishes circadian expression of all genes so far tested, in both continuous light and constant darkness (Wang & Tobin, 1998), comparable to the *lhy-1* line. In addition, the expression of endogenous *CCA1* is very low in *cca1-ox* plants, which suggested that this protein feeds back to inhibit its own expression. Loss of *CCA1* function also affected the clock: a line with a T-DNA insertion in the *CCA1* gene displays a period that is about 2 hours shorter than the wild-type for expression of several clock-regulated genes (Green & Tobin, 1999). Again, this phenotype parallels that of the null *LHY* background. Thus, CCA1 and LHY appeared to be paralogues that played an important role in the plant circadian oscillator, which has been confirmed by analysis of the double mutant (see Section 1.3.1).

A combination of reverse genetics and quantitative trait locus (QTL) analysis of natural genetic variation has demonstrated that the *pseudo-response regulator* (PRR) paralogues of TOC1 play accessory roles in the oscillator. Attention was brought to this family by the identification of TOC1 (PRR1) as a core clock component (Somers *et al.*, 1998b; Strayer *et al.*, 2000). In addition to *TOC1/PRR1*, the *Arabidopsis* genome encodes for four other pseudo-response regulators (*PRR3, PRR5, PRR7,* and *PRR9*) (Matsushika *et al.*, 2000; Strayer *et al.*, 2000). The expression of each gene is clock-controlled, and peak levels of each transcript occur at different times throughout the day. Overexpression of each of these *PRR* genes (with the exception of *PRR7*, which has not been tested) has slight but varying effects on the clock, ranging from slightly short period (*PRR9*) to somewhat long period (*PRR3* and *PRR5*) (Sato *et al.*, 2002; Matsushika *et al.*, 2002a; Murakami *et al.*, 2004). As with the overexpression lines, loss of function for each *PRR* has a limited effect on rhythms (see below).

Complementing these data is the QTL analysis that suggests an important role for PRR7 in natural genetic variation of period length. Analysis of leaf movement rhythms in Columbia-Landsberg *erecta* RILs identified five QTL that affect either period, phase or amplitude (Michael *et al.*, 2003a). Present within a region on the top of chromosome V, which contributes to the circadian period, is the gene for *PRR7*. To verify that PRR7, as well as the other PRR genes, contributes to the circadian period, T-DNA insertion lines for *PRR3-PRR9* were identified and analyzed for alterations in leaf movement rhythms. Consistent with the QTL analysis, seedlings lacking PRR7 either are long period by 1–2 hours (Yamamoto *et al.*, 2003; Michael *et al.*, 2003a) or display a 3–6 hours advanced phase (Kaczorowski & Quail, 2003). On the other hand, loss of either *PRR5* or *PRR3* precipitates a modest short period phenotype (Yamamoto *et al.*, 2003; Michael *et al.*, 2003a). *PRR9* mutants have either a phase or slightly long-period phenotype (Ito *et al.*, 2002; Farre *et al.*, 2005). Clearly, methods other than direct forward screens have been a rich source of core and accessory clock components. Future work in the same vein is likely to produce additional insights into the molecular players in the plant clock (see Section 1.5).

1.3   How the components were placed in the plant clock

Once individual molecular components of the clock were identified, the next task was to link them together into a cogent model. The first testable model for the *Arabidopsis* oscillator arose from several seminal studies that applied hypothesis-based molecular analysis to the available circadian mutants.

1.3.1 CCA1, LHY, and TOC1 contribute to a feedback loop at the core of the oscillator

Given the similarity in their sequences, as well as their phenotypes, LHY and CCA1 appeared to be redundant factors that shared overlapping functions in the heart of the *Arabidopsis* circadian clock. Confirmation of this came with the construction of lines lacking both CCA1 and LHY activity (Alabadi *et al.*, 2002; Mizoguchi *et al.*, 2002). In these lines, expression of clock-controlled genes is rhythmic in driven conditions (i.e. alternating light/dark cycles), but the clock is strongly compromised in both continuous light and continuous darkness. For all genes assayed, rhythmic expression is maintained in the *cca1 lhy* double mutant for only two cycles after release into free-running conditions. After this point, the rhythms rapidly dampen and eventually become arrhythmic. This phenotype clearly demonstrates that LHY and CCA1 are vital for sustained oscillations. However, the function of these two transcription factors must be partially redundant with other unknown protein(s), since rhythms are not completely abolished in the *cca1 lhy* background.

Based on the accumulated evidence from studies of the clocks in all genera, the prototypical oscillator relies on the transcriptional feedback loop, where negative factors feed back to inhibit the action of positive factors (Harmer *et al.*, 2001; Young & Kay, 2001; Van Gelder *et al.*, 2003). An important feature of this simple clock is temporal separation of the activities of proteins with opposing function. Based on this criterion, CCA1/LHY and TOC1 immediately presented themselves as potential actors within such a system. CCA1 and LHY are expressed in the morning (Schaffer *et al.*, 1998; Wang & Tobin, 1998), whereas TOC1 accumulates to maximal levels in the evening (Strayer *et al.*, 2000). The importance of these genes in the circadian system is emphasized by the fact that miss-expression of either has global effects on all circadian activities (Schaffer *et al.*, 1998; Wang & Tobin, 1998; Makino *et al.*, 2002; Matsushika *et al.*, 2002b; Mas *et al.*, 2003a).

Taking this into account, Kay and co-workers closely examined *TOC1* expression in either *cca1-ox* or *lhy-1* seedlings and found that *TOC1* expression is constitutively low in both backgrounds (Alabadi *et al.*, 2001). This was not a general property of the arrhythmic state of these lines, as *TOC1* levels in the arrhythmic *elf3* mutant are constitutively high, not low. Analysis of the *TOC1* promoter revealed a key regulatory region that is responsible for circadian expression of this gene. Within this portion of the promoter lies a nine-nucleotide *cis* element known as the evening element (EE) (Harmer *et al.*, 2000), which is closely related to the CCA1/LHY binding site (Wang *et al.*, 1997). The EE motif was initially recognized through bioinformatic analysis as a sequence that is over-represented in genes with an evening phase (see Section 1.5.2). In the context of the *CCR2* promoter, mutation of the EE vastly reduces circadian-driven expression from this promoter. CCA1 and LHY were found to specifically recognize and bind the EE in the *TOC1* promoter (Alabadi *et al.*, 2001). This supported the idea that CCA1 and LHY are needed to repress *TOC1* transcription in the morning hours.

Important confirmation that CCA1 and LHY act to repress morning expression of *TOC1*, as well as other evening genes, came from analysis of gene expression in the *cca1 lhy* background (Alabadi *et al.*, 2002; Mizoguchi *et al.*, 2002). When these mutant seedlings are grown in light/dark periods of equal duration, peak expression of *TOC1* is shifted to the morning. The same behavior is displayed by other normally evening phased genes, including *GI*, *CCR2*, and *FLAVIN-BINDING KELCH REPEAT F-BOX 1* (*FKF1*) (Alabadi *et al.*, 2002; Mizoguchi *et al.*, 2002; Imaizumi *et al.*, 2003). On the other hand, genes that peak in the morning do not show a change in the timing of their peak expression in the absence of CCA1 and LHY. The promoters of *TOC1*, *GI*, *CCR2*, and *FKF1* all harbor EE (Harmer *et al.*, 2000; Imaizumi *et al.*, 2003); therefore, LHY and CCA1 likely bind to these elements to repress transcription of these genes, which restricts expression of *TOC1* and the others to the evening.

To determine if the regulation observed for *TOC1* was reciprocal, CCA1 and LHY expression was examined in *TOC1* loss-of-function alleles (Alabadi *et al.*, 2002). In a strong *TOC1* loss-of-function background, both *CCA1* and *LHY* expression was significantly reduced relative to message levels in wild type. In addition, the circadian period for both these genes was 2–3 hours shorter in the *toc1* mutant, comparable to that observed for *CAB2* and *CCR2* expression. These data suggested that TOC1 in some way participated in activating the expression of these two genes; however, TOC1 lacks obvious DNA binding motifs. Since regions of TOC1 show significant similarity with domains found in CONSTANS-like transcriptional regulators (Strayer *et al.*, 2000), the strongest possibilities were that it either acts with an accessory transcription factor or influences the activity of a downstream activator. The model that grew out of this work was that CCA1 and LHY make up the negative arm of a transcriptional feedback loop at the core of the clock, and TOC1 contributes either directly or indirectly to the activation of *CCA1* and *LHY* expression (see Section 1.4).

With the accumulation of more recent data, a direct role for TOC1 in the activation of *CCA1* and *LHY* expression appears to be less likely than originally suggested by the initial genetic analysis. For example, overexpression of *TOC1* from the strong, constitutive CaMV35S promoter does not result in elevated expression of either gene; in fact, expression of both *LHY* and *CCA1* is very low in this background (Makino *et al.*, 2002). Similarly, *CCA1* and *LHY* message levels are drastically reduced in *ztl* mutants (Somers *et al.*, 2004) or in plants with impaired SCF function (Harmon, F.G. & S.A. Kay, unpublished observations), where TOC1 protein levels are expected to be elevated because the SCF^{ZTL} complex is not present to target TOC1 to the 26S proteasome. These findings suggest that a response to either elevated or diminished TOC1 activity is for expression of *CCA1* and *LHY* to be squelched for reasons that presently remain unclear. Undoubtedly, TOC1 is important for proper expression of *CCA1* and *LHY*, but its influence may very well be part of a larger network of proteins, which all contribute to control expression of these two transcription factors. Additional regulatory loops are therefore likely to act in the plant clock, as have been found in fungi, flies and mammals (Farre *et al.*, 2005).

Additional support for this possibility is provided by modeling studies, which have demonstrated that all the experimental data cannot be accounted for by the single-feedback loop model (Locke *et al.*, 2005).

1.3.2   The interplay between TOC1 and ZTL sets circadian period

A major step forward in the understanding of how period is established in the *Arabidopsis* clock came with the demonstration of TOC1 degradation through its interaction with ZTL (Mas *et al.*, 2003b). Previous observations had established that the absence of TOC1 shortens the period (Somers *et al.*, 1998b; Strayer *et al.*, 2000), whereas moderate overexpression of *TOC1* from its native promoter (*TOC1* minigene (TMG)) leads to a longer period (Mas *et al.*, 2003a). A similar long-period phenotype is observed when seedlings lack the function of ZTL (Somers *et al.*, 2000). To establish that this inverse relationship stems from a genetic interaction between TOC1 and ZTL, double *ztl toc1* and *ztl* TMG mutants were analyzed for their circadian phenotype (Mas *et al.*, 2003b). The *ztl toc1* double mutant exhibits the same short period phenotype of *toc1* lines; therefore, the *ztl* phenotype requires TOC1 protein. Furthermore, increased TOC1 expression from the TMG transgene exacerbates the long-period phenotype of the *ztl* mutant and is stronger than the TMG alone.

Given that ZTL is an F-box protein, this genetic evidence supports the idea that at least one function of ZTL is to target TOC1 protein for degradation by the ubiquitin/26S proteasome system (Mas *et al.*, 2003b). Consistent with this model, TOC1 and ZTL physically interact *in planta*. In addition, TOC1 protein is subject to proteasome-dependent degradation, and TOC1 is more stable in a *ztl* background. Subsequent work by Somers and co-workers has shown that graded increases in ZTL expression cause a commensurate shortening in free running period (Han *et al.*, 2004). In addition, ZTL was shown to participate in an SCF complex (Han *et al.*, 2004), the presumed role of which is to turnover TOC1 protein. The dramatic effect that manipulation of these two genes has on the free-running period clearly demonstrate that the delicate balance of TOC1 and ZTL activity is vital in the mechanism that sets the circadian period. The major unanswered question is how TOC1 transmits its effects on period, since an activity has yet to be ascribed to the protein.

1.3.3   PRR7 and PRR9 form a feedback loop that affects CCA1 and LHY expression

Interestingly, a recent report has found that seedlings lacking both PRR7 and PRR9 have a significantly stronger circadian phenotype than each of the single mutants (Farre *et al.*, 2005). Although *prr7* mutants are slightly long period and a *prr9* mutant shows changes in phase or a 1.5 hour increase in period, the *prr7 prr9* double mutant has a period that is 6–10 hours longer than wild type. Analysis of gene expression in the single mutants and the *prr7 prr9* background revealed that levels of *CCA1* and *LHY* transcripts are elevated in *prr7* and to a much greater degree in the double mutant. This finding provides a simple explanation for the long-period phenotype observed in the absence of both PRR7 and PRR9, since elevated levels of either CCA1 or LHY cause arrhythmia in constant conditions and a lagging phase in light/dark conditions. Furthermore, these data suggest that PRR7 and PRR9 act together to repress expression of *CCA1* and *LHY*.

Investigation into the factors involved in control of *PRR7* and *PRR9* gene expression showed that CCA1 and LHY drive transcription of *PRR9* and, to a lesser extent, *PRR7* (Farre *et al.*, 2005). The reciprocal effects of these two groups of proteins on the expression of each other raise the exciting possibility that PRR7-PRR9 participate in a feedback loop with CCA1 and LHY, which is analogous to those present in the fly (i.e. VRILLE/PDP1 and dCLK, (Cyran *et al.*, 2003)) and mammalian (i.e. REV-ERBα/RORa and MOP3/BMAL1; Preitner *et al.*, 2002; Sato *et al.*, 2004) clocks. It remains to be determined where the two other PRRs fit into the *Arabidopsis* circadian system, but these proteins may well contribute to similar feedback loops.

1.4   Current framework for understanding the *Arabidopsis* clock

Figure 1.1 shows a model for the *Arabidopsis* circadian system that incorporates all the known components that contribute to the molecular clockworks of the plant. Light signals mediated through the phytochrome (phys) and cryptochrome (crys) photoreceptors are integrated to entrain the clock to the prevailing light/dark regime (Devlin & Kay, 2000; Mas *et al.*, 2000; Somers *et al.*, 1998a). The sensitivity to *red light reduced* 1 (SRR1) protein appears to mediate light input into the clock (Staiger *et al.*, 2003), as does *de-etiolated* 1 (DET1) (Millar *et al.*, 1995b), but the precise mechanism of each remains to be determined. In the core of the oscillator lie the opposing actions of CCA1/LHY and TOC1 to create a negative feedback loop: the negative arm is composed of CCA1 and LHY acting at the *TOC1* promoter, whereas the positive arm requires TOC1 in either a direct or indirect fashion. At dawn, the levels of *CCA1* and *LHY* transcript are approaching their peak, which has been initiated by the positive work of TOC1 and its presumed molecular partners in either a direct or indirect fashion. In the early and mid morning, CCA1 and LHY repress *TOC1* expression through their binding to the EE. Acting at other EEs, the combination of CCA1 and LHY also obviate the expression of other 'evening genes' including *GI*, *ELF4*, *PRR5* and *PRR3*. These two transcription factors also work as positive factors to activate the expression of morning-expressed genes, including *PRR9* and *PRR7*. Casein kinase II (CKII) promotes the activity of CCA1, since phosphorylation is necessary for CCA1 to bind DNA in plant extracts and for normal circadian function (Sugano *et al.*, 1998, 1999; Daniel *et al.*, 2004).

As the day progresses, the fading influence of TOC1 and accumulating levels of PRR9 and PRR7 shut down expression of the two Myb-like transcription factors. It is also likely that an activator(s) of EE genes, which is antagonistic to the negative effects of CCA1 and LHY, is also present throughout the day. Consequently, *TOC1*

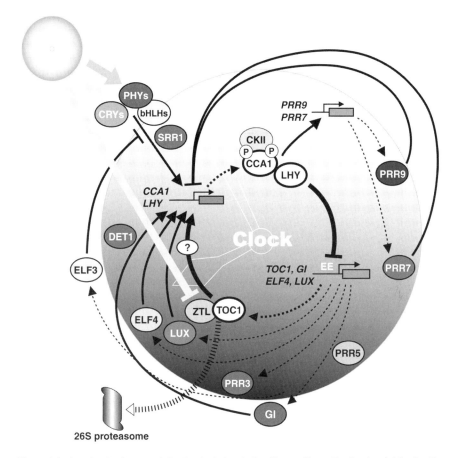

Figure 1.1 A molecular framework for the *Arabidopsis* circadian oscillator. See Section 1.4 for details. Genes are indicated by gray boxes with the names shown to the left. Proteins are depicted by ovals, with the name of the protein indicated inside. The oval with a '?' inside, which bisects the line for TOC1 action, indicates the possibility of an intermediate factor that participates in the activation of *CCA1* and *LHY* expression. Dashed lines indicate transcription/translation. Solid lines show protein activity: lines ending in arrows depict stimulatory function, whereas lines terminating with a perpendicular dash show the negative action of the corresponding proteins.

expression picks up in the evening, along with that of the other evening element genes. ZTL protein is also on the rise, but its degradative potential is inhibited by the presence of light and possibly the absence of other factors such as phosphorylation of TOC1.

Once dusk has fallen, ZTL gains the capacity to direct TOC1 protein to the 26S proteasome, which removes TOC1 from the cell. Before it has been removed, TOC1 participates in some way to activate *CCA1* and *LHY* transcription, which appears to be vital for the establishment of the period of the clock. TOC1 may be

assisted in this role by transcription factors such as the myb-related LUX (Hazen
et al., 2005b) or one of the several basic helix–loop–helix (bHLH) proteins of the
phytochrome-interacting factor (PIF) and PIF3-like (PIL) families, which appear to
interact with TOC1 (Yamashino *et al.*, 2003). Alternatively, intermediate factor(s)
may lie in between the action of TOC1 and stimulation of *CCA1/LHY* transcription.
Expression of *CCA1* is reinforced by ELF4, which has peak expression during this
time period. ELF3 is also available throughout the evening to inhibit errant light
signals that could potentially reset the clock. Proteins that lack obvious functions,
but are likely to take part in the events in the evening, are PRR5, PRR3, and GI.
A more rigorous mathematical approach to modeling the potential roles of these
components, as has been initiated for fly and mammalian clocks (Ueda *et al.*, 2001;
Rand *et al.*, 2004), will be required to provide experimental predictions that go
beyond the purely intuitive approaches used to date.

1.5   What may pave the way to greater understanding of the clock

It should be clear from the discussion above that much of the knowledge of the plant
circadian system stems from genetic analysis and subsequent molecular studies. In
the future, additional effort must be placed on understanding the function of the
proteins that are now in hand, as well as extending the list of circadian clock genes.
In many cases, this requires a shift in attention from genetic analysis to the behavior
of individual proteins or multiprotein complexes. This type of analysis will require
a new set of tools, some of which are outlined in the following sections.

1.5.1   Forward genetics

Forward genetic screens have played a vital role in the identification of plant clock
components; however, all of the genes found through direct forward genetic screens
so far have been based on mutagenesis and screening for perturbed expression of
the same *cab2::luc* reporter gene. Recently, it has become clear that this screen is
approaching saturation, as our group has found multiple alleles of *toc1*, *elf3*, *elf4*,
and *lux* following forward genetic screens in continuous light conditions (Hazen
et al., 2005a). In addition, Millar and co-workers have identified several alleles of
ztl in screens for mutants under continuous darkness (A.J. Millar, personal communication). Limiting genetic dissection of the clock to this single reporter construct
clearly limits the number of loci that can be found through forward genetics. Therefore, future genetic screens should be aimed at finding mutants that affect alternative
reporter constructs. The advantages of this approach are that different reporters will
yield novel mutants, particularly those factors that contribute directly to its expression, as well as genes that contribute to the whole of the oscillator. In addition,
if different clocks do exist in individual cells or tissues, then an alternate reporter
that is specific to that cell type or tissue is necessary to dissect the makeup of each
clock.

Fertile ground for the discovery of additional circadian mutants is likely to be found through the screening of sensitized lines (i.e. plant lines with pre-existing circadian defects) for extragenic mutants. In particular, mutant lines representing weak alleles of clock components and those with weak phenotypes can be mutagenized and screened for stronger clock phenotypes. This approach has the potential to highlight at least three categories of proteins, those that: (1) act together with the mutant protein, (2) are upstream or downstream from the mutant protein in a pathway, or (3) on their own would produce weak phenotypes due to redundancy and, therefore, would be missed in the wild-type background.

Since some redundancy in the function of clock components has been found in the *Arabidopsis* clock, it makes sense to employ mutant lines in forward genetic screens to avoid the possibility of missing clock genes with overlapping functions. A prime example of this in the plant oscillator is the synergistic effect observed with the loss of both CCA1 and LHY. Single knockouts of either gene produce a modest circadian phenotype (Green & Tobin, 1999), but the clock in the double mutant is profoundly compromised (Alabadi *et al.*, 2002; Mizoguchi *et al.*, 2002). Had *CCA1* and *LHY* not been connected through similar overexpression phenotypes and amino acid similarity, it is easy to imagine that screening a population of mutagenized *cca1* seedlings for an enhanced circadian phenotype would have yielded LHY. Taking these factors into account, future genetic screens should be more targeted, either by the use of alternative reporter constructs or by searching for novel mutants in sensitized lines.

Another area where additional investigation should be focused is the makeup of clocks in different tissues. The vast majority of work on the plant circadian system has examined whole seedlings. This approach has been useful in getting an overall picture of the clock, but most likely greatly simplifies the actual plant circadian system. It is abundantly clear that each tissue and cell type in the plant has a clock (Millar, 1998) and these clocks are functionally independent (Thain *et al.*, 2000). For example, the free-running circadian periods for the two reporter genes, *CAB2* and *chalcone synthase* (*CHS*) differ by approximately 1 hour (Thain *et al.*, 2002). A comparable period differential exists for *CAB2* and *PHYB* genes (Hall *et al.*, 2002). In the case of *CAB2* and *CHS*, these two genes are expressed in different portions of the seedling (mesophyll and root, respectively); however, *toc1* and *det1* mutants shorten the rhythms of each to a similar degree (Thain *et al.*, 2002). Therefore, the timepieces that drive the expression of *CAB2* and *CHS* must contain at least some common components. Even with different circadian periods, *CAB2* and *PHYB* expression respond similarly to changes in light environment and mutant backgrounds (Hall *et al.*, 2002), suggesting that these two genes also are controlled by different clocks that are alike in composition. Finally, the clocks in the same organ can be entrained to maintain different phases (Thain *et al.*, 2000), indicating that the clocks in the same region of the plant are autonomous. Therefore, independent clocks, which may be tissue-specific, seem to exist in plants.

An additional complexity in *Arabidopsis* seedlings is the intriguing possibility of separate oscillators that are preferentially entrained by either light or temperature. Upon comparison of *CAB2* and *catalase 3* (*CAT3*) entrainment to light and temperature, McClung and co-workers observed that the phase angle of *CAB2* entrainment is set by light-dark cycles, whereas that of *CAT3* is more sensitive to temperature cycles (Michael *et al.*, 2003b). A phase response curve to examine the response of expression of *CAT3*, *CAB2* and *TOC1* to cold pulses throughout the circadian cycle revealed that the phase of *CAT3* expression is significantly more sensitive to temperature pulses than either *CAB2* or *TOC1* expression, suggesting that the oscillator driving *CAT3* is different from the *TOC1* oscillator. Dissection of these independent, potentially separate, oscillators will require identification of tissue- and temperature-specific promoters, which can be used in luciferase reporter constructs for forward genetic screens.

1.5.2 Genomics and functional genomics

Since a major means of circadian regulation is at the level of transcription, the use of microarrays to study gene expression is particularly useful for study of the circadian clock (Sato *et al.*, 2003) (see Chapter 6). As a tool, data from expression microarrays is useful to develop lists of candidate genes based on expression profile, location of expression, and/or predicted protein domains, which can then be used in functional genomics projects. Profiling also has the power to pinpoint the subset of genes that are the direct targets of a particular transcription factor under study. In addition, alignment of upstream regions of co-regulated genes is likely to produce *cis*-elements that are important for control of this category of expression. Lastly, development of technologies that couple chromatin-immunoprecipitation (ChIP) with hybridization of DNA to gene arrays will be a facile means to identify the promoters regions bound by a particular protein of interest.

Functional genomics holds the promise to identify clock components based on their structure, expression pattern, and/or site of expression. Genomic approaches can be used to develop a list of candidates that can be subsequently analyzed for circadian function with reverse genetic techniques like RNA interference or T-DNA insertions. This method has yet to yield plant clock components, but a systematic analysis of candidate genes is likely to pinpoint proteins that participate in the clock.

A circadian question that is tailored for bioinformatic study is mapping of the signaling pathways that feed out from the oscillator. In particular, microarrays can be used to find the gene targets of key circadian regulators. For example, microarrays were used to divine direct targets of the heterodimeric complex of the *Drosophila* bHLH transcription factors dCLOCK (dCLK) and dCYCLE (dCYC), which make up the positive arm of the fly feedback loop (Young & Kay, 2001; Panda *et al.*, 2002b; Hardin, 2004). Several groups compared the expression profiles of a variety of existing circadian mutants, each of which had predictable effects on the abundance of dCLK and dCYC, to pinpoint genes that were up- or down-regulated in a manner consistent with control by dCLK::dCYC activity (McDonald & Rosbash, 2001; Claridge-Chang *et al.*, 2001; Lin *et al.*, 2002; Ceriani *et al.*, 2003). Profiling of these mutants identified several known clock targets, as well as previously unrecognized targets. A similar study with mouse liver, aimed at finding targets of mCLOCK and mMOP3/BMAL1, which are the mammalian counterparts of dCLK and dCYC (Panda *et al.*, 2002b), examined gene expression in a *Clock*$^{-/-}$ mutant background at the time when activity of this complex is expected to be highest (Panda *et al.*, 2002a). Affected genes included several that displayed rhythmic expression, all of which harbored E-box sequences (i.e. the DNA site recognized by mCLK/MOP3) within their promoter regions. Therefore, these genes represented bona fide targets of mCLK/MOP3, and this validates the microarray analysis.

The application of similar approaches to circadian biology in plants has been limited (Harmer *et al.*, 2000; Schaffer *et al.*, 2001), but should be considered given the advantages available to investigators working with *Arabidopsis*. Namely, the capacity to utilize transgenic plant lines in which the activity of a given protein can be induced by pharmacological means either by (1) allowing the entry of a glucocordicoid fusion protein into the nucleus (Lloyd *et al.*, 1994; Simon *et al.*, 1996; Kang *et al.*, 1999) or (2) the use of an ethanol-inducible transcriptional fusion (Caddick *et al.*, 1998; Salter *et al.*, 1998; Roslan *et al.*, 2001). With precise control over protein function, it is possible to get a direct readout of the genes immediately influenced by the action of the protein, as well as those that lie further downstream. Therefore, well-planned microarray experiments, which take advantage either of circadian mutants or controlled expression of a given factor, should serve to aid in the discovery of genes whose expression is directly influenced by a given clock component.

A concrete example of the utility of combining genome-wide transcription profiling and genome sequence information is the identification of the evening element (EE) by sequence alignment of similarly regulated promoters (Harmer *et al.*, 2000). As discussed above, the EE is a nine-nucleotide *cis*-element that is important for the evening-phased expression of genes such as *CCR2* and *TOC1*. Initial recognition of this promoter element as being a vital piece of circadian control came from the alignment of the 1500 bp upstream sequences of all the 453 cycling genes represented on the first *Arabidopsis* Affymetrix GeneChip. An absolutely conserved sequence (AAAATATCT) was found 46 times in the promoters of 31 genes. Furthermore, the vast majority of these genes were evening phased. Subsequent to this computational analysis, the EE was shown to be important for circadian expression of both the *CCR2* and *TOC1* genes (Harmer *et al.*, 2000; Alabadi *et al.*, 2001). In the future, similar applications of profiling data may yield regulatory motifs that contribute to circadian-driven gene expression of genes in specific tissues or elements that are unique to separate oscillators.

Finally, an important task that must be accomplished after the identification of a clock component is to determine its function, so as to incorporate its action into the circadian system. If the protein in question is a transcription factor, then the most obvious question is the identity of its gene targets. An emerging technique that directly indexes the regions of DNA bound by a transcription factor is chromatin immunoprecipitation (ChIP) (Kuo & Allis, 1999; Orlando, 2000). ChIP is a biochemical technique where the protein of interest, along with its DNA target, is purified from cell extracts by immunoprecipitation. Prior to purification, cells are treated with a fixative to crosslink protein–DNA complexes. After reversal of the crosslinking, the identity of the promoter region accompanying the purified protein is typically determined using PCR with primers directed towards candidate promoters. In terms of circadian biology, ChIP has been successfully used to identify target genes of the mCLK/MOP3 complex (Lee *et al.*, 2001). Although this approach is very powerful, its resolution is limited to those promoter regions that can be predicted based on the identity of the protein under study.

A far more powerful use of this method is to hybridize the isolated genomic DNA fragments to microarrays representing the entire genomic sequence (ChIP-chip), which provides a direct readout of all the promoter sequences present in the mixture. In yeast, where ChIP-chip is currently more technologically feasible, this method has identified genes under the control of chromatin remodeling by histone deacetylation (Robyr & Grunstein, 2003), new origins of replication (Wyrick *et al.*, 2001), and previously unrecognized binding sites for cell-cycle transcription factors (Ren *et al.*, 2000; Simon *et al.*, 2001). Although the array technology suitable for ChIP-chip with more complex eukaryotic genomes, like *Arabidopsis*, remains under development, this technique holds the potential to give a genome-wide picture of every gene that is under control of a given factor.

1.5.3   Transient functional assays

As expected, based on the fly and mouse clock, the emerging picture of the plant clock is quite complex, with the expression of individual factors likely being regulated by multiple interlocking loops. Therefore, assigning a molecular function to a specific protein may require a simplified approach where the candidate is assayed in isolation from the entire circadian system. This is particularly true if the protein is expected to directly influence the expression of a given gene.

One way to accomplish this is through expression analysis with transient transfection of cells using a combination of constructs designed to report the behavior of the protein of interest. The basis of this approach is to measure the activity of the promoter under study, usually by monitoring a transcriptional luciferase fusion, following the addition of different combinations of possible transcription factors or effector proteins. In practice, this requires the co-transformation of the reporter construct with vectors encoding the factors to be studied into either cells in culture or whole seedlings. Recently, this approach has been successfully applied to *Arabidopsis* seedlings to demonstrate that the bHLH transcription factor PIF1 activates transcription in a light-dependent manner (Huq *et al.*, 2004).

As far as circadian biology is concerned, a functional genomics strategy incorporating transient assays was successful in isolating an activator of Mop3/Bmal expression (Sato *et al.*, 2004). After identification of transcripts that are circadian regulated in multiple tissues by transcriptional profiling of several major organs, these candidates were tested for the capacity to stimulate expression of a Bmal1 transcriptional reporter consisting of a promoter fusion with luciferase (Bmal::luc).

Cultures of HeLa cells were cotransfected with a mixture of the reporter construct and an expression vector harboring the cDNA of the candidate genes. The transfected cells were then evaluated for changes in luciferase activity from the Bmal1::luc fusion. A cDNA encoding for the retinoic acid-related orphan receptor, RORa, specifically and potently stimulated the Bmal1 reporter construct. Subsequent work with RORa demonstrated that this transcription factor is a key player in the mammalian core oscillator where it makes up a feedback loop that stimulates expression of Bmal1.

Conceptually, the use of transient assays is not limited to investigating only the function of DNA binding proteins: proteins that directly influence transcription factors, such as kinases and phophatases, can also be assessed using this method. Thus, development of robust transient assay systems in plants holds the promise to draw direct connections between clock components, as well as directly assess the function of a particular protein.

1.5.4   Characterizing protein modification

Another area of the plant clock that has not received the attention it deserves is the influence of post-translational modification on the activity of clock proteins. Given the clear role for this aspect of protein function in the fungal and animal clocks (Young & Kay, 2001; Harms *et al.*, 2004), this is an area that requires more study in plants.

In the *Drosophila* clock, phosphorylation is an important mechanism to generate a delay in the activity of the heterodimeric complex of *period* (PER) and *timeless* (TIM), which represents the negative arm of the feedback loop (Young & Kay, 2001). Phosphorylation of PER by DOUBLETIME (DBT), the fly homologue of casein kinase 1ε, influences both its stability and the subcellular localization of the PER-TIM complex (Kloss *et al.*, 1998; Price *et al.*, 1998). PER accumulation is limited in the absence of TIM, because PER is degraded after phosphorylation by DBT. A rise in TIM and subsequent complex formation with PER abrogates this degradation. In addition to its role in PER stability, phosphorylation is critical for timing of PER-TIM complex entry into the nucleus: translocation of the PER-TIM complex into the nucleus is attendant to phosphorylation of the proteins by several kinases. Thus, a combination of these two regulatory steps is essential for proper rhythm generation in clocks, which emphasizes the importance of understanding the protein modification and dynamics in the plant clock.

In *Arabidopsis*, further investigation in this vein is likely to be fruitful with CCA1, LHY, and TOC1. It is clear that CCA1 is phosphorylated by CKII and this modification is needed for CCA1 to bind the *Lhcb* promoter in plant extracts (Sugano *et al.*, 1998, 1999). In addition, a form of CCA1 that cannot be phosphorylated appears to lack critical circadian functions of the wild-type protein (Daniel *et al.*, 2004). It is not yet clear if these effects are all mediated through CKII, or whether other kinases are involved. Nevertheless, phosphorylation plays an important role in CCA1 activity and its contribution must be more thoroughly investigated. CKII also interacts with and phosphorylates LHY *in vitro*, but the functional significance of this finding remains to be determined (Sugano *et al.*, 1999).

TOC1 interacts with the F-box protein ZTL and this leads to its degradation by the 26S proteasome. Typically, the target of an F-box protein must be phosphorylated prior to complex formation (Vierstra, 2003; Petroski & Deshaies, 2005). Whether this is the case in the TOC1-ZTL interaction remains to be determined, but this is another case where closer investigation should reveal important aspects of plant clock function. Clearly, more work needs to be put into understanding the role of post-translational modification in the function of clock proteins in plants. Finally, there is no information on whether the proteins that participate in the *Arabidopsis* clock are subject to differential compartmentalization like the PER-TIM complex.

1.6 Conclusion

A complementary combination of forward genetics and hypothesis-based molecular studies led to the formulation of the CCA/LHY-TOC1 feedback loop model. As the list of components that contribute to the clock grows longer and the shape of the central oscillator comes into focus, this model is likely to be, and should be, significantly modified to fit the more recently accumulated data. Furthermore, the substantial progress that has been achieved in the last decade, both in the understanding of the plant clock and the techniques available to plant biologists, has made it such that the plant circadian field can move beyond simple mutant identification and analysis. It will be exciting to watch the circadian field expand in the next decade to fill in the gaps that now exist in our present understanding of the plant oscillator.

References

Alabadi, D., Oyama, T., Yanovsky, M.J., Harmon, F.G., Mas, P. & Kay, S.A. (2001) Reciprocal regulation between TOC1 and LHY/CCA1 within the *Arabidopsis* circadian clock. *Science*, **293**, 880–883.

Alabadi, D., Yanovsky, M.J., Mas, P., Harmer, S.L. & Kay, S.A. (2002) Critical role for CCA1 and LHY in maintaining circadian rhythmicity in *Arabidopsis*. *Curr. Biol.*, **12**, 757–761.

Aschoff, J. (1979) Circadian rhythms: influences of internal and external factors on the period measured in constant conditions. *Z. Tierpsychol.*, **49**, 225–249.

Caddick, M.X., Greenland, A.J., Jepson, I., Krause, K.P., Qu, N., Riddell, K.V., Salter, M.G., Schuch, W., Sonnewald, U. & Tomsett, A.B. (1998) An ethanol inducible gene switch for plants used to manipulate carbon metabolism. *Nat. Biotechnol.*, **16**, 177–180.

Ceriani, M.F., Hogenesch, J.B., Straume, M. & Kay, S.A. (2003) Genome-wide expression analysis in Drosophila reveals genes controlling circadian behavior. *Cell. Mol. Neurobiol.*, **23**, 223.

Christie, J.M., Salomon, M., Nozue, K., Wada, M. & Briggs, W.R. (1999) LOV (light, oxygen, or voltage) domains of the blue-light photoreceptor phototropin (nph1): binding sites for the chromophore flavin mononucleotide. *Proc. Natl. Acad. Sci. USA*, **96**, 8779–8783.

Claridge-Chang, A., Wijnen, H., Naef, F., Boothroyd, C., Rajewsky, N. & Young, M.W. (2001) Circadian regulation of gene expression systems in the Drosophila head. *Neuron*, **32**, 657–671.

Covington, M.F., Panda, S., Liu, X.L., Strayer, C.A., Wagner, D.R. & Kay, S.A. (2001) ELF3 modulates resetting of the circadian clock in *Arabidopsis*. *Plant Cell*, **13**, 1305–1315.

Cyran, S.A., Buchsbaum, A.M., Reddy, K.L., Lin, M.C., Glossop, N.R., Hardin, P.E., Young, M.W., Storti, R.V. & Blau, J. (2003) vrille, Pdp1, and dClock form a second feedback loop in the Drosophila circadian clock. *Cell*, **112**, 329–341.

Daniel, X., Sugano, S. & Tobin, E.M. (2004) CK2 phosphorylation of CCA1 is necessary for its circadian oscillator function in *Arabidopsis*. *Proc. Natl. Acad. Sci. USA*, **101**, 3292–3297.

Devlin, P.F. & Kay, S.A. (2000) Cryptochromes are required for phytochrome signaling to the circadian clock but not for rhythmicity. *Plant Cell*, **12**, 2499–2510.

Doyle, M.R., Davis, S.J., Bastow, R.M., McWatters, H.G., Kozma-Bognar, L., Nagy, F., Millar, A.J. & Amasino, R.M. (2002) The ELF4 gene controls circadian rhythms and flowering time in *Arabidopsis thaliana*. *Nature*, **419**, 74–77.

Dunlap, J.C. (1990) Closely watched clocks: Molecular analysis of circadian rhythms in *Neurospora* and *Drosophila*. *TIG*, **6**, 159–165.

Farre, E.M., Harmer, S.L., Harmon, F.G., Yanovsky, M.J. & Kay, S.A. (2005) Overlapping and distinct roles of PRR7 and PRR9 in the *Arabidopsis* circadian clock. *Current Biology*, **15**, 47–54.

Fowler, S., Lee, K., Onouchi, H., Samach, A., Richardson, K., Coupland, G. & Putterill, J. (1999) GIGANTEA: a circadian clock-controlled gene that regulates photoperiodic flowering in *Arabidopsis* and encodes a protein with several possible membrane-spanning domains. *EMBO J.*, **18**, 4679–4688.

Froehlich, A.C., Liu, Y., Loros, J.J. & Dunlap, J.C. (2002) White collar-1, a circadian blue light photoreceptor, binding to the frequency promoter. *Science*, **297**, 815–819.

Glossop, H.R.J. & Hardin P.E. (2002) Central and peripheral circadian oscillator mechanism in files and mammals. *J. Cell Science* **115**:3369–3377.

Green, R.M. & Tobin, E.M. (1999) Loss of the circadian clock-associated protein 1 in *Arabidopsis* results in altered clock-regulated gene expression. *Proc. Natl. Acad. Sci. USA*, **96**, 4176–4179.

Ha, S.-B. & An, G. (1988) Identification of upstream regulatory elements involved in the developmental expression of the *Arabidopsis thaliana cab1* gene. *Proc. Natl. Acad. Sci.*, **85**, 8017–8021.

Hall, A., Bastow, R.M., Davis, S.J., Hanano, S., McWatters, H.G., Hibberd, V., Doyle, M.R., Sung, S., Halliday, K.J., Amasino, R.M. & Millar, A.J. (2003) The TIME FOR COFFEE (TIC) gene maintains the amplitude and timing of *Arabidopsis* circadian clocks. *Plant Cell*, **15**, 2719–2729.

Hall, A., Kozma-Bognar, L., Bastow, R.M., Nagy, F. & Millar, A.J. (2002) Distinct regulation of CAB and PHYB gene expression by similar circadian clocks. *Plant J.*, **32**, 529–537.

Han, L., Mason, M., Risseeuw, E.P., Crosby, W.L. & Somers, D.E. (2004) Formation of an SCF complex is required for proper regulation of circadian timing. *Plant J.*, **40**, 291–301.

Hardin, P.E. (2004) Transcription regulation within the circadian clock: the E-box and beyond. *J. Biol. Rhythms*, **19**, 348–360.

Harmer, S.L., Hogenesch, J.B., Straume, M., Chang, H. S., Han, B., Zhu, T., Wang, X., Kreps, J.A. & Kay, S.A. (2000) Orchestrated transcription of key pathways in *Arabidopsis* by the circadian clock. *Science*, **290**, 2110–2113.

Harmer, S.L., Panda, S. & Kay, S.A. (2001) Molecular bases of circadian rhythms. *Annu. Rev. Cell Dev. Biol.*, **17**, 215–253.

Harms, E., Kivimae, S., Young, M.W. & Saez, L. (2004) Posttranscriptional and posttranslational regulation of clock genes. *J. Biol. Rhythms*, **19**, 361–373.

Hazen, S.P., Borevitz, J.O., Harmon, F.G., Pruneda-Paz J.L., Schultz T.F., Yanovsky J.J., Liljegren S.J., Ecker J.R. & Kay S.A. (2005a) Rapid Array Mapping of Circadian Clock and Developmental Mutations in *Arabidopsis*. *Plant Physiol.*, **138**:990–997.

Hazen, S.P., Schultz T.F., Pruneda-Paz J.L., Borevitz J.O., Ecker J.R. & Kay S.A. (2005b) LUX AR-RHYTHMO encodes a Myb domain protein essential for circadian rhythms. *Proc. Natl. Acad. Sci. USA*, **102**:10387–10392.

He, Q., Cheng, P., Yang, Y., Wang, L., Gardner, K.H. & Liu, Y. (2002) White collar-1, a DNA binding transcription factor and a light sensor. *Science*, **297**, 840–843.

Hicks, K.A., Albertson, T.M. & Wagner, D.R. (2001) EARLY FLOWERING3 encodes a novel protein that regulates circadian clock function and flowering in *Arabidopsis*. *Plant Cell*, **13**, 1281–1292.

Hicks, K.A., Millar, A.J., Carre, I.A., Somers, D.E., Straume, M., MeeksWagner, D.R. & Kay, S.A. (1996) Conditional circadian dysfunction of the *Arabidopsis* early-flowering 3 mutant. *Science*, **274**, 790–792.

Huala, E., Oeller, P.W., Liscum, E., Han, I.S., Larsen, E. & Briggs, W.R. (1998) *Arabidopsis* NPH1: a protein kinase with a putative redox-sensing domain. *Science*, **278**, 2120–2123.

Huq, E., Al Sady, B., Hudson, M., Kim, C., Apel, K. & Quail, P.H. (2004) Phytochrome-interacting factor 1 is a critical bHLH regulator of chlorophyll biosynthesis. *Science*, **305**, 1937–1941.

Huq, E., Tepperman, J.M. & Quail, P.H. (2000) GIGANTEA is a nuclear protein involved in phytochrome signaling in *Arabidopsis*. *Proc. Natl. Acad. Sci. USA*, **97**, 9789–9794.

Imaizumi, T., Tran, H.G., Swartz, T.E., Briggs, W.R. & Kay, S.A. (2003) FKF1 is essential for photoperiodic-specific light signaling in *Arabidopsis*. *Nature*, **426**, 302–306.

Ito, S., Matsushika, A., Yamada, H., SATO, S., Kato, T., Tabata, S., Yamashino, T. & Mizuno, T. (2002) Characterization of the APRR9 pseudo-response regulator belonging to the APRR1/TOC1 quintet in *Arabidopsis thaliana*. *Plant Cell Physiol.*, **44**, 1237–1245.

Kaczorowski, K.A. & Quail, P.H. (2003) *Arabidopsis* PSEUDO-RESPONSE REGULATOR7 (PRR7) is a signaling intermediate in phytochrome-regulated seedling deetiolation and phasing of the circadian clock. *Plant Cell*, **15**, 2654–2665.

Kang, H.G., Fang, Y. & Singh, K.B. (1999) A glucocorticoid-inducible transcription system causes severe growth defects in *Arabidopsis* and induces defense-related genes. *Plant J.*, **20**, 127–133.

Kenigsbuch, D. & Tobin, E.M. (1995) A region of the *Arabidopsis Lhcb1*3* promoter that binds to CA-1 activity is essential for high expression and phytochrome regulation. *Plant Physiol.*, **108**, 1023–1027.

Kloss, B., Price, J.L., Saez, L., Blau, J., Rothenfluh, A., Wesley, C.S. & Young, M.W. (1998) The Drosophila clock gene double-time encodes a protein closely related to human casein kinase I epsilon. *Cell*, **94**, 97–107.

Kuo, M.H. & Allis, C.D. (1999) In vivo cross-linking and immunoprecipitation for studying dynamic Protein: DNA associations in a chromatin environment. *Methods*, **19**, 425–433.

Lee, C., Etchegaray, J.P., Cagampang, F.R.A., Loudon, A.S.I. & Reppert, S.M. (2001) Posttranslational mechanisms regulate the mammalian circadian clock. *Cell*, **107**, 855–867.

Lin, Y., Han, M., Shimada, B., Wang, L., Gibler, T.M., Amarakone, A., Awad, T.A., Stormo, G.D., Van Gelder, R.N. & Taghert, P.H. (2002) Influence of the period-dependent circadian clock on diurnal, circadian, and aperiodic gene expression in Drosophilamelanogaster. *Proc. Nat. Acad. Sci.*, **99**, 9562–9567.

Liu, X.L., Covington, M.F., Fankhauser, C., Chory, J. & Wanger, D.R. (2001) ELF3 encodes a circadian clock-regulated nuclear protein that functions in an *Arabidopsis* PHYB signal transduction pathway. *Plant Cell*, **13**, 1293–1304.

Lloyd, A.M., Schena, M., Walbot, V. & Davis, R.W. (1994) Epidermal cell fate determination in *Arabidopsis*: Patterns defined by a steroid-inducible regulator. *Science*, **266**, 436–439.

Locke, J.C.W., Millar, A.J. & Turner, M.S. (2005) Modeling genetic networks with noisy and varied experimental data: the circadian clock in *Arabidopsis thaliana*. *J. Theor. Biol.*, **234**, 383–393.

Makino, S., Matsushika, A., Kojima, M., Oda, Y. & Mizuno, T. (2001) Light response of the circadian waves of the APRR1/TOC1 quintet: When does the quintet start singing rhythmically in *Arabidopsis*? *Plant Cell Physiol.*, **42**, 334–339.

Makino, S., Matsushika, A., Kojima, M., Yamashino, T. & Mizuno, T. (2002) The APRR1/TOC1 quintet implicated in circadian rhythms of *Arabidopsis thaliana*: I. Characterization with APRR1-overexpressing plants. *Plant Cell Physiol.*, **43**, 58–69.

Mas, P., Alabadi, D., Yanovsky, M.J., Oyama, T. & Kay, S.A. (2003a) Dual role of TOC1 in the control of circadian and photomorphogenic responses in *Arabidopsis*. *Plant Cell*, **15**, 223–236.

Mas, P., Devlin, P.F., Panda, S. & Kay, S.A. (2000) Functional interaction of phytochrome B and cryptochrome 2. *Nature*, **408**, 207–211.

Mas, P., Kim, W.Y., Somers, D.E. & Kay, S.A. (2003b) Targeted degradation of TOC1 by ZTL modulates circadian function in *Arabidopsis thaliana*. *Nature*, **426**, 567–570.

Matsushika, A., Imamura, A., Yamashino, T. & Mizuno, T. (2002a) Aberrant expression of the light-inducible and circadian-regulated APRR9 gene belonging to the circadian-associated APRR1/TOC1 quintet results in the phenotype of early flowering in *Arabidopsis thaliana*. *Plant Cell Physiol.*, **43**, 833–843.

Matsushika, A., Makino, S., Kojima, M. & Mizuno, T. (2000) Circadian waves of expression of the APRR1/TOC1 family of pseudo-response regulators in *Arabidopsis thaliana*: insight into the plant circadian clock, *Plant Cell Physiol.*, **41**, 1002–1012.

Matsushika, A., Makino, S., Kojima, M., Yamashino, T. & Mizuno, T. (2002b) The APRR1/TOC1 quintet implicated in circadian rhythms of *Arabidopsis thaliana*: II. Characterization with CCA1-overexpressing plants, *Plant Cell Physiol.*, **43**, 118–122.

McDonald, M.J. & Rosbash, M. (2001) Microarray analysis and organization of circadian gene expression in Drosophila 36. *Cell*, **107**, 567–578.

McWatters, H.G., Bastow, R.M., Hall, A. & Millar, A.J. (2000) The ELF3 zeitnehmer regulates light signaling to the circadian clock. *Nature*, **408**, 716–720.

Michael, T.P., Salome, P.A. & McClung, C.R. (2003) Two *Arabidopsis* circadian oscillators can be distinguished by differential temperature sensitivity. *Proc. Natl. Acad. Sci. USA*, **100**, 6878–6883.

Millar, A.J. (1998). The cellular organization of circadian rhythms in plants: not one but many clocks. In *Biological Rhythms and Photoperiodism in Plants* (eds P.J. Lumsden & A.J. Millar) pp. 51–68. Oxford, Bios.

Millar, A.J., Carre, I.A., Strayer, C.A., Chua, N.H. & Kay, S.A. (1995a) Circadian clock mutants in *Arabidopsis* identified by luciferase imaging. *Science*, **267**, 1161–1163.

Millar, A.J. & Kay, S.A. (1991) Circadian control of CAB gene-transcription and messenger-RNA accumulation in *Arabidopsis*. *Plant Cell*, **3**, 541–550.

Millar, A.J., Short, S.R., Chua, N.H. & Kay, S.A. (1992) A novel circadian phenotype based on firefly luciferase expression in transgenic plants. *Plant Cell*, **4**, 1075–1087.

Millar, A.J., Straume, M., Chory, J., Chua, N.H. & Kay, S.A. (1995b) The regulation of circadian period by phototransduction pathways in *Arabidopsis*. *Science*, **267**, 1163–1166.

Mizoguchi, T., Wheatley, K., Hanzawa, Y., Wright, L., Mizoguchi, M., Song, H.R., Carre, I.A. & Coupland, G. (2002) LHY and CCA1 are partially redundant genes required to maintain circadian rhythms in *Arabidopsis*. *Dev. Cell*, **2**, 629–641.

Murakami, M., Yamashino, T. & Mizuno, T. (2004) Characterization of circadian-associated APRR3 pseudo-response regulator belonging to the APRR1/TOC1 quintet in *Arabidopsis thaliana*. *Plant Cell Physiol.*, **45**, 645–650.

Orlando, V. (2000) Mapping chromosomal proteins in vivo by formaldehyde-crosslinked-chromatin immunoprecipitation. *Trends Biochem. Sci.*, **25**, 99–104.

Panda, S., Antoch, M.P., Miller, B.H., Su, A.I., Schook, A.B., Straume, M., Schultz, P.G., Kay, S.A., Takahashi, J.S. & Hogenesch, J.B. (2002a) Coordinated transcription of key pathways in the mouse by the circadian clock. *Cell*, **109**, 307–320.

Panda, S., Hogenesch, J.B. & Kay, S.A. (2002b) Circadian rhythms from flies to human. *Nature*, **417**, 329–335.

Panda, S., Poirer, G.G. & Kay, S.A. (2002c) tej defines a role for poly(ADP-ribosyl)ation in establishing period length of the *Arabidopsis* circadian oscillator. *Dev. Cell*, **3**, 51–61.

Park, D.H., Somers, D.E., Kim, Y.S., Choy, Y.H., Lim, H.K., Soh, M.S., Kim, H.J., Kay, S.A. & Nam, H.G. (1999) Control of circadian rhythms and photoperiodic flowering by the *Arabidopsis* GIGANTEA gene. *Science*, **285**, 1579–1582.

Petroski, M.D. & Deshaies, R.J. (2005) Function and regulation of cullin-RING ubiquitin ligases. *Nat. Rev. Mol. Cell. Biol.*, **6**, 9–20.

Preitner, N., Damiola, F., Molina, L.L., Zakany, J., Duboule, D., Albrecht, U. & Schibler, U. (2002) The orphan nuclear receptor REV-ERB alpha controls circadian transcription within the positive limb of the mammalian circadian oscillator. *Cell*, **110**, 251–260.

Price, J.L., Blau, J., Rothenfluh, A., Abodeely, M., Kloss, B. & Young, M.W. (1998) Double-time is a novel Drosophila clock gene that regulates PERIOD protein accumulation. *Cell*, **94**, 83–95.

Rand, D.A., Shulgin, B.V., Salazar, D. & Millar, A.J. (2004) Design principles underlying circadian clocks. *J. R. Soc. Interface*, **1**, 119–130.

Ren, B., Robert, F., Wyrick, J.J., Aparicio, O., Jennings, E.G., Simon, I., Zeitlinger, J., Schreiber, J., Hannett, N., Kanin, E., Volkert, T.L., Wilson, C.J., Bell, S.P. & Young, R.A. (2000) Genome-wide location and function of DNA binding proteins. *Science*, **290**, 2306–2309.

Robyr, D. & Grunstein, M. (2003) Genomewide histone acetylation microarrays. *Methods*, **31**, 83–89.

Rosbash, M. & Hall, J.C. (1989) The molecular biology of circadian rhythms. *Neuron*, **3**, 387–398.

Roslan, H.A., Salter, M.G., Wood, C.D., White, M.R.H., Croft, K.P., Robson, F., Coupland, G., Doonan, J., Laufs, P., Tomsett, A.B. & Caddick, M.X. (2001) Characterization of the ethanol-inducible alc gene-expression system in *Arabidopsis thaliana*. *Plant J.*, **28**, 225–235.

Salome, P.A. & McClung, C.R. (2004) The *Arabidopsis thaliana* clock. *J. Biol. Rhythms*, **19**, 425–435.

Salter, M.G., Paine, J.A., Riddell, K.V., Jepson, I., Greenland, A.J., Caddick, M.X. & Tomsett, A.B. (1998) Characterization of the ethanol-inducible alc gene expression system for transgenic plants. *Plant J.*, **16**, 127–132.

Sato, E., Nakamichi, N., Yamashino, T. & Mizuno, T. (2002) Aberrant expression of the *Arabidopsis* circadian-regulated APRR5 gene belonging to the APRR1/TOC1 quintet results in early flowering and hypersensitiveness to light in early photomorphogenesis. *Plant Cell Physiol.*, **43**, 1374–1385.

Sato, T.K., Panda, S., Kay, S.A. & Hogenesch, J.B. (2003) DNA arrays: Applications and implications for circadian biology. *J. Biol. Rhythms*, **18**, 96–105.

Sato, T.K., Panda, S., Miraglia, L.J., Reyes, T.M., Rudic, R.D., McNamara, P., Naik, K.A., FitzGerald, G.A., Kay, S.A. & Hogenesch, J.B. (2004) A functional genomics strategy reveals Rora as a component of the mammalian circadian clock. *Neuron*, **43**, 527–537.

Schaffer, R., Landgraf, J., Accerbi, M., Simon, V., Larson, M. & Wisman, E. (2001) Microarray analysis of diurnal and circadian-regulated genes in *Arabidopsis*. *Plant Cell*, **13**, 113–123.

Schaffer, R., Ramsay, N., Samach, A., Corden, S., Putterill, J., Carré, I.A. & Coupland, G. (1998) The late elongated hypocotyl mutation of *Arabidopsis* disrupts circadian rhythms and the photoperiodic control of flowering. *Cell*, **93**, 1219–1229.

Simon, I., Barnett, J., Hannett, N., Harbison, C.T., Rinaldi, N.J., Volkert, T.L., Wyrick, J.J., Zeitlinger, J., Gifford, D.K., Jaakkola, T.S. & Young, R.A. (2001) Serial regulation of transcriptional regulators in the yeast cell cycle. *Cell*, **106**, 697–708.

Simon, R., Igeno, M.I. & Coupland, G. (1996) Activation of floral meristem identity genes in *Arabidopsis*. *Nature*, **384**, 59–62.

Somers, D.E., Devlin, P.F. & Kay, S.A. (1998a) Phytochromes and cryptochromes in the entrainment of the *Arabidopsis* circadian clock. *Science*, **282**, 1488–1490.

Somers, D.E., Kim, W.Y. & Geng, R. (2004) The F-box protein ZEITLUPE confers dosage-dependent control on the circadian clock, photomorphogenesis, and flowering time. *Plant Cell*, **16**, 769–782.

Somers, D.E., Schultz, T.F., Milnamow, M. & Kay, S.A. (2000) ZEITLUPE encodes a novel clock-associated PAS protein from *Arabidopsis*. *Cell*, **101**, 319–329.

Somers, D.E., Webb, A.A., Pearson, M. & Kay, S.A. (1998b) The short-period mutant, toc1-1, alters circadian clock regulation of multiple outputs throughout development in *Arabidopsis thaliana*. *Development*, **125**, 485–494.

Staiger, D., Allenbach, L., Salathia, N., Fiechter, V., Davis, S.J., Millar, A.J., Chory, J. & Fankhauser, C. (2003) The *Arabidopsis* SRR1 gene mediates phyB signaling and is required for normal circadian clock function. *Genes Dev.*, **17**, 256–268.

Strayer, C., Oyama, T., Schultz, T.F., Raman, R., Somers, D.E., Mas, P., Panda, S., Kreps, J.A. & Kay, S.A. (2000) Cloning of the *Arabidopsis* clock gene TOC1, an autoregulatory response regulator homolog. *Science*, **289**, 768–771.

Sugano, S., Andronis, C., Green, R.M., Wang, Z.Y. & Tobin, E.M. (1998) Protein kinase CK2 interacts with and phosphorylates the *Arabidopsis* circadian clock-associated 1 protein. *Proc. Natl. Acad. Sci. USA*, **95**, 11020–11025.

Sugano, S., Andronis, C., Ong, M.S., Green, R.M. & Tobin, E.M. (1999) The protein kinase CK2 is involved in regulation of circadian rhythms in *Arabidopsis*. *Proc. Natl. Acad. Sci. USA*, **96**, 12362–12366.

Sun, L., Doxsee, R.A., Harel, E. & Tobin, E.M. (1993) CA-1, a novel phosphoprotein, interacts with the promoter of the *cab*140 gene in *Arabidopsis* and is undetectable in *det*1 mutant seedlings. *Plant Cell*, **5**, 109–121.

Thain, S.C., Hall, A. & Millar, A.J. (2000) Functional independence of circadian clocks that regulate giant gene expression. *Curr. Biol.*, **10**, 951–956.

Thain, S.C., Murtas, G., Lynn, J.R., McGrath, R.B. & Millar, A.J. (2002) The circadian clock that controls gene expression in *Arabidopsis* is tissue specific. *Plant Physiol.*, **130**, 102–110.

Ueda, H.R., Hagiwara, M. & Kitano, H. (2001) Robust oscillations within the interlocked feedback model of Drosophila circadian rhythm 81. *J. Theor. Biol.*, **210**, 401–406.

Van Gelder, R.N., Herzog, E.D., Schwartz, W.J. & Taghert, P.H. (2003) Circadian rhythms: in the loop at last. *Science*, **300**, 1534–1535.

Vierstra, R.D. (2003) The ubiquitin/26S proteasome pathway, the complex last chapter in the life of many plant proteins. *Trends Plant Sci.*, **8**, 135–142.

Wager-Smith K. & Kay S.A. (2000) Circadian rhythm genetics: from flies to mice to humans. *Nat. Genetics* **26**:23–27.

Wang, Z.Y., Kenigsbuch, D., Sun, L., Harel, E., Ong, M.S. & Tobin, E.M. (1997) A Myb-related transcription factor is involved in the phytochrome regulation of an *Arabidopsis* Lhcb gene. *Plant Cell*, **9**, 491–507.

Wang, Z.Y. & Tobin, E.M. (1998) Constitutive expression of the CIRCADIAN CLOCK ASSOCIATED 1 (CCA1) gene disrupts circadian rhythms and suppresses its own expression. *Cell*, **93**, 1207–1217.

Wyrick, J.J., Aparicio, J.G., Chen, T., Barnett, J.D., Jennings, E.G., Young, R.A., Bell, S.P. & Aparicio, O.M. (2001) Genome-wide distribution of ORC and MCM proteins in S. cerevisiae: high-resolution mapping of replication origins. *Science*, **294**, 2357–2360.

Yamamoto, Y., Sato, E., Shimizu, T., Nakamich, N., SATO, S., Kato, T., Tabata, S., Nagatani, A., Yamashino, T. & Mizuno, T. (2003) Comparative genetic studies on the APRR5 and APRR7 genes belonging to the APRR1/TOC1 quintet implicated in circadian rhythm, control of flowering time, and early photomorphogenesis. *Plant Cell Physiol.*, **44**, 1119–1130.

Yamashino, T., Matsushika, A., Fujimori, T., SATO, S., Kato, T., Tabata, S. & Mizuno, T. (2003) A link between circadian-controlled bHLH factors and the APRR1/TOC1 quintet in *Arabidopsis thaliana*. *Plant Cell Physiol.*, **44**, 619–629.

Yanovsky, M.J. & Kay, S.A. (2003) Living by the calendar: how plants know when to flower. *Nat. Rev. Mol. Cell Biol.*, **4**, 265–275.

Young, M.W. & Kay, S.A. (2001) Time zones: A comparative genetics of circadian clocks. *Nat. Rev. Genet.*, **2**, 702–715.

Zagotta, M.T., Hicks, K.A., Jacobs, C.I., Young, J.C., Hangarter, R.P. & Meeks-Wagner, D.R. (1996) The *Arabidopsis* ELF3 gene regulates vegetative photomorphogenesis and the photoperiodic induction of flowering. *Plant J.*, **10**, 691–702.

2 Pseudo-response regulator genes 'tell' the time of day: multiple feedbacks in the circadian system of higher plants

Shigeru Hanano and Seth J. Davis

2.1 Introduction

The circadian system that generates an approximately 24-hour rhythm has evolved as an adaptation to the environmental changes of light and temperature caused by the Earth's rotation (Pittendrigh, 1993; Schultz & Kay, 2003). Many organisms – both prokaryotes and eukaryotes, ranging from cyanobacteria to mammals – have a circadian system. In plants, physiological phenomena of photoperiodism and circadian rhythms are widely known (Thomas & Vince-Prue, 1997). Using their circadian system, plants measure both day-length and seasonal change. This allows these organisms to fine-tune their development with the change of environment conditions. Different plant species have adjusted this to different times. For example, numerous species of plants bloom during different months of the year and at a different time of the day. One of the first uses of this was made by Linnaeus in his 'flower clock' (Fig. 2.1). It 'tells' the time of the day by monitoring when a set of different plant species open their flowers within the daylight period (Linnaeus, 1751; Johnson *et al.*, 1998).

Circadian systems regulate many biological rhythms in plants, including photosynthetic activity, leaf movement, stomatal aperture and the petal opening and closing rhythms important in pollination. However, the molecular mechanism(s) of how plants entrain to their environmental cues and adjust the phase of their rhythms to the day–night cycle is unknown. Recently, pseudo-response regulator (*PRR*) genes have been isolated from the model plant *Arabidopsis thaliana* as candidates within the clock oscillator(s) (Strayer *et al.*, 2000; Makino *et al.*, 2000). Interestingly, mRNA from each of the *PRR* genes oscillates rhythmically, with several hour intervals between them. Perhaps, this 'tells' the time of the day like the 'flower clock' (Matsushika *et al.*, 2000). Mutations in any of the *PRR* genes cause changes in normal circadian processes (see Sections 2.6.1–2.6.5). The *PRR* genes may either be directly involved in the circadian oscillator or function in input to the clock; these functions are not exclusive. Current knowledge of the redundancy and function of *PRR* genes is reviewed in this chapter.

2.2 History of the circadian system

After centuries of wide-ranging physiological and anatomical analyses, genetic approaches have proven an attractive way to identify genes involved in the circadian

Figure 2.1 The Carl-von-Linné flower clock (Linnaeus's flower clock). Each plant species blooms at the distinct time of day (*Source*: Linnaeus, 1751).

system. A screen for mutants, or for natural-allelic variants, can bring us the relationship between genes and physiology. During the past 30 years, genetic approaches have provided us with many clock mutants, and their cognate genes, from a diverse array of organisms. In the early 1970s, the first clock mutant *period* (*per*) was isolated from *Drosophila melanogaster* (Konopka & Benzer, 1971). Since then, scientists have identified more clock mutants, ranging from Drosophila to other species: such as *timeless* (*tim*) from Drosophila, *frequency* (*frq*) from *Neurospora crassa*, *kai* from cyanobacteria, *clock* from mouse, and so on (Feldman & Hoyle, 1973; Sehgal *et al.*, 1994; Kondo *et al.*, 1994).

In 1984, the *PER* gene was cloned (Bargiello *et al.*, 1984; Zehring *et al.*, 1984). This was the first behavioral gene isolated from any animal, and its isolation indicated that a molecular understanding of time measurement was possible. Other clock genes, *kai* in cyanobacteria, *frequency* (*FRQ*) in Neurospora, *timeless* (*TIM*) in Drosophila, *CLOCK* and *BMAL* in mammals, were also identified from forward and reverse genetics (Ishiura *et al.*, 1998; McClung *et al.*, 1989; Myers *et al.*, 1995;

King *et al.*, 1997; Tei *et al.*, 1997; Sun *et al.*, 1997; Allada *et al.*, 1998). Interestingly, although clock genes are not conserved across kingdoms (see Chapter 10), these components contribute to positive/negative feedbacks in all organisms studied (Dunlap, 1996; Young, 1998; Iwasaki & Kondo, 2004). It is widely accepted in the clock community that the core oscillator of circadian system can be explained as one or more feedback mechanisms at the transcriptional/post-transcriptional regulation of clock genes.

In plants, despite ancient knowledge of physiological phenomena, the molecular mechanism(s) of the circadian system had not been defined until recently. During the past 20 years, *Arabidopsis thaliana* has become a model plant for genetic and molecular-genetic studies (Somerville & Koornneef, 2002). As a great breakthrough for the clock community, Millar and Kay developed the *promoter::luciferase* system to monitor circadian rhythms, and to screen for clock mutants (Millar *et al.*, 1992, 1995a,b). Arabidopsis plants carrying a firefly *luciferase* (*LUC*) gene fused to a circadian-regulated *chlorophyll a/b-binding protein* (*CAB*) promoter exhibit robust rhythms in bioluminescence (Millar *et al.*, 1992). At least 21 independent mutations have been isolated from EMS-treated *CAB::LUC* seedlings with a low-light (photon-counting) video camera system (Millar *et al.*, 1995a). A semi-dominant mutant termed *timing of cab gene expression* (*toc1*) was one of the first mutants isolated from this screen. It is the mutant most thoroughly characterized, and the gene encoding *TOC1* has been cloned (Strayer *et al.*, 2000). *TOC1* encodes one of the pseudo-response regulators (*PRRs*) (Makino *et al.*, 2000).

2.2.1   The CCA1/LHY- TOC1 model for the Arabidopsis clock

In another approach, genes encoding related MYB-transcription factors, *Circadian Clock Associated 1* (CCA1) and *Late Elongated Hypocotyl* (LHY), were isolated and implicated in the clock mechanism (Wang & Tobin, 1998; Schaffer *et al.*, 1998). *CCA1* was originally cloned as a gene encoding a protein that binds to the phytochrome-regulated promoter region of a *CAB* gene (Wang *et al.*, 1997). Constitutive expression of the *CCA1* gene causes plants to express multiple arrhythmic phenotypes (Wang & Tobin, 1998). *lhy* was originally isolated as a day-length-insensitive late-flowering mutant (Schaffer *et al.*, 1998). The original allele of the *lhy* mutant caused the overproduction of LHY protein. This abolished the circadian rhythm of leaf movement and *CAB::LUC* gene-expression rhythms (Schaffer *et al.*, 1998). Both CCA1 and LHY were thus shown to contribute to the circadian system in Arabidopsis. The photoperiodic insensitivity of the original mutants is thought to be a consequence of a defect in the circadian clock. The gene expressions of *CCA1/LHY* and *TOC1* oscillate rhythmically with different phase from each other, with a dawn peak for *CCA1/LHY* and a dusk peak for *TOC1*. It was reported that CCA1 and LHY can bind to the promoter region of *TOC1* gene and this represses *TOC1* transcription (Alabadí *et al.*, 2001). Thus, a CCA1/LHY-TOC1 feedback-loop model has been proposed as a central oscillator of angiosperms (see Chapter 1).

2.3   An overview of the phosphorelay two-component system in Arabidopsis

TOC1 contains an atypical response-receiver domain and two putative transcriptional motifs: a basic motif conserved within the CONSTANS family and an acidic domain (Strayer *et al.*, 2000; Makino *et al.*, 2000). The atypical receiver domain is similar to that present in the classical Arabidopsis response regulators (ARRs), which are involved in Histidine (His) to Asparatic acid (Asp) phosphorelay two-component systems (Hwang *et al.*, 2002). However, the TOC1/PRR family lacks the conserved Asp (converted to glutamic acid) on the receiver domain. The Asp is critical for the phosphorelay as a phosphor-accepting site. In fact, there is no evidence of phosphotransfer event in the receiver domain of PRRs. However, the phosphorelay systems are known to contribute to phytohormone signalling, including cytokinin and ethylene signalling (Sweere *et al.*, 2001). Recently, we have found that these phytohormones affect various aspects of circadian system (Hanano *et al.*, unpublished). One of the classical response regulators ARR4 indeed modulates circadian rhythms via light signalling (Hanano *et al.*, unpublished). Both the typical ARRs and atypical PRRs can thus be implicated in various aspects of the clock network. For this reason, we present a brief overview of the classical phosphorelay two-component system.

The two-component systems are phosphorelay signalling mechanisms involved in sensing a variety of environmental stimuli, and are evolutionarily conserved from prokaryotes to eukaryotes (Fig. 2.2) (Hwang *et al.*, 2002). In Arabidopsis, many components in this system play important roles for plant signal transduction, including responses to phytohormones, stress responses and light signalling. The phosphorelay is carried out through His-protein kinases (AHKs), His-containing phosphotransfer proteins (AHPs), and response regulators (ARRs). Most of AHKs are hybrid-type His-kinases, which contain response receiver domain at their C-termini, except for ERS1 and ERS2 (see below). At the first step, AHKs sense the environmental input signal and transduce this via autophosphorylation of His on their transmitter domain. In the hybrid-type system, the phosphate of His is transferred to the Asp on the receiver domain at the C-terminal in AHKs. The phosphate on the receiver domain in AHKs is received to the His on AHPs, and then is relayed to the Asp on the receiver domain in ARRs via AHPs (Fig. 2.2).

2.3.1   AHK and AHP genes involved in the phosphorelay system of Arabidopsis

A genome sequence for Arabidopsis was determined at the end of 2000 (Arabidopsis Genome Initiative, 2000). This sequence information provided us with insight into all genes involved in a phospho-relay system. In Arabidopsis, a total of 11 sensor *AHK* genes, *ETR1, ETR2, ERS1, ERS2, EIN4, AHK1, AHK2, AHK3, AHK4, CKI1* and *CKI2*, have been found. Although *ERS1* and *ERS2* lack a response receiver domain, the others contain this domain at their C-terminus, and thus are

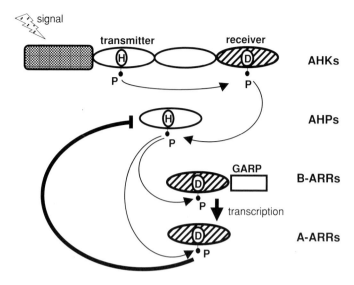

Figure 2.2 Typical two-component system in *Arabidopsis*. H: Histidine, D: Asparatic acid, P: Phosphate residue and GARP: GARP DNA-binding motif (B-motif). The AHKs receive environmental signals at their N-terminus, and then the H on transmitter domain of AHKs is auto-phosphorylated. The phosphate is transferred to the D on receiver domain in AHKs, relayed to the H on AHPs, and then received to the D on ARRs. In cytokinin signalling, the B-type ARRs are phosphorylated by AHPs and this activates transcription of A-type *ARR* genes. The A-type ARRs are thought to be negative-acting factors in cytokinin signalling.

presumed to carry out the hybrid-type AHK-AHP-ARR phospho-relay. Five (ETR1, ETR2, ERS1, ERS2 and EIN4) are ethylene receptors (Hua & Meyerowitz, 1998); a further three (AHK2, AHK3 and AHK4/CRE1/WOL) function in the cytokinin response (Kakimoto, 2003; Higuchi *et al.*, 2004; Nishimura *et al.*, 2004). These three, AHK2, AHK3 and AHK4 bind cytokinin, and are likely receptors. One of the histidine kinases, AHK1, is assumed to be an osmosensor (Urao *et al.*, 1999). Another histidine kinase CKI1 has also been implicated as a cytokinin receptor, as *Arabidopsis* callus overproducing *CKI1* exhibits a *cytokinin independent* cell-division and greening-phenotype (Kakimoto, 1996). However, CKI1 has no ability to bind to cytokinin. Probably, CKI1 overproduction activates the cytokinin signal of AHK2, AHK3 or AHK4. Recently, *CKI1* has been shown to be essential for megagametogenesis (Pischke *et al.*, 2002). The remaining histidine kinase, *CKI2*, has high similarity to *CKI1*; however, the function of this gene has not been reported.

Interestingly, in cyanobacteria, two histidine kinases, SasA and CikA, are known to affect these circadian systems (Iwasaki *et al.*, 2000; Schmitz *et al.*, 2000). The histidine kinase SasA has been identified as a protein that interacts with the clock component KaiC. *SasA* is necessary to sustain robust circadian rhythms and might function in entrainment. Another histidine kinase, *CikA*, is required for resetting the clock in response to light. The *CikA* encodes histidine kinase domain and pseudo-response receiver motif. Perhaps such molecules are at the basal origin of the higher plant clock. AHPs are thought to act downstream of AHKs (Tanaka *et al.*, 2004). There are five *AHP* genes in Arabidopsis genome. All *AHP* genes encode a His-containing phosphotransfer domain, which is similar to that found in the yeast Hpt *Spy1*. The phosphotransfer function of AHPs has been confirmed by complementation of yeast Hpt *ypd1* mutants and *in vitro* phosphorylation experiments with bacteria histidine kinase and ARRs (Suzuki *et al.*, 2002). In recent experiments, *AHP1* and *AHP2* were defined as likely components of cytokinin signalling. These AHPs are shuttled from the cytoplasm to the nucleus in response to cytokinin (Hwang & Sheen, 2001). Plants overproducing *AHP2* show a hypersensitive response in both a hypocotyl and a root-elongation assay (Suzuki *et al.*, 2002). There are currently no reports regarding the function of the other AHPs. In particular, it is of interest how only five AHPs act downstream of the nine AHKs.

2.3.2   Arabidopsis response regulators

Response regulators are encoded by no fewer than 22 genes in Arabidopsis (Hwang *et al.*, 2002; Mizuno, 2004). All response regulators encode a receiver domain at the N-terminus. The *ARR*s are classified into two distinct sub-types based on their primary structures. Eleven A-type response regulators do not seem to have a functional domain, besides the receiver domain. Another eleven B-type response regulators contain functional domains, such as nuclear localization signals, B-motifs (also termed as GARP-DNA binding motif), or transcription activator proline/glutamine-rich domains. These appendages are localized at the C-terminus. It is noteworthy that the B-motif is similar to the MYB-DNA-binding domain in mammals (Riechmann *et al.*, 2000). These B-type response regulators could act as nuclear-located transcription factors. The B-type ARRs (ARR1 and ARR2) activate transcription of A-type *ARR*s, and are reported to function in cytokinin signalling (Sakai *et al.*, 2001; Hwang & Sheen, 2001). The expressions of A-type *ARR*s are rapidly induced by cytokinin via the B-type ARRs (Brandstatter & Kieber, 1998). Most of A-type ARRs are also known to be located in the nucleus, except for ARR16 (Kiba *et al.*, 2002). The double and higher order *arr* mutants show increasing sensitivity to cytokinin in root-elongation responses and in hormone-regulated shoot formation (To *et al.*, 2004). Collectively, it is concluded that these A-type ARRs play important roles in the response of cytokinin signalling.

The typical phosphorelay system in Arabidopsis should consist of 11 *AHK*s→ 5 *AHP*s→22 *ARR*s, as detected from the genome sequence, but pathway-related sequences exist. Interestingly, there are factors similar in sequence to members of *AHK*s and *ARR*s. For example, the photoreceptor phytochrome genes (*PhyA~PhyE*) have sequence similarity to *AHK*s (Krall & Reed, 2000). However, the phytochromes do not contain a conserved His in their 'kinase' domain. Also, the cyanobacteria, and other gram-negative bacteria, have phytochrome-like His kinases (e.g. *CPH1*) (Yeh *et al.*, 1997). CPH1 has sequence similarity with phytochrome in its N-terminus and can bind appropriate chromophores. In addition, its kinase activity is light responsive. It has been concluded that phytochromes evolved from His kinases, such as CPH1 in cyanobacteria (Vierstra & Davis, 2000). A further connection implicating such an evolutionary history of plant phytochromes is the finding that the AHK-related phytochromes can work together with an ARR to connect signalling (Sweere *et al.*, 2001).

2.4 Pseudo-response regulators

The *pseudo-response regulators* (*PRR*s) are related to *ARR*s, and thus are similar to the components in the typical phosphorelay system, but the *PRR* genes are atypical of classical response regulators (Fig. 2.3) (Strayer *et al.*, 2000; Makino *et al.*, 2000). The PRRs lack the conserved asparatic acid (converted to glutamic acid) on their receiver domain. This has led to the suggestion that these proteins are non-functional in AHK phosphorelays. TOC1/PRR1, PRR3, PRR5, PRR7 and PRR9 are classified as a sub-group of the PRRs (the TOC1/PRR family). These five PRRs encode an atypical response-receiver domain (pseudo-receiver domain) at the N-terminus and two putative transcriptional motifs at C-terminus: a basic motif conserved within the

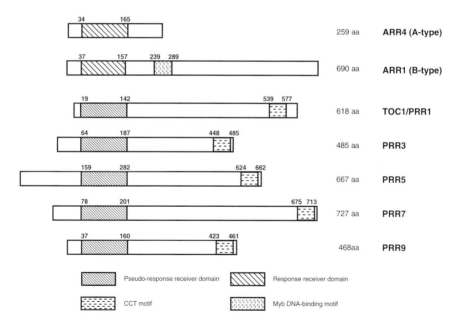

Figure 2.3 Protein structure of TOC1/PRR family in comparison with ARR4 (A-type ARR) and ARR1 (B-type ARR). The sequence information is available from the public database TAIR (http://www.arabidopsis.org/) and MIPS (http://mips.gsf.de/). The protein structures are predicted by Pfam through the web site Motif on GenomeNet (http://motif.genome.jp/). Pseudo-response-receiver domain, response receiver domain, CCT motif (CONSTANS-like motif) and Myb DNA-binding motif (B-motif or GARP-motif) are shown as indicated.

CONSTANS family and an acidic domain. For those PRRs that have, to date, been analyzed, each protein is located in the nucleus. The CONSTANS-motif seems to be responsible for this nuclear-localization (Makino *et al.*, 2000). The members of TOC1/PRR family function in the circadian clock, as described below. The function of the other PRRs is currently unknown. PRR2 has a B-motif which is similar to that found in the B-type response regulator ARR10. The B-motif is related to the MYB-related domain, which thus implicates ARR2 as a supposed transcription factor. The mRNA of *PRR2* is accumulated constitutively over the day, unlike the *TOC1/PRR* family members, perhaps indicating that PRR2 is not involved in clock function.

2.5   Circadian regulation of TOC1/PRR family: CCA1/LHY-TOC1 feedback model and circadian wave form of the TOC1/PRR family

TOC1 encodes a pseudo-response regulator, as described above. *TOC1* mRNA oscillates with the peak at dusk, though the oscillations damp rapidly in constant darkness (Strayer *et al.*, 2000). Makino *et al.* (2000) and Matsushika *et al.* (2000) also identified *TOC1/PRR1* and the homologues *PRR9*, *PRR7*, *PRR5* and *PRR3* (of the *TOC1/PRR* family) from database searches of the Arabidopsis genome. This occurred during their comprehensive cloning of genes involved in two-component system. Interestingly, the mRNAs from these five *PRR* genes oscillate sequentially after dawn with approximately 2–4 hour intervals between the successive family members, under constant light conditions (Fig. 2.4) (Matsushika *et al.*, 2000). *PRR9* transcription is rapidly induced by light (red light) (Makino *et al.*, 2001). The *PRR9* mRNA accumulates after dawn, and then the others follow in the order *PRR7*, *PRR5*, *PRR3* and *TOC1/PRR1* (Matsushika *et al.*, 2000). *TOC1/PRR* family gene expression is in a circadian wave form, which predicts the time of the day. The orchestrated model was the first proposed molecular mechanism acting as a central oscillator. Currently, this model is not favoured. Still, elements of the model are likely to be correct.

   CCA1 and *LHY* are believed to be other components in the central oscillator. The CCA1 and LHY proteins have similar function in light signalling, circadian rhythms and photoperiodic regulation of flowering time. This is striking, given their sequence similarity (Wang & Tobin, 1998; Schaffer *et al.*, 1998). This has led to the early prediction that these genes have partially redundant functions. Plants overproducing either *CCA1* or *LHY* lack detectable circadian rhythms (with one exception, described below) and both genotypes express a late-flowering phenotype. The *cca1* and *lhy* loss-of-function mutations, or *LHY*-RNA interference plants (*lhy*-R), have a shortening-period phenotype, but these plants still maintain rhythms (Green & Tobin, 1999; Mizoguchi *et al.*, 2002; Alabadí *et al.*, 2002). The *lhy11-1 cca1-1* or the *cca1 lhy*-R double mutants cause near arrhythmia of leaf-movement rhythms (Mizoguchi *et al.*, 2002; Alabadí *et al.*, 2002). This can be explained as LHY and CCA1 function redundantly in the circadian system, as previously predicted.

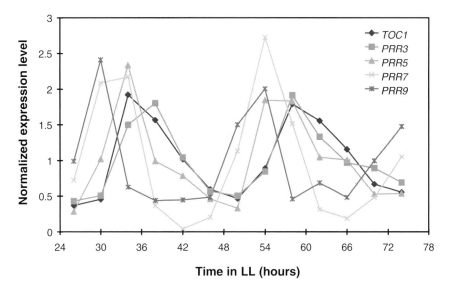

Figure 2.4 The mRNA of the *PRR* genes oscillates with an approximately 24 h period under constant light conditions. The gene-expression profiles are available from the public microarray database Genevestigator https://www.genevestigator.ethz.ch/ (Zimmermann *et al.*, 2004). Each expression data set is normalized to an arbitrary average level. The circadian experiments were performed by Edwards, K. D. and Millar, A. J. (unpublished).

2.5.1 Regulation of TOC1 expression

The rhythmic peak of *TOC1* mRNA in constant light is relatively broad compared to that of the other *PRR*s, and it dampens rapidly in constant darkness (Matsushika *et al.*, 2000). Promoter-deletion analyses of the *TOC1* promoter sequence fused to a *luciferase* gene resulted in the identification of an approximately 200 nt DNA fragment necessary for clock-regulated expression (Alabadí *et al.*, 2001). This DNA fragment contains a so-called evening element (EE; AAAATATCT) known in the promoters of many clock-regulated genes showing a peak at dusk (Harmer *et al.*, 2000). Alabadí *et al.* (2001) showed that LHY and CCA1 can bind to this DNA sequence in a gel-shift assay. As *TOC1* expression is reduced in plants overproducing *LHY* or *CCA1* (*LHY*-ox and *CCA1*-ox, respectively), LHY and CCA1 are believed to regulate *TOC1* expression negatively after binding to the EE in this promoter region. The mRNAs from the *LHY* and *CCA1* genes accumulate in the morning and their proteins are expressed during the day. Later in the evening, as LHY and CCA1 protein levels are reduced, transcription of the *TOC1* gene is activated. In a strong *toc1* allele, such as *toc1-2*, the mRNA level of *LHY* and *CCA1* is reduced. TOC1 is thus believed to act as a positive regulator of *LHY* and *CCA1* transcription. Thereby, the first feedback model of a plant clock was proposed (Alabadí *et al.*, 2001). A combination of genetic, molecular and biochemical analyses lent support to such a model. This model is being refined through current experimentation.

TOC1 is regulated at the post-translational level. The TOC1 protein is targeted by ZEITLUPE (ZTL) for proteolysis (Más *et al.*, 2003b). *ZTL* encodes a Period-ARNT-Sim (PAS)-like Light, Oxygen or Voltage (LOV) domain, an F-box and six kelch repeats (Somers *et al.*, 2000). The LOV domain binds a flavin mononucleotide chromophore, and undergoes light-induced photochemistry. This could suggest that ZTL function is activated by light (Imaizumi *et al.*, 2003). The LOV domain is similar to two-component sensor domains. F-box proteins interact with Skp1/Cullin/F-box (SCF) type E3 ubiquitin ligases, which are involved in proteosome-specific protein degradation. Recently, the LOV domain of ZTL has been shown to interact with TOC1 protein during light-dependent protein degradation (Table 2.2) (Más *et al.*, 2003b). The levels of *TOC1* mRNA and protein are negatively correlated to the pace of the clock. The interaction between TOC1 and ZTL is abolished in *ztl1-1* or *ztl1-3* null alleles, resulting in an apparent constitutive accumulation of TOC1 protein, and thus, a delay in periodicity. TOC1 protein degradation targeted by ZTL appears responsible for regulation of circadian period.

There are a total of three genes in the *ZTL* family [*ZTL* and its homologues *Flavin-binding kelch repeat F box 1* (*FKF1*) and *LOV kelch protein 2* (*LKP2*)] (Somers *et al.*, 2000; Nelson *et al.*, 2000; Schultz *et al.*, 2001). Plants overproducing *LKP2* show arrhythmic phenotypes in many outputs of circadian rhythms (Schultz *et al.*, 2001), whereas the circadian phenotype of the *fkf1* mutant is similar to that of wild-type plants (Nelson *et al.*, 2000; Imaizumi *et al.*, 2003). This could place LKP2 close, or central, to the oscillator, and these results can also be interpreted as that FKF1 is not part of the clock. Recently, LKP2 has been also reported to interact with TOC1 and PRR5 in the yeast two-hybrid system (Yasuhara *et al.*, 2004). But, further analyses of LKP2 in regulation of TOC1 and PRR5 still remain to be done, and thus a direct connection of LKP2 to the core oscillator is not resolved. Anyhow, TOC1 expression is clearly controlled by both transcriptional and post-transcriptional mechanisms to generate the timing of rhythms.

2.5.2   Interactions between TOC1 and other PRR family members

In plants overproducing TOC1/PRR1 (*TOC1*-ox), the mRNA accumulation of *PRR9* is repressed, while the expression of other *PRRs* can be detected. The *TOC1*-ox plant maintains a rhythm for these *PRRs* for at least one cycle (Makino *et al.*, 2002). This could have been assumed to be due to sequential regulation of circadian wave form in TOC1/PRRs family. However, in plants overproducing PRR9 (*PRR9*-ox), the peaks of gene expression from *TOC1, PRR7, PRR5, PRR3, CCA1, LHY* and *Early flowering 3* (*ELF3*) are all advanced. PRR9 therefore is unlikely to repress gene expression of any of the *TOC1/PRR* family, nor other clock genes (Matsushika *et al.*, 2002a). The complete repression of *PRR* genes by the other PRRs has not been detected, except for TOC1/PRR1 repressing *PRR9*.

2.5.3   The role of the evening element in PRR9 regulation

Light input shortens the circadian period and modifies the circadian phase in Arabidopsis (Millar *et al.*, 1995a; Millar & Kay 1996; Anderson *et al.*, 1997; Somers *et al.*, 1998a; Salomé *et al.*, 2002). The expression of *PRR9* is not only circadian-regulated, but also acute-inducible by a light pulse through phytochrome-mediated events, consistent with *CCA1* light induction (Makino *et al.*, 2001; Ito *et al.*, 2003). Recently, it has been reported that the promoter of the *PRR9* gene contains an evening element (EE), even though *PRR9* is light-inducible and is a morning gene (Salomé & McClung, 2004; Farré *et al.*, 2005; Ito *et al.*, 2005). Promoter analysis of *PRR9* identified two functional sequences: a light-responding region and a circadian-controlled region (Ito *et al.*, 2005). Deletion of the light-responding region results in an evening phase of *PRR9* gene expression. The EE may control the circadian pattern, but not affect the phase strongly. In *cca1-1 lhy-R* double mutants, *PRR9* expression is greatly reduced (Farré *et al.*, 2005). Recently, LHY and CCA1 have also been shown to bind the EE element in the *PRR9* promoter (Farré *et al.*, 2005). The peak of these transcription factors corresponds to *PRR9* transcription. Taken together, this suggests that CCA1 acts as a positive regulator of *PRR9* expression, which is different from the LHY and CCA1 function in *TOC1* repression. Interestingly, the PRR9 protein has also been shown to interact with TOC1 in the yeast two-hybrid system (Table 2.2) (Ito *et al.*, 2005). Thus, *PRR9* gene expression is regulated positively by LHY and CCA1 and negatively by TOC1 at the transcriptional level, and then may be modulated by TOC1 at the post-transcriptional level.

2.5.4   Interactions between PRRs and clock components

The expression of *PRR7* follows *PRR9* expression (Fig. 2.4). *PRR7* gene expression is likely to be independent from *PRR9* expression because of their distinct roles in the circadian system, as described below (Farré *et al.*, 2005). *PRR7* has three CCA1 binding sites (CBS) in its promoter, which are also found in *CAB* promoters. Farré *et al.* (2005) showed that LHY and CCA1 can bind to the promoter of *PRR7*, as such binding occurs in the *PRR9* promoter. These transcription factors may act to promote the expression of *PRR9* and *PRR7* in parallel, although *PRR7* is still expressed in *cca1-1 lhy-R*. Perhaps another factor specifically regulates PRR7, while not regulating PRR9.

PRR5 expression follows *PRR7*. Nakamichi *et al.* (2005) described rhythmic expression of *PRR5* in the *prr7-11* background. They concluded that distinct transcriptional mechanisms are involved. *PRR3* rises after *PRR5* accumulates, and then *TOC1/PRR1* is transcribed around dusk. The *PRR3* peak nearly overlaps with *TOC1/PRR1* (Fig. 2.4) (Matsushika *et al.*, 2000). In plants overproducing PRR5 (*PRR5*-ox), the mRNA of *PRR3* and *TOC1/PRR1* accumulates more than is seen in wild-type. In addition, the expression of both *PRR9* and *PRR7* are slightly reduced (Sato *et al.*, 2002). Interestingly, in *prr5-11* and *PRR5*-ox plants, the acute

induction of *PRR9* and *CCA1* is modified (Yamamoto *et al.*, 2003). PRR5 negatively regulates the acute response to light. PRR5 might regulate the transcription
of other *PRR* genes, especially antagonizing PRR9 function. However, in *prr5-1*
and *prr5-11*, *prr7-3* or *prr9-1* null mutants, the *TOC1/PRR1* mRNA accumulation
level is not changed significantly, although period length is changed. In addition,
CCA1 is slightly reduced in *prr5-1* (Eriksson *et al.*, 2003), and both *LHY* and *CCA1*
are slightly increased in *prr7* (Farré *et al.*, 2005). It is inconsistent with the data
that the acute response of *CCA1* is increased in *prr5-11*, and decreased in *PRR5-*
ox (Yamamoto *et al.*, 2003). Collectively, these observations implicate each of the
PRRs in clock regulation.

2.6   A detailed description of signal convergence in the *PRR* genes

2.6.1   *TOC1/PRR1, At5g61380*

The *toc1* mutant was isolated via *luciferase* screening (Millar *et al.*, 1995a). As
mentioned, *toc1* is the best characterized *PRR* mutant, and a shortened circadian
periodicity of various rhythms can be seen (Table 2.1) (Millar *et al.*, 1995a; Somers
et al., 1998a; Dowson-Day & Millar 1999). The original mutant *toc1-1* bears a
consistent short-period phenotype with the *CAB::LUC* marker in LL and in leaf
movement assays (Millar *et al.*, 1995a). *toc1-1* carrying *CCR2::LUC* also expresses
a short-period phenotype in constant darkness (Strayer *et al.*, 2000). *toc1-1* shortens
periodicity by about 2–3 hours over a wide range of red-light intensities and temperatures (Somers *et al.*, 1998b). The effect is independent of the light input to the clock
(Strayer *et al.*, 2000). The period of stomatal conductance and hypocotyl elongation
rhythms is also shortened with approximately 3 hours in *toc1-1* (Somers *et al.*, 1998b;
Dowson-Day & Millar, 1999). This has led to the suggestion that a cell-autonomous
clockwork is active in various organs (Somers *et al.*, 1998b). The *toc1-1* mutation
affects a variety of clock-controlled processes perhaps independent of plant age.
Therefore, TOC1 is interpreted to be a critical component of the central oscillator.

2.6.1.1   *toc1-2 and TOC1 mini-gene mutants*
More recently, a strong allele *toc1-2* was analyzed (Strayer *et al.*, 2000). *toc1-2* is
a recessive allele that expresses a truncated protein via an aberrant mRNA caused
by incorrect splicing. To examine the role of *TOC1* in the circadian system, *TOC1-*
RNA interference (RNAi) plants (*toc1-R*) were also prepared. In *toc1-2*, *LHY* and
CCA1 expression levels are reduced in comparison with *toc1-1* (Alabadí *et al.*,
2001). *CCA1* gene expression is induced by red light; however, the light induction
of *CCA1* is also reduced in *toc1-2* and *toc1-R* (Más *et al.*, 2003a). In *toc1-R*, it
was found, using an RT-PCR assay, that the mRNA levels of other four *PRRs* are
similar to that observed in wild type (Más *et al.*, 2003a). In various intensities of
red light, both *toc1-R* and *toc1-2* exhibit arrhythmic expression of *CAB::LUC* and
CCR2::LUC. In constant darkness, arrhythmic gene expression of *CCR2::LUC* in
toc1-R and *toc1-2* is also observed. However, in constant blue light, oscillations in

Table 2.1 Circadian and morphological phenotypes of *prr* mutants and *PRR*-ox plants

Genotype	Circadian phenotype	Thermocycle (in LL)	R-acute	Hypocotyl	Flowering
toc1-1	2–3 h short				Early (SD), normal (21 h SD)
toc1-2	Arrhythmic (R/D), 3–4 h short in (B)		Low	Long (R/FR)	
toc1-R	Arrhythmic (R/D), 3–4 h short in (B)				Normal
prr9	Late phase/long period (W, B), normal (R/D)	Fails to entrain		Slightly long	Normal
prr7-1, -2	Arrhythmic (R), early phase (W)			Long (R/FR)	
prr7-3	Long period (R > B), normal (D)	Initially fails to entrain		Long (R)	Late (LD)
prr5	Short period (W, B), normal (R/D/E)	Short period	High	Long	Late
prr3	Mild short period				
TMG	Long period				
TOC1-ox	Long period			Short (R)	Normal
PRR9-ox	Short period			Slightly short (R)	Early (SD)
PRR7-ox	Not reported				
PRR5-ox	Low amplitude long period		Low	Short (R/B)	Early
PRR3-ox	Slightly long period/late phase			Long (R)	Late (LD)
prr9prr5	Normal (W)				Late
prr9prr7	Very long (R, B, W), arrhythmic (D)	Fails to entrain			
prr7prr5	Low amplitude, short			Very long (R)	Very late

R: Red light, FR: far-Red, B: Blue light, W: White light, D: dark.
E: etiolated seedlings after temperature-entrainment, SD: short day, LD: long day.
R-acute: acute response of *CCA1* gene expression induced by Red light.
TMG: *TOC1* mini-gene.

toc1-R and *toc1-2* still remain, albeit with a 3–4 hour shorter period than that in wild type. Thus, TOC1 plays an important role in the integration of red-light signalling to the circadian system, but the *TOC1* loss-of-function by RNAi still shows 3–4 hour shorter rhythms, suggesting that other components compensate the circadian function under blue light.

To investigate the correlations between the amount of *TOC1* transcripts and period length, TOC1 mini-gene (TMG) transgenic plants expressing additional copies of the *TOC1* locus were analyzed (Más *et al.*, 2003a). *TOC1* RNA levels are higher in TMG lines than in wild type, but these plants are still rhythmic. This is unlike what was seen in *TOC1*-ox lines that constitutively expressed *TOC1* (Makino *et al.*, 2002). In TMG plants, the *CAB::LUC* rhythm is longer than that of wild type and correlates with the levels of *TOC1* transcripts, whereas *TOC1*-ox has an arrhythmic phenotype. Thus, increased *TOC1* levels delays the pace of clock, but maintaining

TOC1 rhythmic expression keeps overt rhythmicity. Constitutive *TOC1* expression disrupts the rhythm.

2.6.1.2 TOC1 and phytochrome

Más *et al.* (2003a) measured the hypocotyl length of various *toc1* genotypes. Both *toc1-2* and *toc1-R* have a long-hypocotyl phenotype under constant red and far-red light. In contrast, TMG plants showed short hypocotyl in red light. This implicates TOC1 as a positive element in phytochrome signalling. Under blue light and in constant darkness, the hypocotyl lengths of *toc1-2*, *toc1-R* and TMG plants are similar to that of wild type. However, the phenotypes of hypocotyl length are milder than that of photoreceptor mutants, and independent of light intensity. *TOC1*-ox plants also have shorter petioles than wild-type plants (Makino *et al.*, 2002). This might be consistent with the hypocotyl phenotype of *toc1-2*, *toc1-R* and TMG plants, suggesting the broad function of TOC1 on red-light-dependent cell elongation. The red-light-mediated induction of *CCA1/LHY* expression was also decreased in *toc1-R* and *toc1-2* plants (Más *et al.*, 2003a). YFP::TOC1 appeared exclusively in the nucleus in a distinctive speckled pattern, which is supposed to be co-located with the phytochromes (Strayer *et al.*, 2000; Makino *et al.*, 2000). Taken together, TOC1 is an important factor for regulating hypocotyl elongation controlled by red and far-red.

2.6.1.3 TOC1 and flowering time

The *toc1* mutant has a defect in photoperiodic induction of flowering time in the Landsberg *erecta* background (Somers *et al.*, 1998b). Flowering time of *toc1-1* in a short day of 8 hours-light/16 hours-darkness is earlier than in wild type. Thus, *toc1-1* has a reduced sensitivity to the day length. Under short days, the peak of *CONSTANS* (*CO*) gene expression in *toc1-1* is advanced, and this results in the activation of *FLOWERING TIME* (*FT*). This promotes the photoperiodic pathway of flowering (Yanovsky & Kay, 2002). Importantly, the early-flowering phenotype under short days is not observed under 21 h short day (7 hour-light/14 hour-dark) that matches endogenous period of the *toc1-1*. The phase shift of *CO* expression is supposed to be caused by a clock defect, which is tightly linked to the flowering phenotype. However, *toc1-R* does not have the same effect on the flowering time (Más *et al.*, 2003a). Furthermore, it is curious that *TOC1*-ox plants do not show a strong flowering phenotype even though *GIGANTEA* (*GI*) expression is markedly reduced in expression (Makino *et al.*, 2002). TOC1 may also affect the downstream of *CO* in the photoperiodic pathway or the other regulatory pathways of flowering time, perhaps independent from its role in periodicity and phase of *GI* expression (Hayama & Coupland, 2003).

2.6.2 PRR9, At2g46790

PRR9 expression is induced by light and is circadian-regulated with the peak of the expression rhythm around dawn. *PRR9* gene expression is repressed in *TOC1*-ox plants (Makino *et al.*, 2002). Plants overproducing PRR9 (*PRR9*-ox) have a

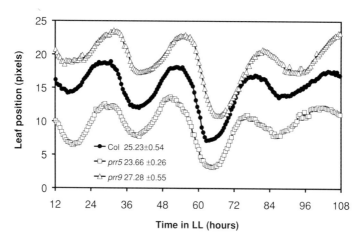

Figure 2.5 Leaf-movement circadian phenotypes of *prr5-1* and *prr9-1*. One-week-old seedlings were entrained to a 12 hour light/12 hour dark cycle and were transferred into constant light conditions. The leaf position was measured every 30 min. The traces of leaf position (pixels) are averages of data from 6–10 leaves. The period length is calculated by BRASS (http://www.amillar.org) with the fast Fourier transform-nonlinear least squares method (FFT-NLLS) (Plautz, 1997).

short-period phenotype (Matsushika *et al.*, 2002a). In *PRR9*-ox plants, *PRR3* and *TOC1/PRR1* gene expression levels are slightly reduced. A *prr9* null mutant was isolated from public T-DNA insertion lines (Eriksson *et al.*, 2003; Michael *et al.*, 2003b). This *prr9* mutant was found to have a long-period phenotype for various circadian rhythms, including gene expression rhythms and leaf movement under constant light (Fig. 2.5; Table 2.1) (Eriksson *et al.*, 2003; Farré *et al.*, 2005). However, Michael *et al.* (2003) did not report this long-period phenotype in a *prr9* mutant, and instead, they found a delayed phase in a cotyledon-movement assay. This may be either explained as that the *prr9* phenotype is dependent upon specific light conditions or by the different outputs measured. The *prr9* mutant has a long-period phenotype under constant blue light (Eriksson *et al.*, 2003), but the period length of *prr9* was similar to that of wild type under red light and constant darkness; in contrast, *toc1-2* has a strong phenotype (arrhythmia) in red and constant darkness. Even under blue-light conditions, gene expression still oscillates in the *prr9* null mutant. PRR9 appears not to be a central clock component, but may integrate blue-light input to the clock, compensating LHY/CCA1-TOC1 feedbacks. Alternatively, PRR9 might function redundantly in the circadian core oscillator with other PRRs, such as PRR7 (described below). TOC1 expression levels are slightly increased in *prr9* and marginally decreased in *PRR9*-ox. This could explain the period phenotype, as the amount of TOC1 delays the pace of circadian clock.

After temperature-entrainment, the *prr9* mutant did not have a period phenotype in etiolated seedlings (Eriksson *et al.*, 2003), yet, they failed to entrain to the thermocycles when grown under constant light (Salomé & McClung, 2005). Such a response points to convergence of temperature and light inputs to the entrainment mechanisms.

prr9 seedlings exhibit a mild long-hypocotyl phenotype under constant red or blue light, which is not dependent on light intensity (Eriksson *et al.*, 2002). The hypocotyl length is slightly short in *PRR9*-ox under red-light, but the phenotype does not mimic that seen in *TOC1*-ox lines (Sato *et al.*, 2002). *PRR9*-ox plants flower earlier than wild-type, especially under short-day condition, whereas *prr9* does not show this flowering difference (Matsushika *et al.*, 2002a; Hanano *et al.*, unpublished). From these results, one could conclude that PRR9 functions in light signalling and/or a photoperiodic pathway. As many genes exhibit advanced expression in *PRR9*-ox, if the phase of *CO* expression is also regulated by PRR9-mediated clock, this could be the cause of early flowering in this genotype.

2.6.3   PRR7, At5g02810

prr7-1 and *prr7-2* were originally isolated from a genetic screen for de-etiolation defective seedlings under red and far-red light conditions (Kaczorowski & Quail, 2003). Both *prr7-1* and *prr7-2* have stop codons in the pseudo-receiver domain. Each bears long-hypocotyl phenotypes under constant red or far-red light conditions (Table 2.1). PRR7::GUS fusion protein is localized in the nucleus. In *prr7* mutants, *CCA1* and *LHY* gene expression are not rhythmic under constant red light. These transcription factors are constitutively expressed in *prr7* plants, which has led to the suggestion that PRR7 represses the gene expression of *CCA1* and *LHY* around dusk. Under constant white light, *CCA1*, *LHY* and *TOC1* all oscillate rhythmically with earlier phase in *prr7* than in wild type (Kaczorowski & Quail, 2003).

prr7-3 (also known as *prr7-11*) was isolated from T-DNA insertion lines and assayed for various phenotypes (Yamamoto *et al.*, 2003; Farré *et al.*, 2005). *prr7-3* mutant has a long-period phenotype. This is in contrast to the *prr7-1* and *prr7-2* alleles which show early-phase phenotypes. Farré *et al.* introduced *CCR2::LUC* into the *prr7-3* mutant. The *prr7* mutation lengthens period length of this reporter under red light, and this effect was stronger than that seen under blue light. This inconsistency amongst the *prr7* alleles might be caused by the position of the mutation in *prr7-3*. Here, it is closer to the 5-prime ATG than in *prr7-1* and *prr7-2*, perhaps indicating that *prr7-1* and *prr7-2* are not nulls. Another interesting point is that the *prr7-3* mutant initially fails to entrain to thermocycles (Salomé & McClung, 2005). This failure to follow temperature entrainment may be caused by a weaker gating to temperature in the circadian system. In constant to darkness, the period phenotype is not significant. Additional phenotypes in *prr7-3* are similar to previous alleles, in that all have long-hypocotyl phenotypes under red light. The *prr7-3* flowers later than wild type when grown under long-day conditions. This has led to the suggestion that *PRR7* is involved in the photoperiodic pathway of flowering (Nakamichi *et al.*, 2005). Finally, *PRR7* is located at the top of chromosome 5, which is linked to QTL known as *ANDANTE (AND)*, *ANOTHER ANDANTE (AAN)* and *TAU5A* (Swarup *et al.*, 1999; Michael *et al.*, 2003b). Each of these is a period-length circadian QTL

(Michael *et al.*, 2003b). However, whether PRR7 exhibits sequence variation to account for the variation is not known.

2.6.4   *PRR5, At5g24470*

PRR5 overexpressing lines (*PRR5*-ox) and *prr5* null mutants were isolated and assayed for clock responses. *PRR5*-ox has short hypocotyls when grown under red light or blue light, and this genotype has an early-flowering phenotype (Sato *et al.*, 2002). The flowering phenotype of this overexpressor appears photoperiod-dependent, but it is seen even in long-day conditions. In *PRR5*-ox, many gene-expression rhythms have lowered amplitude. The gene expression of *TOC1/PRR1* and *PRR3* are constitutively high, but *CCA1, LHY* and *GI* expression are reduced in the *PRR5*-ox. PRR5 suppressed *GI* gene expression. But, in spite of low *GI* levels, the *PRR5*-ox line flowered early. This is quite surprising because the flowering phenotype is inconsistent with the current flowering model. Perhaps, the peak of the *GI* gene expression in *PRR5*-ox is earlier than in wild type. Such an early phase, even if low in amplitude, could generate early flowering (Hayama & Coupland, 2003). In addition, *LHY* and *CCA1* gene expression levels are relatively lower in *PRR5*-ox, compared to in wild type. The early-flowering phenotype of *PRR5*-ox could be because of the lowered level of these transcription factors.

   prr5-1, prr5-3, prr5-11 and *prr5-12* have been isolated from T-DNA inser-tion populations and assayed for phenotypes (Table 2.2) (Eriksson *et al.*, 2003; Yamamoto *et al.*, 2003; Michael *et al.*, 2003). The circadian period under constant white light in *prr5* is 1.0–2.4 h shorter than in wild type (Fig. 2.5). The *prr5* phe-notype is not observed under constant red light, in darkness, nor in etiolated plants entrained to temperature cycles. However, under constant white light, *prr5* failed to re-entrain to temperature cycles (Salomé & McClung, 2005). Interestingly, the failure of temperature re-entrainment is larger than that expected from the change in the free-running cycle, which is always modest (approximately 1 hour short pe-riod). This could be explained in that thermocycles contain a cool-temperature phase and this is what *prr5* mutants fail to detect. It will be interesting to test whether the circadian phenotype of *prr5* is stronger at lower mean temperatures. Formally, PRR5

Table 2.2 Possible protein partners of PRRs

Protein	Interaction	Posible function
TOC1/PRR1	ZTL, LKP2	Proteolysis
	PRR9	
	PIF3, PIL family	bHLH DNA-binding
	ABI3	ABA signalling
PRR9	TOC1/PRR1	
PRR7	Not reported	
PRR5	LKP2	Proteolysis
	WNK1	Kinase
PRR3	WNK1	Kinase

could play a role in either temperature entrainment of cotyledon movement, in clock accuracy at lowered temperatures, or PRR5 might have a role in temperature compensation. *prr5* mutants also have a long hypocotyl and late flowering phenotype (Eriksson *et al.*, 2003; Yamamoto *et al.*, 2003).

2.6.5   *PRR3, At5g60100*

PRR3 has also been studied using both loss-of-function and gain-of-function approaches (Table 2.1). A *prr3-1* mutant was isolated from T-DNA insertion populations. It has a mild short-period phenotype (Michael *et al.*, 2003b). *PRR3*-ox plants under constant light have either a slightly long-period or a late-phase phenotypes (Murakami *et al.*, 2004). From these data, it has been concluded that PRR3 functions in the circadian system, but its effects are relatively weak. *PRR3*-ox also has a long hypocotyl under red light and it is late flowering under long-day conditions. This is opposite to the effects seen in *PRR5*-ox. *PRR3* and *PRR5* genes are expressed at similar times of the day and have some effects on the circadian system in the same direction, whereas the morphological phenotype of *PRR3*-ox and *PRR5*-ox are antagonised. These observations indicate a complex mechanism exists that intersects between the circadian system and various morphological phenomena.

2.7   Phylogenetic and epistatic relations between the PRRs

All *PRRs* in the *TOC1* family are expressed in a circadian manner with approximately 2–4 hour intervals between each successive wave of gene expression. Each has distinct genetic effects on the circadian system. The morning expressed genes *PRR9* and *PRR7* advance the pace of the clock, while the evening expressed genes *PRR5*, *PRR3* and *TOC1/PRR1* delay it. In a phylogenetic tree, *PRR9* and *PRR5* are located in the same branch, *PRR3* and *PRR7* cluster together in a separate branch, and *TOC1/PRR1* is in a third (Fig. 2.6) (Matsushika *et al.*, 2000). With regard to the clock, *PRR9* and *PRR5* seem to act in blue-light input, rather than red, while *TOC1/PRR1* and *PRR7* function in the red-light input to the clock, rather than blue (Eriksson *et al.*, 2003; Más *et al.*, 2003a; Farré *et al.*, 2005). TOC1/PRR1 represses *PRR9* gene expression and physically interacts with PRR9 protein (Makino *et al.*, 2002; Ito *et al.*, 2005).

To investigate the genetic interaction amongst the *PRRs*, the double mutants *prr9prr5*, *prr9prr7* and *prr7prr5* have been generated and analyzed (Table 2.1) (Eriksson *et al.*, 2003; Farré *et al.*, 2005; Salomé & McClung, 2005; Nakamichi *et al.*, 2005). Both *prr9* and *prr5* have a strong phenotype under blue light, but in the opposite direction. If there is genetic interaction between them, one of the effects could remain in the double mutant, whereas if they function in distinct pathways the period phenotype could be additive. The period length of the *prr9prr5* for leaf movement under constant light condition is similar to that of wild type,

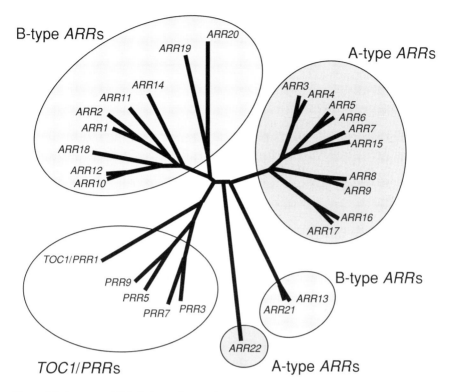

Figure 2.6 An unrooted N-J phylogenetic tree of response regulators and the *TOC1/PRR* family of Arabidopsis. The receiver domains of response regulators and pseudo-response receiver domains of TOC1/PRR proteins are aligned on the web site CLUSTALW on GenomeNet http://clustalw. genome.jp/.

which could be interpreted as that PRR9 and PRR5 function in distinct pathways (Eriksson *et al.*, 2003). PRR7 may function in red light input to the clock, whereas PRR9 acts in a blue light pathway, as described above. Each single mutant did not affect the circadian system in constant dark. Under red or blue light conditions, *prr9prr7* double mutants have a very long period phenotype, which might mean that both PRR9 and PRR7 have distinct functions in the integration of light input to the circadian system (Farré *et al.*, 2005). The *prr9prr7* double mutant also has defects in temperature entrainment when grown under constant light (Salomé & McClung, 2005). Probably, these mutations affect the resetting signals of environmental stimuli. It is very curious that the double mutants harbour a strong phenotype (arrhythmic) in constant darkness, whereas each single mutant seems to be same as wild type in the constant dark condition. This implies genetic redundancy. The genetic PRR9 function of the clock in darkness is currently mysterious, as *PRR9* expression levels are low under these conditions (Makino *et al.*, 2001).

2.8   Partners of PRR proteins

2.8.1   PIF3-like family

Recently, in addition to CCA1, LHY and ZTL, many partner proteins interacting with TOC1/PRR family have been reported (Table 2.2). There is no proof that TOC1 protein acts on the *LHY* or *CCA1* promoter directly. This has led to the suggestion that other factors co-function in the induction of *LHY* and *CCA1* transcription. Candidate genes include *PHYTOCHROME-INTERACTING FACTOR 3* (*PIF3*) and the related *PIF3-like* (*PIL*) gene family (Ni *et al.*, 1998; Martinez-Garcia *et al.*, 2000; Yamashino *et al.*, 2003). PIF3 is one of the seven PIF3-like (PIL) proteins in the 162-member bHLH protein group. The induction of *CCA1* and *LHY* transcription was proposed to be mediated via PIF3 (Martinez-Garcia *et al.*, 2000). The red-light photoreceptor phyB activates PIF3, which binds to the promoter of CCA1 and LHY. In addition, in the yeast two-hybrid system, TOC1 physically binds to PIF3 and PIL1 (Makino *et al.*, 2002). In further two-hybrid analyses, TOC1 also interacts with all PIL family members, except for HFR1 (Yamashino *et al.*, 2003). HFR1 is involved in phyA-mediated far-red-light signalling, whereas the PIF3 and PIF4 members of the PIL family play a role in phyB-mediated red-light signalling (Fairchild, *et al.*, 2000; Huq & Quail, 2002). This might be reasonable if TOC1 integrates red-light signalling to the clock (phyB would be activated, bind PIF3 and/or PIF4; this complex would interact with TOC1 to induce *LHY* and *CCA1* expression). However, *PIF3* anti-sense lines do not have detectable circadian phenotypes (Oda *et al.*, 2004). In addition, *pif3* null mutants do not have alterations in *LHY* nor *PRR9* mRNA accumulation (Monte *et al.*, 2004). The *pil1* null mutant also has no phenotype in circadian-controlled gene expression (Yamashino *et al.*, 2003). The gene expression levels of *PIL4* and *PIL6* genes are inducible by light and the expression of these genes is under clock control with a morning peak. The mRNA levels of these *PIL4* and *PIL6* genes are reduced in *TOC1*-ox. However, a *pil6* mutant also does not have clock phenotypes. *pil6-1* studies have supported a negative role for this factor in red light-mediated photomorphogenesis (Fujimori *et al.*, 2004). PIL5 is also known as a phytochrome-binding protein that regulates seed germination (Oh *et al.*, 2004), perhaps excluding this factor from clock function. To date, none of the *PIF3* or *PIL* genes have been reported to regulate the induction of the *CCA1* and *LHY*. *PIF3* and *PIL* family seem to function in phytochrome signalling rather than in the circadian system. Why these bHLH transcription factors bind to the PRRs is not clear.

2.8.2   ZTL family

TOC1 interacts with LKP2 and ZTL (Más *et al.*, 2003b; Yasuhara *et al.*, 2004). As described above, ZTL regulates the protein levels of TOC1 via protein degradation. As alterations in *LKP2* levels also generate circadian phenotypes, LKP2 may function in the similar pathway to degrade TOC1. LKP2 also interacts with PRR5, in addition to TOC1/PRR1, but does not appear to bind to other PRRs (Yasuhara *et al.*,

2004). The binding of LKP2 is only observed for TOC1/PRR1 and for PRR5. It is interesting to note that only TOC1 and PRR5 function to repress the period of the clock. Perhaps this is because LKP2 functions in a similar way as ZTL. This suggestion of protein degradation is consistent with known phenotypes in mutations of these genes. Analysis of a loss-of-function alleles in *lkp2* will assist such a hypothesis.

2.8.3   Other interactions

In yeast two-hybrid analysis, PRR3 associated with the protein kinase, WNK1 (Murakami-Kojima *et al.*, 2002). The kinase domain of WNK1 has high similarity to those of CTR1/EDR1 (Raf)-family of MAPKKKs. CTR1 has been shown to function as a negative regulator of the ethylene-response pathway (Kieber *et al.*, 1993). In *in-vitro* phosphorylation experiments, PRR3 is phosphorylated by WNK1. The *WNK1* gene is transcribed in a circadian manner with a peak phase similar to that of *PRR3*. WNK1 also interacts with PRR5. It is not known if PRR5 is phosphorylated by WNK1. There are nine known members of WNK1 family in Arabidopsis. *WNK2*, *WNK4* and *WNK6* are also circadian regulated (Nakamichi *et al.*, 2002). The molecular-genetic effects of the *WNK* family have yet to be reported.

The *TOC1/PRR1* gene has also been cloned as ABI3 (ABA insensitive 3)-interacting protein 1 (Kurup *et al.*, 2000). Restated, this means that TOC1 protein interacts with ABI3, a factor involved in ABA signalling. ABA is a known phytohormone that regulates the stress response, such as responses to drought and cold. TOC1 may also play a role for the integration of ABA signalling to the clock. Functional analysis of ABI3 with regard to the circadian system remains.

2.9   A brief overview of PRRs in other plants

Recently, five *PRR* orthologues have been identified from rice (*OsPRRs*) (Murakami *et al.*, 2003). In a phylogenetic tree, *OsPRR1* is the *TOC1/PRR1* orthologues, *OsPRR59* and *OsPRR95* are on the same branch as *PRR5* and *PRR9*, and *OsPRR37* and *OsPRR73* cluster together with *PRR7* and *PRR3*. All five *OsPRRs* are expressed in a rhythmic expression pattern, perhaps suggesting that all higher plants use five *PRR* genes in their clock. *OsPRR73* and *OsPRR37* have a circadian peak around the middle of the day. The mRNA of *OsPRR95* and *OsPRR59* and *OsPRR1* accumulate later in the evening. However, none of the *OsPRR* genes are induced by light nor are any expressed around morning. Recall that PRR9 in Arabidopsis is expressed at dawn. Interestingly, *OsPRR73* and *OsPRR37* expression under short-day conditions seems 'spiky.' This could imply transcript repression in darkness. The difference of expression patterns between the long-day plant Arabidopsis and the short-day plant rice could suggest that *PRR* genes function in photoperiodism, such as sensing the diurnal light environment, or perhaps, these genes function as outputs from the clock in the flowering pathway. In rice, *Early heading date1* (*Ehd1*) has been isolated as a flowering-time QTL. This gene confers early induction of flowering in short days (Doi *et al.*, 2004). *Ehd1* encodes a B-type response regulator, which consists of an authentic response receiver domain and GARP DNA-binding motif. The *Ehd1* shows diurnal expression pattern with dual peaks in the late night and in middle of the day in short day conditions. Although *Ehd1* functions in induction of *FT*-like gene expression, *Ehd1* does not seem to affect the circadian system (Doi *et al.*, 2004). This gene has high similarities to *ARR12* in Arabidopsis. A function for Arabidopsis *ARR12* in circadian control or floral-induction has not been reported.

Lower plants and algae lack obvious *PRRs* (Mittag *et al.*, 2005). Where did the *PRR* genes in higher plants come from evolutionarily? Some histidine kinases in bacteria contain pseudo-response receivers, as described above. Probably, higher plant *PRR* genes evolved from two-hybrid sensor histidine kinases originally from bacteria, perhaps cyanobacteria. From there a further division of function of these hybrid kinases led to a separation of *AHKs*, phytochromes and *PRRs*.

2.10   Perspective

The TOC1/PRR family plays important roles within the circadian system. The CCA1/LHY-TOC1/PRR1 feedback loop has been proposed to function as a central oscillator (Fig. 2.7A) (Alabadí *et al.*, 2001; Eriksson & Millar, 2003). Recently, evidence for CCA1/LHY-PRR9/PRR7 positive regulation has also been considerable to compensate the TOC1/PRR1 feedbacks (Farré *et al.*, 2005). However, the CCA1/LHY-TOC1/PRR family feedbacks are still unclear. In *TOC1*-ox, the transcripts of *LHY* and *CCA1* still accumulate rhythmically, albeit with a low amplitude (Makino *et al.*, 2002). Though the expression of *TOC1* damps rapidly in constant dark, *LHY* and *CCA1* still oscillate rhythmically at low levels (Makino *et al.*, 2000; Doyle *et al.*, 2002). Even in the strong allele *toc1-2* or *TOC1*-RNA interference plants (*toc1-R*), *LHY* and *CCA1* are still rhythmic under constant light (Alabadí *et al.*, 2001). The rhythmic expression of *CAB2* still remains under constant blue or white-light conditions (Más *et al.*, 2003a). In these genotypes grown under blue light, other factors must compensate the feedback model. In constant darkness, *LHY, CCA1* and *TOC1* expression is damped. So how does a robust rhythm of many genes, such as *CCR2* and *CAT3*, continue? Even though *PRR9* is strongly repressed in wild type grown in constant darkness, where do the strong clock phenotypes of *prr9prr7* double mutant come from? These questions point to multiple-feedback loops waiting to be discovered (Fig. 2.7B).

CCA1-ox loses rhythmic mRNA accumulation from all five *PRR* genes (Matsushika *et al.*, 2002b). In *CCA1*-ox, *TOC1* is expressed constitutively and with a low amplitude. In plants overproducing both CCA1 and PRR5, *TOC1* clearly oscillates (Fujimori *et al.*, 2005). The expression level of *PRR9* is not induced drastically in *CCA1*-ox (Matsushika *et al.*, 2002a), which is not consistent with the model that CCA1 and LHY positively regulate the expression of *PRR9* and *PRR7* (Farré *et al.*, 2005). These inconsistencies can be explained if other MYB-related transcription

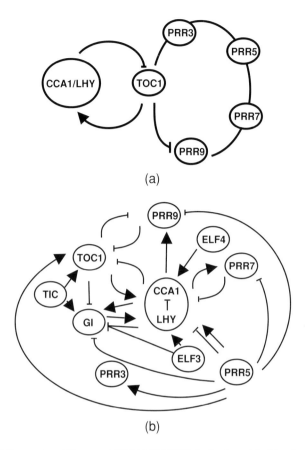

Figure 2.7 (a) One version of the current CCA1/LHY-TOC1 feedback model of the circadian-clock system in Arabidopsis (edited from Eriksson and Millar, 2003). (b) One possible multiple-feedback model. Each arrow is predicted from the gene expressions in respective mutants and overexpressers. The strengths of genetic interaction are more or less, and are not indicated. An arrow indicates that a given gene positively regulates its target. A stopped arrow indicates negative regulation of the target.

factors, such as *EARLY-PHYTOCHROME-RESPONSIVE 1* (*EPR1*) (Kuno *et al.*, 2003), function redundantly in the expression of *PRR* genes.

The expression of *LHY* and *CCA1* does not correctly correspond to TOC1 protein accumulation. There is also no proof that TOC1 protein acts on the *LHY* or *CCA1* promoter. These facts suggest that other factors must function in the induction and repression of *LHY* and *CCA1* transcription. Candidate genes include *PHYTOCHROME-INTERACTING FACTOR 3* (*PIF3*) and *PIF3-like* (*PIL*) family (Ni *et al.*, 1998; Martinez-Garcia *et al.*, 2000; Yamashino *et al.*, 2003). However, there is no evidence that the PIF3/PIL family contributes to the circadian system. These factors could have a specific function on the integration between TOC1 and phytochrome signalling, but not on the circadian system. Or, perhaps, redundancy masks a lack of clock phenotype in *pif3*. Analysis of other *PIL* genes and multiple mutations of this family should lend an answer to this ongoing question regarding TOC1 induction of *LHY* and *CCA1*. There is also a time lag between *TOC1* mRNA and protein accumulation. This could be explained by the protein degradation of TOC1 targeted by ZTL (Más *et al.*, 2003b). However, even though TOC1 protein is quite stable in the *ztl-1* mutant, it is curious that a rhythm of *TOC1* gene expression persists. How can *ztl*, which mimics *TOC1*-ox at the TOC1 expression level, be rhythmic?

2.10.1 Limitations of the current model

Further confusions exist in the current feedback model (Eriksson & Millar, 2003). In *cca1* or *lhy-12* single mutants, the phases of the evening genes *GI* and *TOC1* are shifted resulting in early expression. In the *cca1 lhy-12* double mutant, *TOC1* expression level is high and fairly constitutive, and *GI* mRNA levels are low (Mizoguchi *et al.*, 2002). In *TOC1*-ox, mRNA accumulation of *GI* is also reduced (Makino *et al.*, 2002). In *gi*, *LHY* and *CCA1* levels are low (Park *et al.*, 1999; Fowler *et al.*, 1999). Interestingly, even though *CCA1* and *LHY* are repressed in *gi* mutants, *GI* gene expression still keeps a robust rhythm in constant dark (Park *et al.*, 1999). Cotyledon-movement rhythms are robust in *gi* (Tseng *et al.*, 2004). Are these rhythms present in *gi* independent from CCA1 and LHY? Even though TOC1 induces *CCA1* and *LHY* gene expression in the model, TOC1 represses the gene expression of *GI* known to induce *CCA1* and *LHY* gene expression (Park *et al.*, 1999). If TOC1 repressed *GI* expression, why are *CCA1* and *LHY* transcribed? Where should we put GI in the model? The flowering phenotypes of the *prr* genotypes, such as *prr5* and *prr7prr5*, are also inconsistent with the current flowering model of photoperiodic induction. Even though mRNA level of *LHY* and *CCA1* are low and though the *GI* phase is advanced, why do these lines flower late?

How does ELF3 fit into the confusions? The *elf3* mutation disrupts circadian rhythms under constant light conditions, but not in constant darkness (Hicks *et al.*, 1996). The *ELF3* gene is supposed to play a role in the circadian-gating mechanism (McWatters *et al.*, 2000). In *elf3*, *GI* gene expression is arrhythmic (constitutively high) in constant light, which may be caused by the gating disruption (Park *et al.*, 1999). In contrast, *LHY* expression is rapidly repressed in *elf3* in constant light (Schaffer *et al.*, 1998). In the *lhy* mutant overproducing *LHY*, *GI* and *CCA1* gene expression are repressed. In *CCA1*-ox, *GI* mRNA accumulates to a high level. In *gi-3*, both *CCA1* and *LHY* are repressed (Fowler *et al.*, 1999). Taken together, ELF3 activates *LHY* gene expression and represses *GI* gene expression (probably via LHY). LHY also represses *CCA1*. CCA1 activates *GI* gene expression. GI activates *LHY* and *CCA1* gene expression. Curiously, *lhy* overproducing LHY abolishes the circadian rhythms of various outputs such as *GI*, *CCA1* and *CCR2* gene expression and leaf movement, but does not disrupt the robust rhythm of *ELF3* gene expression (Hicks *et al.*, 2001). Is the rhythm of *ELF3* gene expression regulated by an oscillator independent of *LHY* function?

Where do we place ELF3, ELF4 and TIC in the feedback model? In *elf3-1* and *tic*, *TOC1* expression is constitutively high or at least averages to a normal level, as expected from *LHY* and *CCA1* repression (Alabadí *et al.*, 2001; Hall *et al.*, 2003). However, maximal *TOC1* RNA levels only accumulate to half of that seen in wild type (Hall *et al.*, 2003). In *tic*, *TOC1* and *GI* transcripts are also low even though *CCA1* and *LHY* accumulate to levels similar to that seen in wild type. In the *tic elf3* double mutant, both *CCA1* and *TOC1* expression rhythms have low amplitude, which causes an arrhythmic phenotype. TIC and ELF3 may thus conduct the LHY/CCA1-TOC1 feedback loop. In addition, the *elf4* mutation has a defect in clock accuracy and disrupts rhythmic expression of *CCA1* (Doyle *et al.*, 2002). However, the expression of *PRR* genes is unknown. Analyses of *PRR* genes in *elf3*, *elf4* and *tic* mutants should be carried out in these lines.

Mutations in each of the *TOC1/PRR* family members, or overexpression of any of these genes, alters circadian period (Table 2.1) (Matsushika *et al.*, 2002a; Sato *et al.*, 2002; Eriksson *et al.*, 2003b; Yamamoto *et al.*, 2003; Ito *et al.*, 2003; Murakami *et al.*, 2004; Michael *et al.*, 2003; Farré *et al.*, 2005; Salomé & McClung, 2005). The *TOC1/PRR* family affects the circadian system in distinct ways, and probably at distinct times of the day. Interestingly, the morning genes *PRR9* and *PRR7* advance the pace of clock, but *PRR5*, *PRR3* and *TOC1/PRR1* delay it. The *PRRs* may compose the multiple oscillators (Fig. 2.7B). How this would be integrated is not clear from present data. *TOC1/PRR1* is unusual amongst the *PRR* gene family. Mutations in the other four lead to moderate effects, compared to the strong effect seen in *toc1* mutants.

The current CCA1/LHY-TOC1 feedback model seems to be too simplistic to account for the clock system. In contrast to the feedback models presented for other organisms, no fully characterized null mutant of Arabidopsis completely disrupts robust rhythms of all output markers. Multiple oscillators can be suggested with the redundancy of clock genes, such as *TOC1/PRR* family and *CCA1/LHY* family. In such a scenario, the proposal for a *TOC1, CCA1/LHY* feedback loop would remain, and multiple loops would be added to this causing feedback regulation overlaying feedback regulation.

In plants, multiple oscillators can be detected at various physiological levels. Plants have distinct clocks in different cell layers and tissues. These share components, and probably these differences reflect the non-autonomy of the plant clock (Thain *et al.*, 2000, 2002). But this is yet one more complexity in deciphering the plant clock. In addition, at least two different circadian clocks, distinguishable by their sensitivity to environmental light or temperature signals, regulate the transcription of genes (Michael *et al.*, 2003a). So, multiple cells have multiple clocks with multiple interconnecting feedback loops. Why is this all so complicated? In an organism hypersensitive to environmental perception, which is clearly the case for higher land plants, such a complex interconnecting loop seems plausible, if not likely.

Other possible functions for the *TOC1/PRR* family exist in the integration of environmental input to the clock. For example, TOC1 interacts with ABI3. Does TOC1 have a role in ABA signalling, or does ABI3 mediate ABA inputs to the clock? This is unknown. Additionally, *PRR9* and *PRR5* function in the circadian system under blue light rather than red. Do *PRR9* and *PRR5* mediate the light input to the clock? *PRR9*, *PRR7* and *PRR5* fail to entrain to thermocycles. Do these *PRRs* play a role in resetting by the temperature rather than in central oscillation? Like *CikA* in cyanobacteria, the *TOC1/PRR* family may also integrate the environmental signals to the circadian system. Here, the balance of PRR proteins would fine-tune the circadian system.

Plants are sessile organisms always exposed to ever-changing environments. This is true at the level of an individual, and over population times that occur during species migration. For this, circadian fitness can be important for plants to adapt to their environments (Green *et al.*, 2002). In fact, *PRR7* and *TOC1/PRR1* are located in QTL (Swarup *et al.*, 1999; Michael *et al.*, 2003b). This could be interpreted as that allelic variation has been captured over micro-evolutionary times to improve population survival. These loci might act in Arabidopsis as a primary source of natural variation, allowing modest complementary positive and negative effects on the circadian system to enhance fitness in local environments (Michael *et al.*, 2003). The redundancy of clock genes in plants can be explained by their adaptation to environments during land plants evolution. Many functions of the *TOC1/PRR* family are still unclear. Particularly, is *TOC1* (and each member of *TOC1/PRR* family) really a component of central oscillators? Or, are *TOC1/PRR* family connecting input signal to the circadian system, or are they composing outer-feedbacks? However, it is clear that these *PRRs* are the genes that 'tell the time of day' in the fashion of Linnaeus' flower clock. Studies on the *PRRs* should thus 'open us up' to the mechanism(s) of circadian clocks in plants.

Acknowledgements

Many thanks to Malgorzata Domagalska and Elsebeth Kolmos for proof reading and for helpful comments, and to Harriet McWatters for her editorial assistance. We also would like to thank Anthony Hall and Harriet McWatters for the opportunity, and the time, to write this chapter.

References

Alabadí, D., Oyama, T., Yanovsky, M.J., Harmon, F.G., Más, P. & Kay S.A. (2001) Reciprocal regulation between *TOC1* and *LHY/CCA1* within the Arabidopsis circadian clock. *Science*, **293**, 880–883.
Alabadí, D., Yanovsky, M.J., Más, P., Harmer, S.L. & Kay, S.A. (2002) Critical role for CCA1 and LHY in maintaining circadian rhythmicity in Arabidopsis. *Curr. Biol.*, **12**, 757–761.
Allada, R., White, N.E., So, W.V., Hall, J.C. & Rosbash, M. (1998) A mutant Drosophila homolog of mammalian Clock disrupts circadian rhythms and transcription of period and timeless. *Cell*, **93**, 791–804.
Anderson, S.L., Somers, D.E., Millar, A.J., Hanson, K., Chory, J. & Kay, S.A. (1997) Attenuation of phytochrome A and B signaling pathways by the Arabidopsis circadian clock. *Plant Cell*, **9**, 1727–1743.

Arabidopsis Genome Initiative. (2000) Analysis of the genome sequence of the flowering plant *Arabidopsis thaliana. Nature*, **408**, 796–815.

Bargiello, T.A., Jackson, F.R. & Young, M.W. (1985) Restoration of circadian behavioral rhythms by gene transfer in Drosophila. *Nature*, **312**, 752–754.

Brandstatter, I. & Kieber, J.J. (1998) Two genes with similarity to bacterial response regulators are rapidly and specifically induced by cytokinin in Arabidopsis. *Plant Cell*, **10**, 1009–1019.

Doi, K., Izawa, T., Fuse, T., Yamanouchi, U., Kubo, T., Shimatani, Z., Yano, M. & Yoshimura, A. (2004) *Ehd1*, a B-type response regulator in rice, confers short-day promotion of flowering and controls FT-like gene expression independently of Hd1. *Genes Dev.*, **18**, 926–936.

Dowson-Day, M.J. & Millar, A.J. (1999) Circadian dysfunction causes aberrant hypocotyl elongation patterns in Arabidopsis. *Plant J.*, **17**, 63–71.

Doyle, M.R., Davis, S.J., Bastow, R.M., McWatters, H.G., Kozma-Bognár, L., Nagy, F., Millar, A.J. & Amasino, R.M. (2002) The *ELF4* gene controls circadian rhythms and flowering time in *Arabidopsis thaliana. Nature*, **419**, 74–77.

Dunlap, J.C. (1996) Genetics and molecular analysis of circadian rhythms. *Annu. Rev. Genet.*, **30**, 579–601.

Eriksson, M.E. & Millar, A.J. (2003) The circadian clock. A plant's best friend in a spinning world. *Plant Physiol.*, **132**, 732–738.

Eriksson, M.E., Hanano, S., Southern, M.M., Hall, A. & Millar, A.J. (2003) Response regulator homologues have complementary, light-dependent functions in the Arabidopsis circadian clock. *Planta*, **218**, 159–162.

Fairchild, C.D., Schumaker, M.A. & Quail, P.H. (2000) HFR1 encodes an atypical bHLH protein that acts in phytochrome A signal transduction. *Genes Dev.*, **14**, 2377–2391.

Farré, E.M., Harmer, S.L., Harmon, F.G., Yanovsky, M.J. & Kay, S.A. (2005) Overlapping and distinct roles of PRR7 and PRR9 in the Arabidopsis circadian clock. *Curr. Biol.*, **15**, 47–54.

Feldman, J.F. & Hoyle, M.N. (1973) Isolation of circadian clock mutants of Neurospora crassa. *Genetics*, **75**, 605–613.

Fowler, S., Lee, K., Onouchi, H., Samach, A., Richardson, K., Morris, B., Coupland, G. & Putterill, J. (1999) *GIGANTEA*: a circadian clock-controlled gene that regulates photoperiodic flowering in *Arabidopsis* and encodes a protein with several possible membrane-spanning domains. *EMBO J.*, **18**, 4679–4688.

Fujimori, T., Sato, E., Yamashino, T. & Mizuno, T. (2005) PRR5 (PSEUDO-RESPONSE REGULATOR 5) Plays Antagonistic Roles to CCA1 (CIRCADIAN CLOCK-ASSOCIATED 1) in *Arabidopsis thaliana. Biosci. Biotechnol. Biochem.*, **69**, 426–430.

Fujimori, T., Yamashino, T., Kato, T. & Mizuno, T. (2004) Circadian-controlled basic/helix-loop-helix factor, PIL6, implicated in light-signal transduction in *Arabidopsis thaliana. Plant Cell Physiol.*, **45**, 1078–1086.

Green, R.M. & Tobin, E.M. (1999) Loss of the circadian clock-associated protein 1 in Arabidopsis results in altered clock-regulated gene expression. *Proc. Natl. Acad. Sci. USA*, **96**, 4176–4179.

Green, R.M., Tingay, S., Wang, Z.Y. & Tobin, E.M. (2002) Circadian rhythms confer a higher level of fitness to Arabidopsis plants. *Plant Physiol.*, **129**, 576–584.

Hall, A., Bastow, R.M., Davis, S.J., Hanano, S., McWatters, H.G., Hibberd, V., Doyle, M.R., Sung, S., Halliday, K.J., Amasino, R.M. & Millar, A.J. (2003) The *TIME FOR COFFEE* gene maintains the amplitude and timing of Arabidopsis circadian clocks. *Plant Cell*, **15**, 2719–2729.

Hanano, S., Domagalska, M.A., Davis, S.J. (2005) Phytohormone Symphony regulater the plant circadian clock, unpublished research.

Harmer, S.L., Hogenesch, J.B., Straume, M., Chang, H.S., Han, B., Zhu, T., Wang, X., Kreps, J.A. & Kay, S.A. (2000) Orchestrated transcription of key pathways in Arabidopsis by the circadian clock. *Science*, **290**, 2110–2113.

Hayama, R. & Coupland, G. (2003) Shedding light on the circadian clock and the photoperiodic control of flowering. *Curr. Opin. Plant Biol.*, **6**, 13–19.

Hicks, K.A., Millar, A.J., Carré, I.A., Somers, D.E., Straume, M., Meeks-Wagner, D.R. & Kay, S.A. (1996) Conditional circadian dysfunction of the Arabidopsis early-flowering 3 mutant. *Science*, **274**, 790–792.

Hicks, K.A., Albertson, T.M. & Wagner, D.R. (2001) Early flowering3 encodes a novel protein that regulates circadian clock function and flowering in Arabidopsis. *Plant Cell*, **13**, 1281–1292.

Higuchi, M., Pischke, M.S., Mähönen, A.P., Miyawaki, K., Hashimoto, Y., Seki, M., Kobayashi, M., Shinozaki, K., Kato, T., Tabata, S., Helariutta, Y., Sussman, M.R. & Kakimoto, T. (2004) *In planta* functions of the Arabidopsis cytokinin receptor family. *Proc. Natl. Acad. Sci. USA*, **101**, 8821–8826.

Hua, J. & Meyerowitz, E.M. (1998) Ethylene responses are negatively regulated by a receptor gene family in *Arabidopsis thaliana. Cell*, **94**, 261–271.

Huq, E. & Quail, P.H. (2002) PIF4, a phytochrome-interacting bHLH factor, functions as a negative regulator of phytochrome B signaling in Arabidopsis. *EMBO J.*, **21**, 2441–2450.

Hwang, I., Chen, H.C. & Sheen, J. (2002) Two-component signal transduction pathways in Arabidopsis. *Plant Physiol.*, **129**, 500–515.

Hwang, I. & Sheen, J. (2001) Two-component circuitry in Arabidopsis cytokinin signal transduction. *Nature*, **413**, 383–389.

Imaizumi, T., Tran, H.G., Swartz, T.E., Briggs, W.R. & Kay, S.A. (2003) FKF1 is essential for photoperiodic-specific light signaling in Arabidopsis. *Nature*, **426**, 302–306.

Ishiura, M., Kutsuna, S., Aoki, S., Iwasaki, H., Andersson, C.R., Tanabe, A., Golden, S.S., Johnson, C. H. & Kondo, T. (1998) Expression of a gene cluster *kaiABC* as a circadian feedback process in cyanobacteria. *Science*, **281**, 1519–1523.

Ito, S., Matsushika, A, Yamada, H, Sato, S., Kato, T., Tabata, S., Yamashino, T. & Mizuno, T. (2003) Characterization of the APRR9 pseudo-response regulator belonging to the APRR1/TOC1 quintet in *Arabidopsis thaliana. Plant Cell Physiol.*, **44**, 1237–1245.

Ito, S., Nakamichi, N., Matsushika, A., Fujimori, T., Yamashino, T. & Mizuno, T. (2005) Molecular dissection of the promoter of the light-induced and circadian-controlled *APRR9* gene encoding a clock-associated component of *Arabidopsis thaliana. Biosci. Biotechnol. Biochem.*, **69**, 382–390.

Iwasaki, H. & Kondo, T. (2004) Circadian timing mechanism in the prokaryotic clock system of cyanobacteria. *J. Biol. Rhythms*, **19**, 436–444.

Iwasaki, H., Williams, S.B., Kitayama, Y., Ishiura, M., Golden, S.S. & Kondo, T. (2000) A kaiC-interacting sensory histidine kinase, SasA, necessary to sustain robust circadian oscillation in cyanobacteria. *Cell*, **101**, 223–233.

Johnson, C.H., Knight, M., Trewavas, A., & Kondo, T. (1998) A clockwork green: circadian programs in photosynthetic organisms. In: *Biological Rhythms and Photoperiodism in Plants* (eds P. J. Lumsden & A.J. Millar), pp. 1–34, BIOS Scientific Publishers Limited, Oxford.

Kaczorowski, K.A. & Quail, P.H. (2003) Arabidopsis pseudo-response regulator 7 is a signaling intermediate in phytochrome-regulated seedling deetiolation and phasing of the circadian clock. *Plant Cell*, **15**, 2654–2665.

Kakimoto, T. (1996) CKI1, a histidine kinase homolog implicated in cytokinin signal transduction. *Science*, **274**, 982–985.

Kakimoto, T. (2003) Perception and signal transduction of cytokinins. *Annu. Rev. Plant Biol.*, **54**, 605–627.

Kiba, T., Yamada, H. & Mizuno, T. (2002) Characterization of the ARR15 and ARR16 response regulators with special reference to the cytokinin signaling pathway mediated by the AHK4 histidine kinase in roots of *Arabidopsis thaliana. Plant Cell Physiol.*, **43**, 1059–1066.

Kieber, J.J., Rothenberg, M., Roman, G., Feldmann, K.A. & Ecker, J.R. (1993) *CTR1*, a negative regulator of the ethylene response pathway in Arabidopsis, encodes a member of the raf family of protein kinases. *Cell*, **72**, 427–441.

King, D.P., Zhao, Y., Sangoram, A.M., Wilsbacher, L.D., Tanaka, M., Antoch, M.P., Steeves, T.D., Vitaterna, M.H., Kornhauser, J.M., Lowrey, P.L., Turek, F.W. & Takahashi, J.S. (1997) Positional cloning of the mouse circadian clock gene. *Cell*, **89**, 641–653.

Kondo, T., Tsinoremas, N.F., Golden, S.S., Johnson, C.H., Kutsuna, S. & Ishiura, M. (1994) Circadian clock mutants of cyanobacteria. *Science*, **266**, 1233–1236.

Konopka, R.J. & Benzer, S. (1971) Clock mutants of Drosophila melanogaster. *Proc. Natl. Acad. Sci. USA*, **68**, 2112–2116.

Krall, L. & Reed, J.W. (2000) The histidine kinase-related domain participates in phytochrome B function but is dispensable. *Proc. Natl. Acad. Sci. USA*, **97**, 8169–8174.

Kuno, N., Møller, S.G., Shinomura, T., Xu, X., Chua, N.H., Furuya, M. (2003) The novel MYB protein early-phytocrome-responsive1 is a component of a slave circadian oscillator in Arabidopsis. *Plant Cell*, **15**, 2476–2488.

Kurup, S., Jones, H.D. & Holdsworth, M.J. (2000) Interactions of the developmental regulator ABI3 with proteins identified from developing Arabidopsis seeds. *Plant J.*, **21**, 143–155.

Linnaeus, C. (1751) *Philosophia Botanica*. Godofr. Kiesewetter, Stockholm.

Makino, S., Kiba, T., Imamura, A., Hanaki, N., Nakamura, A., Suzuki, T., Taniguchi, M., Ueguchi, C., Sugiyama, T. & Mizuno, T. (2000) Genes encoding pseudo-response regulators: insight into His-to-Asp phosphorelay and circadian rhythm in *Arabidopsis thaliana*. *Plant Cell Physiol.*, **41**, 791–803.

Makino, S., Matsushika, A., Kojima, M., Oda, Y. & Mizuno, T. (2001) Light response of the circadian waves of the APRR1/TOC1 quintet: when does the quintet start singing rhythmically in Arabidopsis? *Plant Cell Physiol.*, **42**, 334–339.

Makino, S., Matsushika, A., Kojima, M., Yamashino, T. & Mizuno, T. (2002) The APRR1/TOC1 quintet implicated in circadian rhythms of *Arabidopsis thaliana*: I. Characterization with APRR1-overexpressing plants. *Plant Cell Physiol.*, **43**, 58–69.

Martinez-Garcia, J.F., Huq, E. & Quail, P.H. (2000) Direct targeting of light signals to a promoter element-bound transcription factor. *Science*, **288**, 859–863.

Más, P., Alabadí, D., Yanovsky, M.J., Oyama, T. & Kay, S.A. (2003a) Dual role of TOC1 in the control of circadian and photomorphogenic responses in Arabidopsis. *Plant Cell*, **15**, 223–236.

Más, P., Kim, W.Y., Somers, D.E. & Kay, S.A. (2003b) Targeted degradation of TOC1 by ZTL modulates circadian function in *Arabidopsis thaliana*. *Nature*, **426**, 567–570.

Matsushika, A., Imamura, A., Yamashino, T. & Mizuno, T. (2002a) Aberrant expression of the light-inducible and circadian-regulated *APRR9* gene belonging to the circadian-associated APRR1/TOC1 quintet results in the phenotype of early flowering in *Arabidopsis thaliana*. *Plant Cell Physiol.*, **43**, 833–843.

Matsushika, A., Makino, S., Kojima, M. & Mizuno, T. (2000) Circadian waves of expression of the APRR1/TOC1 family of pseudo-response regulators in *Arabidopsis thaliana*: insight into the plant circadian clock. *Plant Cell Physiol.*, **41**, 1002–12.

Matsushika, A., Makino, S., Kojima, M., Yamashino, T. & Mizuno, T. (2002b) The APRR1/TOC1 quintet implicated in circadian rhythms of *Arabidopsis thaliana*: II. Characterization with CCA1-overexpressing plants. *Plant Cell Physiol.*, **43**, 118–122.

McClung, C.R., Fox, B.A. & Dunlap, J.C. (1989) The Neurospora clock gene frequency shares a sequence element with the Drosophila clock gene period. *Nature*, **339**, 558–562.

McWatters, H.G., Bastow, R.M., Hall, A. & Millar, A.J. (2000) The *ELF3 zeitnehmer* regulates light signaling to the circadian clock. *Nature*, **408**, 716–720.

Michael, T.P., Salomé, P.A. & McClung, C.R. (2003a) Two Arabidopsis circadian oscillators can be distinguished by differential temperature sensitivity. *Proc. Natl. Acad. Sci. USA*, **100**, 6878–6883.

Michael, T.P., Salomé, P. A., Yu, H.J., Spencer, T.R., Sharp, E.L., McPeek, M.A., Alonso, J.M., Ecker, J.R. & McClung, C.R. (2003b) Enhanced fitness conferred by naturally occurring variation in the circadian clock. *Science*, **302**, 1049–1053.

Millar, A.J. & Kay, S.A. (1996) Integration of circadian and phototransduction pathways in the network controlling *CAB* gene transcription in Arabidopsis. *Proc. Natl. Acad. Sci. USA*, **93**, 15491–15496.

Millar, A.J., Carré, I.A., Strayer, C.A., Chua, N.H. & Kay, S.A. (1995a) Circadian clock mutants in Arabidopsis identified by luciferase imaging. *Science*, **267**, 1161–1163.

Millar, A.J., Short, S.R., Chua, N.H. & Kay, S.A. (1992) A novel circadian phenotype based on firefly luciferase expression in transgenic plants. *Plant Cell*, **4**, 1075–1087.

Millar, A.J., Straume, M., Chory, J., Chua, N.H. & Kay, S.A. (1995b) The regulation of circadian period by phototransduction pathways in Arabidopsis. *Science*, **267**, 1163–1166.

Mittag, M., Kiaulehn, S. & Johnson C.H. (2005) The circadian clock in chlamydomonas reinhardtii. What is it for? What is it similar to? *Plant Physiol.*, **137**, 399–409.

Mizoguchi, T., Wheatley, K., Hanzawa, Y., Wright, L., Mizoguchi, M., Song, H.R., Carré, I.A. & Coupland, G. (2002) *LHY* and *CCA1* are partially redundant genes required to maintain circadian rhythms in Arabidopsis. *Dev. Cell*, **2**, 629–641.

Mizuno, T. (2004) Plant response regulators implicated in signal transduction and circadian rhythm. *Curr. Opin. Plant Biol.*, **7**, 499–505.

Monte, E., Tepperman, J.M., Al-Sady, B., Kaczorowski, K.A., Alonso, J.M., Ecker, J.R., Li, X., Zhang, Y. & Quail, P.H. (2004) The phytochrome-interacting transcription factor, PIF3, acts early, selectively, and positively in light-induced chloroplast development. *Proc. Natl. Acad. Sci. USA*, **101**, 16091–16098.

Murakami, M., Ashikari, M., Miura, K., Yamashino, T. & Mizuno, T. (2003) The evolutionarily conserved *Os*PRR quintet: rice pseudo-response regulators implicated in circadian rhythm. *Plant Cell Physiol.*, **44**, 1229–1236.

Murakami, M., Yamashino, T. & Mizuno, T. (2004) Characterization of circadian-associated APRR3 pseudo-response regulator belonging to the APRR1/TOC1 quintet in *Arabidopsis thaliana*. *Plant Cell Physiol.*, **45**, 645–650.

Murakami-Kojima, M., Nakamichi, N., Yamashino, T. & Mizuno, T. (2002) The APRR3 component of the clock-associated APRR1/TOC1 quintet is phosphorylated by a novel protein kinase belonging to the WNK family, the gene for which is also transcribed rhythmically in *Arabidopsis thaliana*. *Plant Cell Physiol.*, **43**, 675–683.

Myers, M.P., Wager-Smith, K., Wesley, C.S., Young, M.W. & Sehgal, A. (1995) Positional cloning and sequence analysis of the Drosophila clock gene, *timeless*. *Science*, **270**, 805–808.

Nakamichi, N., Kita, M., Ito, S., Sato, E., Yamashino, T. & Mizuno, T. (2005) The Arabidopsis pseudo-response regulators, PRR5 and PRR7, coordinately play essential roles for circadian clock function. *Plant Cell Physiol.*, **46**, 609–619.

Nakamichi, N., Murakami-Kojima, M., Sato, E., Kishi, Y., Yamashino, T. & Mizuno T. (2002) Compilation and characterization of a novel WNK family of protein kinases in Arabiodpsis thaliana with reference to circadian rhythms. *Biosci. Biotechnol. Biochem.*, **66**, 2429–2436.

Nelson, D.C., Lasswell, J., Rogg, L.E., Cohen, M.A. & Bartel, B. (2000) *FKF1*, a clock-controlled gene that regulates the transition to flowering in Arabidopsis. *Cell*, **101**, 331–340.

Ni, M., Tepperman, J.M. & Quail, P.H. (1998) PIF3, a phytochrome-interacting factor necessary for normal photoinduced signal transduction, is a novel basic helix-loop-helix protein. *Cell*, **95**, 657–667.

Nishimura, C., Ohashi, Y., Sato, S., Kato, T., Tabata, S. & Ueguchi, C. (2004) Histidine kinase homologs that act as cytokinin receptors possess overlapping functions in the regulation of shoot and root growth in Arabidopsis. *Plant Cell*, **16**, 1365–1377.

Oda, A., Fujiwara, S., Kamada, H., Coupland, G. & Mizoguchi, T. (2004) Antisense suppression of the Arabidopsis PIF3 gene does not affect circadian rhythms but causes early flowering and increases FT expression. *FEBS Lett.*, **557**, 259–264.

Oh, E., Kim, J., Park, E., Kim, J.-I., Kang, C. & Choi, G. (2004) PIL5, a phytochrome-interacting basic helix-loop-helix protein, is a key negative regulator of seed germination in *Arabidopsis thaliana*. *Plant Cell*, **16**, 3045–3058.

Park, D.H., Somers, D.E., Kim, Y.S., Choy, Y.H., Lim, H.K., Soh, M.S., Kim, H.J., Kay, S.A. & Nam, H.G. (1999) Control of circadian rhythms and photoperiodic flowering by the Arabidopsis GIGANTEA gene. *Science*, **285**, 1579–1582.

Pischke, M.S., Jones, L.G., Otsuga, D., Fernandez, D.E., Drews, G.N. & Sussman, M.R. (2002) An Arabidopsis histidine kinase is essential for megagametogenesis. *Proc. Natl. Acad. Sci. USA*, **99**, 15800–15805.

Pittendrigh, C.S. (1993) Temporal Organization: Reflections of a Darwinian clock-watcher. *Annu. Rev. Physiol.*, **55**, 17–54.

Riechmann, J.L., Heard, J., Martin, G., Reuber, L., Jiang, C.-Z., Keddie, J., Adam, L., Pineda, O., Ratcliffe, O.J., Samaha, R.R., Creelman, R., Pilgrim, M., Broun, P., Zhang, J.Z., Ghandehari, D.,

Sherman, B.K. & Yu, G.L. (2000) Arabidopsis transcription factors: genome-wide comparative analysis among eukaryotes. *Science*, **290**, 2105–2110.

Plautz, J.D., Straume, M., Stanewsky, R., Jamison, C.F., Brandes, C., Dowse, H.B., Hall, J.C. & Kay, S.A. (1997) Quantitative analysis of Drosophila period gene transcription in living animals. *J. Biol. Rhythms*, **12**, 204–217.

Sakai, H., Honma, T., Aoyama, T., Sato, S., Kato, T., Tabata, S. & Oka, A. (2001) ARR1, a transcription factor for genes immediately responsive to cytokinins. *Science*, **294**, 1519–1521.

Salomé, P.A., Michael, T.P., Kearns, E.V., Fett-Neto, A.G., Sharrock, R.A. & McClung, C.R. (2002) The *out of phase 1* mutant defines a role for PHYB in circadian phase control in Arabidopsis. *Plant Physiol.*, **129**, 1674–1685.

Salomé, P.A. & McClung, C.R. (2004) The *Arabidopsis thaliana* clock. *J. Biol. Rhythms*, **19**, 425–435.

Salomé, P.A. & McClung, C.R. (2005) Pseudo-response regulator 7 and 9 are partially redundant genes essential for the temperature responsiveness of the Arabidopsis circadian clock. *Plant Cell*, **17**, 791–803.

Sato, E., Nakamichi, N., Yamashino, T. & Mizuno, T. (2002) Aberrant expression of the Arabidopsis circadian-regulated *APRR5* gene belonging to the APRR1/TOC1 quintet results in early flowering and hypersensitiveness to light in early photomorphogenesis. *Plant Cell Physiol.*, **43**, 1374–1385.

Schaffer, R., Ramsay, N., Samach, A., Corden, S., Putterill, J., Carré, I.A. & Coupland, G. (1998) The late elongated hypocotyl mutation of Arabidopsis disrupts circadian rhythms and the photoperiodic control of flowering. *Cell*, **93**, 1219–1229.

Schmitz, O., Katayama, M., Williams, S.B., Kondo, T. & Golden, S.S. (2000) CikA, a bacteriophytochrome that resets the cyanobacterial circadian clock. *Science*, **289**, 765–768.

Schultz, T.F. & Kay, S.A. (2003) Circadian clocks in daily and seasonal control of development. *Science*, **18**, 326–328.

Schultz, T.F., Kiyosue, T., Yanovsky, M., Wada, M. & Kay, S.A. (2001) A role for LKP2 in the circadian clock of Arabidopsis. *Plant Cell*, **13**, 2659–2670.

Sehgal, A., Price, J.L., Man, B. & Young M.W. (1994) Loss of circadian behavioral rhythms and *per* RNA oscillations in the Drosophila mutant *timeless. Science*, **263**, 1603–1606.

Somers, D.E., Devlin, P.F. & Kay, S.A. (1998a) Phytochromes and cryptochromes in the entrainment of the Arabidopsis circadian clock. *Science*, **282**, 1488–1490.

Somers, D.E., Schultz, T.F., Milnamow, M. & Kay, S.A. (2000) *ZEITLUPE* encodes a novel clock-associated PAS protein from Arabidopsis. *Cell*, **101**, 319–329.

Somers, D.E., Webb, A.A., Pearson, M. & Kay, S.A. (1998b) The short-period mutant, *toc1-1*, alters circadian clock regulation of multiple outputs throughout development in *Arabidopsis thaliana. Development*, **125**, 485–494.

Somerville, C. & Koornneef, M. (2002) A fortunate choice: the history of Arabidopsis as a model plant. *Nat. Rev. Genet.*, **3**, 883–889.

Strayer, C., Oyama, T., Schultz, T.F., Raman, R., Somers, D. E., Más, P., Panda, S., Kreps, J.A. & Kay, S.A. (2000) Cloning of the Arabidopsis clock gene *TOC1*, an autoregulatory response regulator homolog. *Science*, **289**, 768–771.

Sun, Z.S., Albrecht, U., Zhuchenko, O., Bailey, J., Eichele, G. & Lee, C.C. (1997) *RIGUI*, a putative mammalian ortholog of the Drosophila *period* gene. *Cell*, **90**, 1003–1011.

Suzuki, T., Ishikawa, K., Yamashino, T. & Mizuno, T. (2002) An Arabidopsis histidine-containing phosphotransfer (HPt) factor implicated in phosphorelay signal transduction: overexpression of AHP2 in plants results in hypersensitiveness to cytokinin. *Plant Cell Physiol.*, **43**, 123–129.

Swarup, K., Alonso-Blanco, C., Lynn, J.R., Michaels, S.D., Amasino, R.M., Koornneef, M. & Millar, A.J. (1999) Natural allelic variation identifies new genes in the Arabidopsis circadian system. *Plant J.*, **20**, 67–77.

Sweere, U., Eichenberg, K., Lohrmann, J., Mira-Rodado, V., Bäurle, I., Kudla, J., Nagy, F., Schäfer, E. & Harter, K. (2001) Interaction of the response regulator ARR4 with phytochrome B in modulating red light signaling. *Science*, **294**, 1108–1111.

Tanaka, Y., Suzuki, T., Yamashino, T. & Mizuno T. (2004) Comparative studies of the AHP histidine-containing phosphotransmitters implicated in His-to-Asp phosphorelay in *Arabidopsis thaliana*. *Biosci. Biotechnol. Biochem.*, **68**, 462–465.

Tei, H., Okamura, H., Shigeyoshi, Y., Fukuhara, C., Ozawa, R., Hirose, M. & Sakaki, Y. (1997) Circadian oscillation of a mammalian homologue of the Drosophila *period* gene. *Nature*, **389**, 512–516.

Thain, S.C., Hall, A. & Millar A.J. (2000) Functional independence of circadian clocks that regulate plant gene expression. *Curr. Biol.*, **10**, 951–956.

Thain, S.C., Murtas, G., Lynn, J.R., McGrath, R.B. & Millar, A.J. (2002) The circadian clock that controls gene expression in Arabidopsis is tissue specific. *Plant Physiol.*, **130**, 102–110.

Thomas, B. & Vince-Prue, D. (1997) *Photoperiodism in Plants* (2nd Ed.). Academic Press, New York.

To, J.P., Haberer, G., Ferreira, F.J., Deruère, J., Mason, M.G., Schaller, G.E., Alonso, J.M., Ecker, J. R. & Kieber, J.J. (2004) Type-A Arabidopsis response regulators are partially redundant negative regulators of cytokinin signaling. *Plant Cell*, **16**, 658–671.

Tseng, T.S., Salomé, P.A., McClung, C.R. & Olszewski, N.E. (2004) SPINDLY and GIGANTEA interact and act in *Arabidopsis thaliana* pathways involved in light responses, flowering, and rhythms in cotyledon movements. *Plant Cell*, **16**, 1550–1563.

Urao, T., Yakubov, B., Satoh, R., Yamaguchi-Shinozaki, K., Seki, M., Hirayama, T. & Shinozaki, K. (1999) A transmembrane hybrid-type histidine kinase in Arabidopsis functions as an osmosensor. *Plant Cell*, **11**, 1743–1754.

Vierstra, R.D. & Davis, S.J. (2000) Bacteriophytochromes: new tools for understanding phytochrome signal transduction. *Semin. Cell Dev. Biol.*, **11**, 511–521.

Wang, Z.-Y., Kenigsbuch, D., Sun, L., Harel, E., Ong, M.S. & Tobin, E.M. (1997) A Myb-related transcription factor is involved in the phytochrome regulation of an Arabidopsis *Lhcb* gene. *Plant Cell*, **9**, 491–507.

Wang, Z.-Y. & Tobin, E.M. (1998) Constitutive expression of the *CIRCADIAN CLOCK ASSOCIATED 1* (*CCA1*) gene disrupts circadian rhythms and suppresses its own expression. *Cell*, **93**, 1207–1217.

Yamamoto, Y., Sato, E., Shimizu, T., Nakamich, N., Sato, S., Kato, T., Tabata, S., Nagatani, A., Yamashino, T. & Mizuno T. (2003) Comparative genetic studies on the *APRR5* and *APRR7* genes belonging to the APRR1/TOC1 quintet implicated in circadian rhythm, control of flowering time, and early photomorphogenesis. *Plant Cell Physiol.*, **44**, 1119–1130.

Yamashino, T., Matsushika, A., Fujimori, T., Sato, S., Kato, T., Tabata, S. & Mizuno, T. (2003) A link between circadian-controlled bHLH factors and the APRR1/TOC1 quintet in *Arabidopsis thaliana*. *Plant Cell Physiol.*, **44**, 619–629.

Yanovsky, M.J. & Kay, S.A. (2002) Molecular basis of seasonal time measurement in Arabidopsis. *Nature*, **419**, 308–312.

Yasuhara, M., Mitsui, S., Hirano, H., Takanabe, R., Tokioka, Y., Ihara, N., Komatsu, A., Seki, M., Shinozaki, K. & Kiyosue, T. (2004) Identification of ASK and clock-associated proteins as molecular partners of LKP2 (LOV kelch protein 2) in Arabidopsis. *J. Exp. Bot.*, **55**, 2015–2027.

Young, M.W. (1998) The molecular control of circadian behavioral rhythms and their entrainment in Drosophila. *Annu. Rev. Biochem.*, **67**, 135–152.

Yeh, K.C., Wu, S.H., Murphy, J.T. & Lagarias, J.C. (1997) A cyanobacterial phytochrome two-component light sensory system. *Science*, **277**, 1505–1508.

Zehring, W.A., Wheeler, D.A., Reddy, P., Konopka, R.J., Kyriacou, C.P., Rosbash, M. & Hall, J.C. (1984) P-element transformation with *period* locus DNA restores rhythmicity to mutant, arrhythmic drosophila melanogaster. *Cell*, **39**, 369–376.

Zimmermann P., Hirsch-Hoffmann M., Hennig L. & Gruissem W. (2004) GENEVESTIGATOR. Arabidopsis microarray database and analysis toolbox. *Plant Physiol.*, **136**, 2621–2632.

3 Multiple and slave oscillators

Dorothee Staiger, Corinna Streitner, Fabian Rudolf
and Xi Huang

3.1 Introduction

Endogenous rhythms with a period of about a day are basic features of almost all living organisms (Dunlap, 1999; Young & Kay, 2001). The underlying pacemaker is the circadian clock, an autonomous oscillator within the cell. Through a series of biochemical reactions, it generates a self-sustained rhythm of approximately 24 hours and imposes this rhythmicity on to physiological, biochemical and molecular events. This endows an organism with the ability to anticipate recurrent changes in its surroundings and thus to adapt to the environmental cycles of day and night and the seasons of the year.

In the model plant *Arabidopsis thaliana*, the cellular oscillator consists of autoregulatory feedback loops through which clock proteins regulate transcription of their own genes (see Chapters 1 and 2). Thus, it conforms to the paradigm of the well-described oscillators in *Drosophila melanogaster*, *Neurospora crassa* and mammals (Dunlap, 1999; McClung *et al.*, 2002; Coupland, 2003). The key players of the *Arabidopsis* core clockwork, however, differ fundamentally from the known components in *D. melanogaster* and mammals on the one hand and *N. crassa* on the other. The lack of sequence conservation implies an independent evolutionary origin of plant clocks (see Chapter 10).

Apart from alternating activation and inhibition by feedback of clock gene transcription, clock proteins are subject to additional layers of control, namely phosphorylation with consequences for DNA-binding activity and, perhaps, nuclear localization, as well as regulated life-span of the proteins (Sugano *et al.*, 1999; Mas *et al.*, 2003; Daniel *et al.*, 2004). Moreover, post-transcriptional steps of mRNA processing are expected to shape the transcript profile across the circadian cycle (So & Rosbash, 1997; Edery, 1999; Staiger & Heintzen, 1999; Staiger et al., 2003a).

Regulation of output rhythms by the central clockwork in large parts relies on rhythmic gene expression. A plethora of mRNAs that undergo circadian oscillations in steady-state abundance specify transcription factors, suggesting a transcriptional regulatory network downstream of the oscillator (Harmer *et al.*, 2000; McClung, 2000; Staiger, 2002). Recently, two negative feedback loops of regulatory proteins have been identified in *Arabidopsis* operating downstream of the cell's core clockwork. These autoregulatory circuits represent the first examples of molecular slave oscillators (Rudolf *et al.*, 2004). One of them is centered around the putative transcription factor EPR1 that closely resembles the clock proteins CCA1 and LHY

(Kuno *et al.*, 2003). Thus, it may utilize a mechanism similar to that of the CCA1-LHY-TOC1 core oscillator for autoregulation and regulation of downstream targets. The key component of the second circadian feedback loop is the RNA-binding protein *AT*GRP7 (*Arabidopsis thaliana* glycine-rich RNA-binding protein, also known as *CCR2* (cold and clock regulated 2)). It affects oscillations of its own transcript as well as of a heterologous transcript at the post-transcriptional level and thus represents a mechanistically different circadian circuitry (Heintzen *et al.*, 1997; Staiger *et al.*, 2003a).

It is not clear at present how many oscillators are used to orchestrate the rhythms of higher plants. In an extreme case, it may be one for each output rhythm whereas in the most economic scenario it may be only one in charge of all rhythms. We discuss below recent work pointing to an intermediate situation of more than one clock ticking in different parts of the plant (Thain *et al.*, 2000) (Section 3.3.2). However, the distribution of labour among different oscillators and their hierarchical organization is not fully understood (Millar, 1998). To provide a framework, the first part of the chapter will review the setup of the mammalian timing system.

3.2   Central and peripheral oscillators in the mammalian timing

Very early on, Aschoff concluded from his experiments on bird and mammalian physiology and behaviour that the circadian system comprises multiple oscillators (Aschoff, 1960). He stressed the importance of endogenous coordination among these individual oscillators in the organism without the help of external 'zeitgebers' ('*timegivers*'), a feature he named 'internal synchronization'. The role of external zeitgebers in accomplishing temporal organization was largely confined to phase-setting.

It had long been accepted that the structure of the best-known circadian systems in mammals or insects was hierarchical with a pacemaker residing in the brain that beats out the rhythm for oscillating body functions. In the hypothalamus of mammals, a pair of specialized organs known as the suprachiasmatic nuclei (SCN) houses the central pacemaker. Observations indicated that most SCN neurones are equipped with a circadian oscillator: in cultured SCN explants from transgenic animals harbouring luciferase (LUC) driven by the promoter of the mouse clock gene Period1 (*mPer1*), LUC rhythms can be recorded from individual neurones (Yamaguchi *et al.*, 2003). Moreover, the SCN neurones displayed rhythms in their electrical activity even when they were dissociated in culture (Welsh *et al.*, 1995). Individual cells exhibited very different periods *ex vivo*, however. Thus, in the SCN tight coupling among these cells is required to accomplish the overall period length communicated to downstream targets.

The SCN neurones are also the targets for entrainment by environmental light processed by the eyes and transmitted via the retinohypothalamic tract. Within the SCN, information about light and dark is decoded and the phase of the central pacemaker is adjusted accordingly. Following this entrainment by external zeitgebers, internal synchronization of output rhythms takes place. Several ways of communication by the SCN neurones are known including secreted polypeptides that reach their targets by diffusion and electric signals (Albrecht, 2004; Kriegsfeld et al., 2004).

A few years ago anatomically separate SCN regions have been assigned to morning and evening electrophysiological activities (Jagota et al., 2000). Likewise, in Drosophila melanogaster it has recently been shown that morning oscillators and evening oscillators are situated in distinct neurones (Grima et al., 2004; Stoleru et al., 2004). Thus, different clocks co-exist within a tissue.

In the 1990s, evidence accumulated that peripheral tissues may indeed contain clocks capable of maintaining circadian rhythmicity to some extent in 'isolation' (Giebultowicz & Hege, 1997; Plautz et al., 1997; Whitmore et al., 1998). This was mainly demonstrated by monitoring clock gene expression in different organs, or by following cyclic LUC activity driven by a clock gene promoter in organs dissected from transgenic animals. These initial experiments revealed that circadian oscillations in the periphery damped out rapidly, and thus the underlying clocks were regarded as slave oscillators controlled by the SCN. Around the same time, Ueli Schibler's team found that rat fibroblasts, an immortalized cell line kept in culture for more than 25 years without contacts to SCN tissue, underwent circadian oscillations in clock gene expression when exposed to a serum shock and even maintained the same phase angles between the transcripts found in the SCN (Balsalobre et al., 1998).

These groundbreaking observations indicated that peripheral oscillators by and large are made up from the same clock proteins as the master oscillator in the SCN. SCN and peripheral clocks listen to different entraining cues: peripheral clocks of mammals are not directly light-sensitive but are adjusted to the day-night-cycle by information processed through the SCN. In addition, food uptake can influence peripheral oscillators directly and which indeed may be their prevalent timing cue (Yamazaki et al., 2000; Le Minh et al., 2001; Stokkan et al., 2001).

Recently, Takahashi and co-workers presented evidence that oscillations in peripheral tissues can be self-sustained (Yoo et al., 2004). This insight was possible through the use of an mPer2::LUC construct designed carefully to retain all regulatory aspects of the authentic mPer2 gene. LUC was fused to the terminal exon of the endogenous mPer2 locus through homologous recombination. In contrast to Per1 promoter fusions used before, this construct contains both proximal and distal regulatory motifs and thus its expression reflects transcriptional as well as post-transcriptional contributions to the regulation of the mPER2 protein. Peripheral tissues from mPer2::LUC knock-in mice in cultures expressed persistent LUC rhythms. Whilst, as expected, surgical SCN lesions in the animals led to arrhythmic locomotor activity; they did not inhibit LUC rhythms in cultured peripheral tissue. However, SCN lesions disrupted phase synchrony from tissue to tissue within individual animals (see Fig. 3.1). Together, these experiments imply that the SCN is not required for persistent mPER2::LUC rhythms in peripheral tissue and thus may synchronize rather than drive rhythms in the periphery (Yoo et al., 2004).

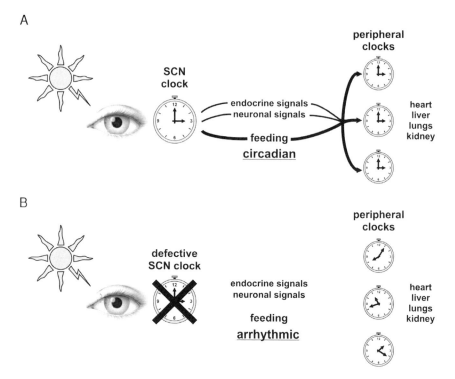

Figure 3.1 The SCN in mammals synchronizes oscillators in peripheral tissues. (A) The SCN is entrained by light. Rhythmic endocrine and neuronal output signals of the SCN clock as well as feeding are the main zeitgebers for the peripheral clocks. (B) In the case of a defective SCN clock, peripheral clocks maintain oscillations but their phases are no longer coordinated (Gachon *et al.*, 2004).

Taking this a step further, through recording of fluorescence rhythms on a single cell basis the Schibler lab uncovered self-sustained clocks in cultured fibroblasts that even continued to tick during cell division and were passed on to the next generation (Nagoshi *et al.*, 2004). Real-time bioluminescence imaging of transgenic cell lines expressing a *Bmal1*::LUC construct showed that progressive amplitude loss with time occurs because the individual molecular oscillators fall out of phase rather than actually damp. Thus, circadian oscillators in cultured fibroblasts have to be considered as robust as oscillators in the SCN.

3.3   Organization of the timing system in plants

Ample evidence suggests that timekeeping in higher plants like in animals relies on multiple clocks in different tissues directing a plethora of rhythmic outputs (Millar, 1998). The organization of the plant circadian system, however, has long been thought to stand apart from the traditional view of a hierarchical system in animals due to the lack of an obvious pacemaker in a single central place (McWatters *et al.*, 2001). The recent change in paradigm blurs the fundamental differences between the SCN oscillator and the peripheral oscillators (Section 3.1): clocks that can function autonomously in peripheral tissue and even in cultured cells rely on the SCN clock largely for synchronization rather than for maintenance of rhythms.

In this section we will review recent approaches to unravel how the circadian clock system is arranged in higher plants.

3.3.1 Are clocks localized in different parts of the plant?

It is well established that individual parts of a higher plant are equipped with clocks, because rhythms in distinct organs have been reported for several decades, many of which persist in organs detached from the intact plant (see Fig. 3.2). Most obviously, petals of numerous flowers open and close at defined times in the circadian cycle. This is caused by an underlying rhythm of cell elongation that persists, for example, in isolated *Kalanchoë* flowers (Engelmann & Johnsson, 1977). The stem of *Chenopodium rubrum* plants grows through rhythms in cell elongation with a parallel rhythm in oleic acid content (Lecharny *et al.*, 1990). Also, elongation of

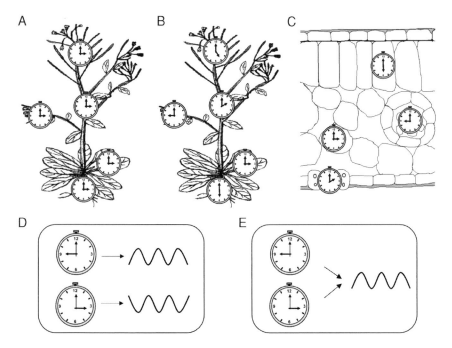

Figure 3.2 A higher plant may have multiple copies of the same clock in different organs (A), different clocks in different organs (B), or within a tissue (C), or even within a cell (D). A single output rhythm can be under the control of two separate oscillators (E).

the *Arabidopsis* hypocotyl goes through alternating phases of rapid growth at subjective dusk and daily growth arrest around subjective dawn that persist in LL. This is reflected in rhythmic changes in cotyledon angles (Engelmann *et al.*, 1992; Dowson-Day & Millar, 1999). Disruption of the morning growth arrest either in mutants defective in ELF3 or in transgenic plants constitutively overexpressing CCA1 leads to an elongated hypocotyl, indicating that this rhythm largely relies on the CCA1 oscillator (Wang & Tobin, 1998; Reed *et al.*, 2000).

Similarly, many plants in the legume and oxalis families move their leaves up and down due to rhythmic swelling and shrinking of motor cells in specialized organs, the pulvini (Satter & Galston, 1981). This is caused by ion fluxes through antagonistic rhythmic opening and closure of potassium channels and concomitant changes in membrane potential in flexor and extensor cells (Kim *et al.*, 1993). Notably, this rhythm persists in protoplasts derived from the pulvinar motor cells for several days in LL (Mayer & Fischer, 1994).

Another example for a circadian rhythm in an isolated cell is the control of stomatal movement. Stomata rhythmically open and close to allow gas exchange through the leaf surface. Guard cells in epidermal peels of *Vicia faba* retain this rhythm, suggesting that they have an independent clock (Gorton *et al.*, 1989).

The design of rhythmic molecular markers provides the possibility to probe clocks virtually in all cell types. For example, petiole explants derived from transgenic CAB::LUC tobacco plants were shown to maintain oscillations in LL (Thain *et al.*, 2002).

Moreover, the observation that *Arabidopsis* cell cultures show oscillations in transcript abundance, provides an access to study the dynamic properties of individual clocks and to probe their mechanisms pharmacologically (Nakamichi *et al.*, 2004).

3.3.2   Does the same type of clock control rhythms in distinct parts of the plant?

The persistency of circadian rhythms in tissue explants and even in single cells reveals the presence of clocks in many different parts of the plant (Section 3.3.1). Does the same type of oscillator control multiple rhythms with distinct temporal and spatial properties or are there specific clocks for a certain tissue or output (see Fig. 3.2)? Clocks can be distinguished by looking at their hands: output rhythms can only maintain different free-running periods if the underlying oscillators have different molecular characteristics. In contrast, a single clock can adjust rhythms to different phases so that processes peaking at separate times of the day may well be controlled by the same molecular oscillator.

Indeed, diverse rhythmic outputs in plants can attain deviating periods under free-running conditions. In *Phaseolus vulgaris*, leaf movement and stomata movement exhibit different periods (Hennessey & Field, 1992). Circadian rhythms with different periods including cell division in the apex and sensitivity to photoperiod in the leaves have been suggested to contribute to flower induction in *Chenopodium rubrum* (King, 1975). In tobacco, *CAB* oscillates with a period different from that of the rhythms in cytosolic free calcium ions (Sai & Johnson, 1999). *PHYB*::LUC oscillations and *CAB*::LUC oscillations free-run with slightly different period lengths in transgenic *Arabidopsis* (Hall *et al.*, 2002).

Recently, McClung's group found two clocks with differential temperature responses in *Arabidopsis* seedlings (Michael *et al.*, 2003). A clock regulating *CAB* expression does not respond well to temperature cycles while responding perfectly to light–dark cycles. It also is not influenced by the steep rise in temperature when seeds are released from stratification. In contrast, another clock regulating *CAT3* responds perfectly well to temperature cycles and the release from stratification. Furthermore, *CAB*::LUC and *CAT3*::LUC rhythms free-run with a different period after entrainment to light–dark cycles, and their phases respond differentially to photoperiods of varying length, again indicating control by distinct oscillators (Michael *et al.*, 2003). Notably, the initial characterization of *CAT3* transcript oscillations in *CCA1* overexpressors also had uncovered a differential influence of this clock component on *CAT3* and *CAB* (Wang & Tobin, 1998). *CAB* and *CAT3* show partially overlapping expression in mesophyll and epidermal cells of cotyledons. The underlying oscillators thus exist side by side in the same tissues and it will be a challenge to determine whether or not they may even be organized within the very same cell. This will require simultaneous recording of reporter proteins with distinct luminescence peaks driven by the *CAB* and *CAT3* promoters, respectively, in the same transgenic plant.

The co-existence of two independent clocks within a single cell has been discovered only once: in the unicellular dinoflagellate *Gonyaulax polyedra* (now renamed *Lingulodinium polyedrum*) two oscillators control endogenous rhythms in bioluminescence (which peaks during the subjective day) and cell aggregation (peaking at the end of the subjective night) (Roenneberg & Morse, 1993). Under constant dim red light, the aggregation rhythm runs with a shorter period than the bioluminescence rhythm. This internal desynchronization of two rhythmic outputs within a single cell pointed to the existence of an 'A' oscillator controlling aggregation and a 'B' oscillator controlling bioluminescence within the very same cell (Roenneberg & Morse, 1993).

While the presence of clocks with different properties in individual parts of the plant is well documented, it is not clear whether these clocks have a distinct molecular architecture. Mutant hunts in *Arabidopsis* have identified a suite of clock proteins. Based on their interdependent expression patterns, they are all arranged in the positive and negative limbs of interconnected feedback loops (McClung *et al.*, 2002; Coupland, 2003). Thus, there is currently no evidence for the existence of more than one basic core clockwork (see Chapter 10).

Some of the rhythms that free-run with different periods in *Arabiodpsis* have been shown to be influenced by the same clock proteins. The *toc1* (also known as *aprr1* or *prr1*) mutation shortens *CAB*::LUC rhythms to 21 h and leaf movement rhythms to 23.3 h (Millar *et al.*, 1995). Parallel measurement of stomatal conductance and CO_2 assimilation in a single *Arabidopsis* plant has shown that the *ztl-1* mutation extends the free-running period of stomatal conductance by 5 hours and the period of CO_2 assimilation by 3.5 hours, respectively (Dodd *et al.*, 2004). The circadian control of stomatal conductance and CO_2 fixation thus is uncoupled. ZTL influences the period of the clock by targeted degradation of the clock protein TOC1 (Mas *et al.*, 2003). Thus, these rhythms appear to be controlled by the same oscillator that controls *CAB* rhythms. Stomatal conductance rhythms previously had been found to have a short period in *toc1* mutant plants, implicating TOC1 function in the guard cell clock (Somers *et al.*, 1998b).

These findings imply that the same clock proteins work in different clocks. However, they do, not provide any evidence for the existence of two or more clocks constructed according to fundamentally different building plans. If the clocks share by and large the same components, tissue-specific or cell-type specific factors may endow them with distinct properties. For example, differences in photoreceptor abundance in clock-containing cells will result in differential input to the clock, and light intensity can have an effect on the pace of the clock (Somers *et al.*, 1998a). Distinct properties may also be provided by clock-associated proteins. The *gi-2* allele of Gigantea, for example, provokes different periods for *CAB*::LUC rhythms and leaf movement (Park *et al.*, 1999).

3.3.3   Do individual clocks interact with each other?

An important aspect of circadian control is the endogenous coordination of physiology and metabolism between individual parts of the plant. To achieve such an 'internal synchronization' one may envisage extensive crosstalk and reciprocal regulation among individual clocks.

Despite the lack of a humoral system, plant tissues are endowed with ways of communication over long distances, for example to report attacks by pathogens to distant parts or to transmit photoperiodic timing from the perceiving leaf to the apex. Obvious candidates for a putative communication among individual clocks would be phytohormones. Indeed, rhythmic hormone levels have been reported including oscillations of cytokinin in carrot and oscillations of auxin in rosette leaves of *Arabidopsis thaliana* (Paasch *et al.*, 1997; Jouve *et al.*, 1999). Notably, transcripts encoding the auxin efflux carriers *PIN3* and *PIN7* are rhythmically expressed (Harmer *et al.*, 2000). Ethylene rhythms were found in several plant species including *Chenopodium rubrum* and Sorghum, but they did not influence growth rhythms or expression of CAB::LUC and PHYB::LUC reporters in *Arabidopsis* (Machackova *et al.*, 1997; Finlayson *et al.*, 1998; Thain *et al.*, 2004). In another investigation gibberellin, brassinolide and abscisic acid had an influence on the period, cytokinin on the phase and auxin on the amplitude of promoter:LUC rhythms (Hanano *et al.*, 2004). Thus, rhythmicity of some phytohormones may play a role within the circadian system.

An early study showed that gibberellic acid (GA_3) phase-shifts leaf movement rhythms in *Gossypium hirsutum* (Viswanathan & Subbaraj, 1983). The availability of *Arabidopsis* mutants impaired in GA biosynthesis and responses will allow researchers to test a potential influence of GA_3 on well-defined rhythmic parameters.

The identification of melatonin, the hormone encoding photoperiodic information in animals, in several plant species was met with some enthusiasm (Kolar *et al.*, 1997). Despite the finding of a higher melatonin concentration during the night in *Chenopodium rubrum*, there is no firm indication as yet of its causal involvement in rhythms or photoperiodism.

The actual degree of autonomy of cellular clocks was determined in an elegant series of experiments by the Millar group employing molecular markers as clock readout (Thain *et al.*, 2000). When the two cotyledons of transgenic tobacco seedlings were entrained to different light–dark regimes, rhythms of *CAB*::LUC stably maintained their respective phase in a subsequent free-run in continuous light. Similarly, when the top and bottom part of a cotyledon were entrained to opposing light–dark cycles, free-running rhythms were maintained in LL, indicating that clocks in different tissues of the same organ show no extensive crosstalk.

One example has been found of two oscillators influencing a single output rhythm (see Fig. 3.2E). Accumulation of a tobacco *CAB* transcript during early development is under control of separate oscillators, only one of which is coupled to phytochrome (Kolar *et al.*, 1995). Through the use of wheat and *Arabidopsis* promoters driving LUC in transgenic tobacco, it has been shown that both oscillators act at the level of transcription (Kolar *et al.*, 1998). This is reminiscent of the 'splitting' phenomenon observed before the identification of morning and evening oscillators in mammals and *Drosophila* (Section 3.2.) (Pittendrigh & Daan, 1976). Activity rhythms in nocturnal rodents can dissociate into two components that free-run with different periods, implicating dual-oscillator control of one output.

3.3.4 Do chloroplasts retain remnants of a clock of an endosymbiontic ancestor?

Chloroplasts share a common ancestor with photoautotrophic cyanobacteria. Thus, does the cell of a higher plant contain traces of an organellar clock in addition to the nucleus-based clock? The circadian clock in prokaryotic cyanobacteria mechanistically differs from the transcription-based autoregulatory feedback loops in higher plants, mammals, *Drosophila* or *Neurospora* (see Chapter 10). Homologues to the cyanobacterial *kai* clock genes have not been identified in fully sequenced plastid genomes (Ditty *et al.*, 2003). So the question arises as to what happened to the endosymbiont's original clock?

A separate clock in plastids may have become superficial. Most *Arabidopsis* genes of cyanobacterial origin have been relocated to the host's nuclear genome. Many of them code for proteins involved in photosynthesis and, if circadian regulation occurs it is via the host's clock.

The wheat *psbD* gene is the first example of a chloroplast gene regulated by the circadian clock at the level of transcription. The promoter is recognized by an *E. coli* type RNA polymerase typical for chloroplasts. The *sigA* factor encoding a

σ factor subunit of the polymerase oscillates in phase with *psbD* mRNA, consistent with the possibility that this nuclear encoded σ factor mediates clock regulation of the chloroplast *psbD* promoter (Morikawa *et al.*, 1999).

3.4   Molecular organization of the clock within a plant cell

The molecular framework of the *Arabidopsis* central oscillator with three cycling components CCA1, LHY, TOC1 that reciprocally regulate each other conceptually is well understood (Alabadi *et al.*, 2002; Mizoguchi *et al.*, 2002) (see also Chapter 1, this volume). Further challenges are to understand the function of the numerous components closely related to either the myb-like factors CCA1 and LHY or the APRR family which includes TOC1 (Andersson *et al.*, 1999; Makino *et al.*, 2001, 2002; Matsushika *et al.*, 2002; Murakami-Kojima *et al.*, 2002; Sato *et al.*, 2002; Nakamichi *et al.*, 2003; Yamashino *et al.*, 2003; Murakami *et al.*, 2004) (see Chapter 2). Moreover, a suite of proteins has been identified contributing to rhythm generation of the core clockwork which have not yet been unambiguously placed in the current picture of the interlocking autoregulatory loops (Fowler *et al.*, 1999; Park *et al.*, 1999; Doyle *et al.*, 2002; Hall *et al.*, 2003; Staiger *et al.*, 2003b). The large-scale screening of *Arabidopsis* mutants with rhythmic LUC reporter genes by a high-throughput real-time bioluminescence monitoring system may reveal additional clock-associated components (Onai *et al.*, 2004).

Fine-tuning of the oscillator to the local environment is achieved through continuous monitoring of the day outside and integration of these signals with the day inside. Resetting adjusts the phase of the clock through altering the level of one of its components (Devlin, 2002; Fankhauser & Staiger, 2002; Millar, 2004).

3.5   Linking gene expression to the clockwork – how output genes are regulated

Apart from regulating their own expression, clock proteins must be able to signal rhythmicity to downstream targets that eventually produce overt rhythms. They may accomplish this by direct control of downstream genes or via cascades of oscillating gene products (Brown & Schibler, 1999). Regulation can occur at the level of transcription (Sections 3.5.1 and 3.5.2), RNA processing (Sections 3.5.3.1–3.5.3.3) and translation (Section 3.5.4). The identification in *Arabidopsis* of oscillating downstream genes that themselves are part of negative autoregulatory feedback loops (Section 3.6) points to the existence of a hierarchical system in the cell with molecular slave oscillators as signalling intermediates (Rudolf *et al.*, 2004).

3.5.1   Transcriptional control of output genes

For most genes that have been investigated in detail, clock regulation has been shown to reside within the promoter, pointing to clock control of transcription leading to mRNA peaks at different times of the day (Nagy *et al.*, 1988; Anderson *et al.*, 1994; Liu *et al.*, 1996; Staiger & Apel, 1999; Staiger *et al.*, 1999). The development of microarrays and Genechips allowed for a comprehensive exploration of genome-wide transcriptome data for rhythmic gene expression, pioneered in *Arabidopsis*. About 6% of the genes were found oscillating at the level of transcript abundance in seedlings (Harmer *et al.*, 2000; Schaffer *et al.*, 2001).

The proportion of oscillatory patterns based on transcriptional regulation was defined on a global basis in an *in vivo* enhancer trapping experiment (Michael & McClung, 2003). Firefly luciferase driven by a minimal promoter that itself is not subject to clock regulation was randomly inserted into the *Arabidopsis* genome and T_2 transgenic plants were assayed for LUC rhythms. Notably, 36% of the LUC-expressing plants showed circadian rhythms with periods between 22 and 28 h. This immediately suggests that clock control at the level of transcription is more prevalent than anticipated from the microarray data and that additional layers of regulation may be imposed at the level of mRNA stability (see Section 3.5.3.1).

In an extreme case, random insertion of a promoter-less luxAB gene in the cyanobacterium *Synechococcus* sp. PCC7942 had uncovered clock regulation of almost every gene (Liu *et al.*, 1995). In line with such a global transcriptional control, a σ like transcription factor has been identified as part of the clock output pathway that affects some, but not all oscillating transcripts (Tsinoremas *et al.*, 1996). It is not yet known to what extent these oscillations manifest themselves in rhythms in steady-state mRNA abundance.

Clock-controlled promoters of output genes can be direct targets for clock proteins and thus autoregulation within the core clock and regulation of downstream genes may share the same molecular basis. Repression of *TOC1* within the oscillator has been suggested to be a consequence of direct CCA1 and LHY binding to its promoter, because recombinant CCA1 and LHY in vitro bind to *TOC1* promoter fragments comprising the evening element (Alabadi *et al.*, 2001). Similarly, transcriptional activation of *LHC* output gene expression around dawn may be due to direct interaction between *cis*-regulatory promoter elements and these myb transcription factors. In fact, CCA1 originally was cloned on the basis of its binding to a CA-rich CAB promoter fragment (Wang *et al.*, 1997).

3.5.2 Identification of cycling transcription factors mediating rhythmic output

Interestingly, transcript profiling on a genome-wide basis revealed transcripts that may themselves play a regulatory role in addition to being clock outputs. Notably, several transcription factors have been identified that may convey clock-control to output targets. For example, the key regulator PAP1 (production of anthocyanin pigment), which shows high amplitude cycling in phase with 23 phenylpropanoid biosynthetic genes, is implicated in the synchronous activation of the entire pathway before dawn (Harmer *et al.*, 2000). Similarly, several cold-induced transcripts undergo circadian oscillations, reaching their highest level around 8 hours after onset

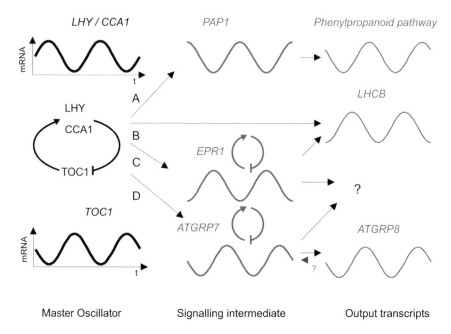

Figure 3.3 A roadmap to clock output. The 24 hour rhythm of the CCA1-LHY-TOC1 oscillator is transmitted to output genes through a rhythmic transcription factor (A) or direct interaction of the clock proteins with promoters (B) or intermediate 'slave' oscillators that act either transcriptionally (C) or post-transcriptionally (D) upon downstream genes. Reproduced with permission from The Biochemist (Rudolf *et al.*, 2004).

of light. The DREB1a/CBF3 transcription factor, that binds to promoter elements of these cold-regulated genes, peaks in the middle of the day, again suggests a role in timing of these transcripts (Harmer *et al.*, 2000).

These examples point to secondary output components mediating circadian rhythmicity within the cell in addition to direct regulation of primary output genes by clock proteins (see Fig. 3.3). A precedent for such an indirect regulation is the mammalian DBP transcription factor that oscillates with an exceptional high amplitude due to circadian control and which is at least partly generated through binding of the pacemaker protein CLOCK to E-box motifs within *Dbp* introns (Ripperger *et al.*, 2000). DBP in turn controls a set of output genes including members of the cytochrome P450 family. Transcriptional cascades of this kind provide a means to direct output rhythms governed by a single oscillator to different phases that are determined by the half-life of the regulators.

3.5.3   Post-transcriptional control

Beyond the regulation of output genes at the level of transcription, several mechanisms operate between the initiation of mRNA transcription and the translation of

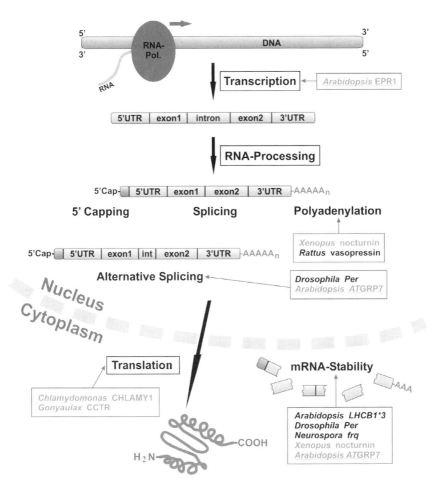

Figure 3.4 Post-transcriptional regulation in the circadian system. Examples for clock output genes or core clock genes are indicated in boxes (black: transcripts which are affected at the indicated level; grey: proteins which do affect processes at the indicated level; grey/underlined: proteins which regulate their own expression creating a negative feedback loop). For details see the text. To complete the picture, EPR1 that autoregulates its expression at the transcriptional level and CHLAMY1/CCTR that presumably are involved in clock regulation of translation are included.

the encoded protein. This post-transcriptional control affects the fate of the mRNA, including its processing, stability, transport and translation, and contributes to the phase and amplitude of its circadian oscillation (see Fig. 3.4). These mechanisms in large parts rely on RNA-binding proteins that interact with pre-mRNAs. Thus, the identification of rhythmically expressed RNA-binding proteins was not a surprise (Carpenter *et al.*, 1994; Heintzen *et al.*, 1994; Mittag *et al.*, 1994; McNeil *et al.*, 1998).

Knowledge of circadian regulation of RNA metabolism in higher plants is rather limited. In this section we will also highlight examples of post-transcriptional regulation impacting on output genes as well as the core clock genes in other organisms, as they might well be precedents for control mechanisms to be active in higher plants.

3.5.3.1 RNA stability

Often post-transcriptional regulation of circadian gene expression is not revealed at first sight. This was initially observed for the *Arabidopsis* transcripts *LHCB1*1* (*CAB2*) and *LHCB1*2* (*CAB3*) that oscillate with a circadian rhythm while the *LHCB1*3* (*CAB1*) transcript is present in nearly equal amounts at all times. Both nuclear run-on transcription assays and the analysis of promoter fusions show that *LHCB1*3* transcription is rhythmic, pointing to regulated mRNA stability that converts oscillating transcription into a nearly constant steady-state *LHCB1*3* mRNA concentration (Millar & Kay, 1991). Similarly, the *CAT3* promoter is rhythmically transcribed in DD whereas *CAT3* mRNA damps to a constitutive high level under the same conditions (Michael & McClung, 2002). Another example for such a 'hidden' post-transcriptional control is the nitrate reductase (NIA2) mRNA in *Arabidopsis* which oscillates in a circadian manner whereas nuclear run-on transcriptional analysis shows a constant rate of *NIA2* gene transcription (Pilgrim *et al.*, 1993). These examples demonstrate opposite regulatory effects on the respective primary transcripts: on the one hand, a rhythmic transcription rate is converted into a constant transcript concentration and on the other hand, oscillations in transcript abundance are generated despite a constant transcription rate.

Arabidopsis cDNA microarrays have been employed to study RNA decay at a genomic scale. This analysis revealed that 1% of the transcripts were unstable with half-lives of less than 1 h (Gutierrez *et al.*, 2002). Of particular interest is the finding that among these highly destabilized transcripts are several which are controlled by the circadian clock. This investigation emphasizes the importance of mRNA stability for shaping transcript oscillations of specific phases and amplitudes in *Arabidopsis*.

A precedent for this is the *Drosophila Per* mRNA which shows an increased stability in the rising phase of the daily oscillation and a decreased stability in the declining phase. This post-transcriptional control presumably is responsible for the cycling *Per* mRNA levels observed in transgenic flies that express a promoterless *Per* gene at more or less the same rate across the circadian cycle (So & Rosbash, 1997).

3.5.3.2 Polyadenylation

Several years ago, polyadenylate tail length was implicated in circadian regulation in rat. The concentration of the neuropeptide vasopressin oscillates in the cerebrospinal fluid. It originates in the SCN where vasopressin mRNAs with differences in poly(A) tail length are found dependent on the time of day (Robinson *et al.*, 1988). This observation suggested an involvement of post-transcriptional regulation in vasopressin expression in addition to the direct transcriptional regulation through the pacemaker components CLOCK-BMAL1 inferred from mice (Jin *et al.*, 1999).

The *nocturnin* mRNA that oscillates in the circadian-clock containing photore-ceptor cells of *Xenopus laevis* retinas has recently been shown to code for a poly(A) specific 3′ exonuclease (Baggs & Green, 2002). Deadenylation is the rate-limiting step in mRNA turnover and significantly affects the half-life of mRNAs. The circadian oscillation of *nocturnin* expression points to deadenylation targets that are involved in rhythmic processes. The selectivity in choice of targets is probably achieved with the help of interaction partners on the protein level (Baggs & Green, 2002). Sequence analysis of *nocturnin* shows that possible homologues in *Arabidopsis* do exist (Baggs & Green, 2002).

3.5.3.3 Splicing

Processing of the *Drosophila Per* pre-mRNA is regulated at the level of alternative splicing, leading to two *Per* mRNAs which differ in the existence of an intron in the 3′ untranslated region (3′ UTR). The cycling of both mRNA types is equal, but flies which only produce the intron-containing *Per* mRNA show a slower accumulation of PER protein, indicating that the intron affects the translation of *Per* mRNA. Further investigations revealed that the splicing of the intron in the 3′ UTR is regulated by the clock and depends on the photoperiod and the temperature. This mechanism provides the ability to directly respond to environmental changes and adjust the circadian expression of the *Per* gene (Majercak *et al.*, 2004).

Alternative splicing also plays an important role in the autoregulatory feedback loop of *Arabidopsis ATGRP7* (Section 3.6.1; see Fig. 3.6).

3.5.3.4 Translational control mediated by a new class of RNA-binding proteins

Circadian rhythmicity of certain proteins is controlled at the translational level in *Lingulodinium* (formerly, *Gonyaulax*) (Morse *et al.*, 1989). An RNA-binding activity, CCTR, specifically interacting with UG repeats in the 3′ UTR of the luciferin binding protein (LBP) has been identified in protein extracts (Mittag *et al.*, 1994). Its abundance oscillates with nadir levels at the phase when *LBP* mRNA is translated, arguing that the RNA-binding activity might act as a translational repressor generating the circadian rhythm in LBP protein concentration.

Recently, purification and cloning of the CCTR analogue in the green alga *Chlamydomonas reinhardtii* has been reported. Target mRNAs to which CHLAMY1 shows the highest binding activity encode proteins of the nitrogen and carbon metabolism (Waltenberger *et al.*, 2001). CHLAMY1 represents a very interesting novel type of heterodimeric RNA binding protein consisting of the C1 subunit with three KH (lysine homology) RNA binding domains and the C3 subunit with three RNA recognition motifs (RRMs). The C3-subunit of CHLAMY1 has the same domain structure as UG- and CUG-binding proteins and members of the CELF (CUG-BP-ETR-3-like factors) family in mammals, which are involved in different post-transcriptional processes like mRNA stability or translational and splicing control (Zhao *et al.*, 2004). The C3 subunit shares significant homology with a

CUG-binding protein from rat. Notably patients suffering from muscular dystrophy which is caused by changes in the repeat length of the CUG motifs have also defects in the circadian system (Zhao et al., 2004).

A survey of the Arabidopsis genome for RNA binding proteins with conserved domains revealed 196 RRM-containing proteins and 26 proteins comprising KH domains (Lorkovic & Barta, 2002). Some transcripts encoding proteins implicated in various aspects of RNA processing are among the cycling transcripts identified on GeneChips (Harmer et al., 2000). Thus, post-transcriptional and translational control may have a wider impact on circadian rhythmicity also in Arabidopsis. In the long run, one may expect bioinformatics-based approaches to add to the identification of target transcripts of RNA-binding proteins. This would, however, require a more refined understanding of the molecular underpinnings of the interaction between certain types of RNA-binding domains with their cognate cis-regulatory motifs. At present, our knowledge is at its infancy, in contrast to corresponding information about binding sequences of DNA-binding modules. This is partly due to the fact that RNA recognition is based on sequence composition as well as secondary and tertiary structure of the RNA, in addition to sequence per se.

3.5.4   RNA-based regulation: The role of an antisense RNA

Recent investigations in Neurospora crassa implicate a naturally occurring antisense RNA in circadian regulation. It is transcribed in opposite direction from the locus of the clock gene frequency (frq). In wild-type fungi, it cycles 180° out of phase with the frq sense transcript and, like frq, is induced by light (Kramer et al., 2003). Interestingly, in mutants in which light induction of the antisense transcript is not possible, the phase of the clock is delayed. Furthermore, light entrainment, mediated by rapid induction of frq mRNA in wild type, is abnormal in these mutant strains. As there is, as yet, no indication for an effect on frq mRNA stability, the molecular basis of this type of regulation remains to be established.

3.5.5   Control by second messengers: do Ca^{2+} ions mediate rhythmic output?

Apart from regulation exerted through transcription factors or RNA-binding proteins, rhythmicity from the core clockwork may also be communicated through second messengers. It was previously observed that CAB expression can be regulated by Ca^{2+} ions (Neuhaus et al., 1993). Because oscillations of free cytosolic Ca^{2+} and CAB::LUC peak simultaneously, the hypothesis was tested whether oscillations in free cytosolic Ca^{2+} ions could mediate CAB transcript rhythms in tobacco calli transgenic for both the calcium reporter apoaequorin and a CAB2::LUC fusion. CAB oscillations persisted in undifferentiated calli in the absence of calcium oscillations, indicating that there is no hierarchical relationship among them (Sai & Johnson, 1999). Whether Ca^{2+} mediates other outputs such as stomatal conductance rhythms remains to be seen.

3.6 Slave oscillators

Transduction of circadian timing within the cell is thought to mainly occur through a linear series of regulatory events that can impart on most steps from gene expression to protein function (see Fig. 3.3A, B). Apart from that, two oscillatory circuits have been described that may serve as signalling intermediates in clock output control (see Fig. 3.3C, D).

3.6.1 The clock-regulated RNA-binding protein ATGRP7 in Arabidopsis thaliana

Transcripts encoding predicted RNA-binding proteins that undergo circadian oscillations in *Arabidopsis thaliana* and *Sinapis alba* with a peak at the end of the day have been described (van Nocker & Vierstra, 1993; Carpenter *et al.*, 1994; Heintzen *et al.*, 1994). The oscillations of the *ATGRP7* (also designated *CCR2*) mRNA are at least in part due to time-of-day specific activity of the promoter (Staiger & Apel, 1999). A minimal clock-responsive element mediating a basal circadian oscillation with peak abundance at the end of the day is located downstream of −112.

Several lines of evidence indicate that this transcriptional regulation is exerted through the CCA1-LHY-TOC1 oscillator. In *Arabidopsis* plants overexpressing CCA1, the *ATGRP7* transcript shows irregular fluctuations with a reduced amplitude in continuous light, whereas overexpression of mutant CCA1 does not abolish *ATGRP7* rhythms, indicating that *ATGRP7* is under negative control of CCA1 (Wang & Tobin, 1998). In plants constitutively overexpressing LHY, *ATGRP7* expression is arrhythmic at trough levels in continuous light (Schaffer *et al.*, 1998). Because mutation of a conserved evening element within the minimal clock-responsive element abolishes *ATGRP7* (also called *CCR2*) rhythms, it is conceivable that CCA1 and/or LHY may directly act on the promoter. Furthermore, *ATGRP7* receives temporal input from TOC1 because in the *toc1* mutant the period of *ATGRP7* oscillations is shortened (Kreps & Simon, 1997; Strayer *et al.*, 2000).

Transcriptional regulation accounts only in part for the observed mRNA oscillation. In transgenic plants overexpressing the *ATGRP7* protein from a constitutive promoter (*ATGRP7*-ox plants), the oscillation of the endogenous *ATGRP7* transcript is strongly depressed, indicating that the *ATGRP7* mRNA is under control by negative feedback (see Fig. 3.5A). This autoregulation is not conferred by the promoter because in *ATGRP7*-ox plants harbouring an *ATGRP7*-β-glucuronidase reporter gene oscillations of the reporter mRNA continue (Staiger & Apel, 1999). Taken together, these observations suggest that *ATGRP7* is part of a feedback loop – a 'slave oscillator' – downstream of the core oscillator that accomplishes negative autoregulation at the post-transcriptional level.

The residual oscillation of the endogenous *ATGRP7* transcript in the *ATGRP7*-ox plants is due to an alternatively spliced transcript that appears at the expense of the mRNA. The strongly reduced half-life of this alternatively spliced *ATGRP7* transcript species may account for its low steady-state abundance (see Fig. 3.5B).

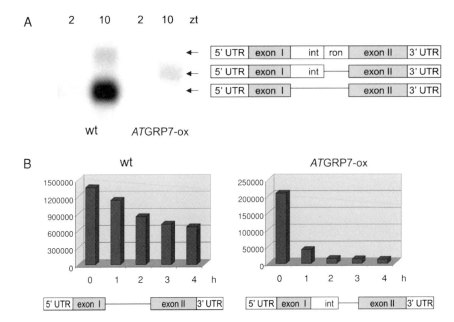

Figure 3.5 (A) Constitutive overexpression of *AT*GRP7 in transgenic *Arabidopsis* leads to the appearance of an alternatively spliced transcript at the expense of the mature mRNA. (B) To determine the half-life of this low-abundant transcript, plants were treated with Cordycepin to inhibit de novo RNA synthesis. Samples were withdrawn at the timepoints indicated. The residual amounts of mRNA in wild-type plants and alternatively spliced transcript in *AT*GRP7-ox plants were quantified (Staiger *et al.*, 2003a).

This implicates a mechanism how *AT*GRP7 contributes to oscillations of its own transcript (see Fig. 3.6). In response to increasing *AT*GRP7 protein accumulation during the circadian cycle, a shift in splice site selection occurs, favouring the production of the alternatively spliced transcript that rapidly decays. Moreover, a premature stop prevents its translation into functional protein so that *AT*GRP7 protein levels decline. Recombinant *AT*GRP7 protein specifically interacts with its own transcript in vitro, suggesting that the shift in splice site selection and downregulation in vivo may be initiated by binding of *AT*GRP7 to its pre-mRNA (Staiger *et al.*, 2003a).

Post-translational modifications may also contribute to *AT*GRP7 cycling. The pronounced oscillation in *AT*GRP7 protein steady-state abundance suggests that the protein has a short half-life and may undergo regulated decay. Similar to known circadian oscillators, the protein may undergo phosphorylation. In plants with increased activity of casein kinase 2 due to overexpression of its regulatory subunit CKB3, *ATGRP7* transcript rhythms are shortened (Daniel *et al.*, 2004). Whether this is exclusively due to the short period of the CCA1 and LHY clock proteins in these plants or whether CK2 also directly acts on *ATGRP7* is not known.

Pre-mRNA

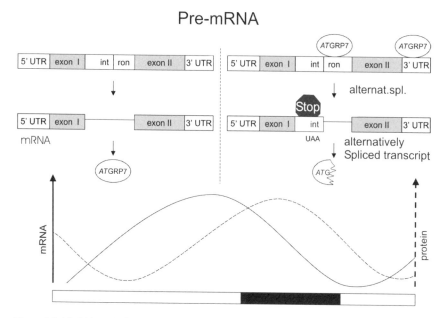

Figure 3.6 Model how *AT*GRP7 influences oscillations of its own transcript. Following transcriptional activation the pre-mRNA is spliced to produce mature mRNA. During the rising phase of *ATGRP7* oscillations, accumulating *AT*GRP7 protein is taken up into the nucleus and presumably by binding to the second part of the intron and/or the 3' UTR provokes a shift in splice site. The alternatively spliced transcript has a reduced half-life and thus the *ATGRP7* transcript level declines. Furthermore, a premature stop codon precludes translation into functional protein.

Apart from the self-regulation, *AT*GRP7 depresses the oscillations of *ATGRP8* encoding a related RNA-binding protein, indicating that *AT*GRP7 regulates other clock-controlled transcripts apart from its own. Furthermore, comparison of transcript profiles between wild-type plants and *AT*GRP7-ox plants revealed several transcripts with differences in steady-state abundance, suggesting that these are targets of *AT*GRP7 regulation (F. R., Neuenschwander, Page, Uchida, Furuya, C. S., X. H., D.S., unpublished observation). Thus, the *AT*GRP7 feedback loop may serve as a signalling intermediate within a clock output pathway. Such a component that itself oscillates with a very high amplitude may be an optimal component of the transduction chain to prevent damping. The negative autoregulation warrants shut-off of the signal in time for the next incoming signal. Furthermore, depending on the half-lives of their components, such slave oscillators may also serve to phase rhythms to different times of the day.

3.6.2 Transcriptional feedback loop as slave oscillator: EPR1

Recently, a second example of a molecular slave oscillator has been found that presumably is based on transcriptional feedback regulation and thus is built more like the core oscillator. A novel MYB protein early-phytochrome-responsive1 (EPR1) was identified by fluorescent differential display based on its rapid induction by red light (Kuno *et al.*, 2003). The transcript closely resembles *CCA1* and *LHY* and undergoes circadian oscillations, peaking a few hours later into the day than *CCA1* RNA. Constitutive overexpression of EPR1 represses accumulation of its own transcript, suggesting that EPR1 protein is part of a negative feedback loop through which it contributes to oscillations of its own transcript. Based on its predicted DNA-binding activity, the negative autoregulation may occur through interfering with transcriptional activation of its own promoter, as with the core oscillator. In addition to the *EPR1* transcript itself, *CAB* oscillations are perturbed in EPR1-ox plants, implicating ERR1 as a regulator of *CAB*. Transcripts of *CCA1* and *LHY* are not affected, whereas *EPR1* is arrhythmic in plants constitutively overexpressing CCA1 and LHY, indicating that EPR1 does not feed back on the central oscillator. Therefore the EPR1 feedback loop operates downstream of the CCA1-LHY-TOC1 oscillator and may confer its 24 hour rhythm upon output genes such as *CAB* (see Fig. 3.3C).

3.7 Summary

Recent advances in our understanding of the clock system in *Arabidopsis* are largely due to the development of molecular hands that can be studied non-invasively. Markers have been found to free-run with different periods, pointing to different underlying clocks (Section 3.3.2 and 3.3.3). Light entrainment in one organ does not affect entrainment of the rest of the plant. This indicates that the cellular clocks are autonomous, lacking extensive crosstalk between clocks even in the same tissue (Section 3.2.3). This high degree of autonomy is in line with the absence of a central pacemaker. Strong coupling between individual oscillators, as in the SCN, favours their accuracy. So the relative independence of individual plant clocks may provide versatility rather than exact timekeeping. Due to their sessile life-style, plants are forced to adjust to environmental changes. As photoautotrophic organisms, they are particularly dependent on light and thus have developed a suite of photoreceptors that allow monitoring of subtle changes in light quality, quantity and duration. These photoreceptors are present in most cells and thus environmental cues can be directly communicated to the cellular clocks via input pathways. Synchronization in the plant may thus be achieved via parallel entrainment, rather than through hierarchical entrainment with a central unit communicating timing information to subsidiary clocks (see Fig. 3.7).

Recently, it became evident that the organization of the plant circadian system with autonomous cellular clock lacking an obvious pacemaker may in fact not differ so much from the organization of the mammalian circadian system. Clocks in peripheral cells are no longer thought to be dependent on the SCN master oscillator for the maintenance of circadian rhythms. Rather, they may be considered as members in an orchestra that would run out of phase with each other if not managed by the SCN conductor (Section 3.2).

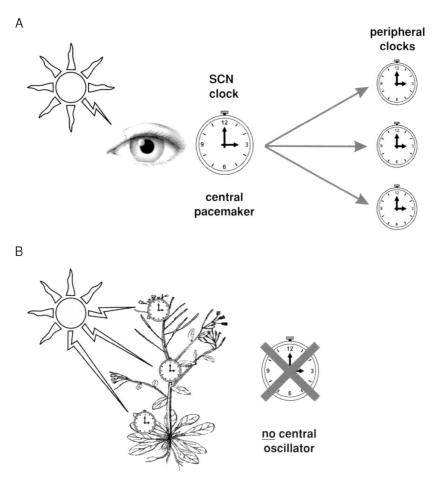

Figure 3.7 (A) In mammals, cellular clocks are entrained to the environmental light–dark cycles predominantly through the master clock in the SCN that processes photic information received from the eyes. (B) In higher plants, parallel entrainment of cellular clocks by photoreceptors residing in the very same cell may be the prevalent way of coordination with the environment.

The molecular basis of different clocks remains to be determined although the same clock proteins are part of different clocks (Section 3.3.2). The investigation of mutants defective in specific clock-associated proteins for their influence on distinct rhythms may help to unravel the basis of oscillators controlling rhythms with different period lengths (Section 3.3.1). Feedback loops operating downstream of the core clockwork add another twist to the hierarchy and broaden the means of output control (Section 3.6). Transcript profiling of mutants lacking EPR1 or *AT*GRP7, respectively, as well as of mutants in the core clock proteins will allow to determine the degree of overlap between their regulatory networks and provide an estimate to which extent the CCA1-LHY-TOC1 core oscillator communicates timing information through these channels.

References

Alabadi, D., Oyama, T., Yanovsky, M.J., Harmon, F.G., Mas, P. & Kay, S.A. (2001) Reciprocal regulation between TOC1 and LHY/CCA1 within the *Arabidopsis* circadian clock. *Science*, **293**, 880–883.

Alabadi, D., Yanovsky, M.J., Más, P., Harmer, S.L. & Kay, S.A. (2002) Critical role for CCA1 and LHY in Maintaining circadian rhythmicity in *Arabidopsis*. *Curr. Biol.*, **12**, 757–761.

Albrecht, U. (2004) The mammalian circadian clock: a network of gene expression. *Front Biosci.*, **9**, 48–55.

Anderson, S.L., Teakle, G.R., Martino-Catt, S.J. & Kay, S.A. (1994) Circadian clock- and phytochrome-regulated transcription is conferred by a 78 bp *cis*-acting domain of the *Arabidopsis* CAB2 promoter. *Plant J.*, **6**, 457–470.

Andersson, C.R., Harmer, S.L., Schultz, T.F. & Kay, S.A. (1999) The Reveille (REV) family of DNA binding proteins and the circadian clock. *Abstract from the 10th International Conference on Arabidopsis Research, Australia.*

Aschoff, J. (1960) Exogenous and endogenous components in circadian rhythms. In: *Cold Spring Harbor Symp. Quant. Biol. XXV: Biological Clocks.* (ed A. Chovnick), pp. 11–28. Cold Spring Harbor Press, Cold Spring Harbor, New York.

Baggs, J.E. & Green, C.B. (2002) Nocturnin, a deadenylase in *Xenopus laevis* retina: a mechanism for posttranscriptional control of circadian-related mRNA. *Curr. Biol.*, **13**, 189–198.

Balsalobre, A., Damiola, F. & Schibler, U. (1998) A serum shock induces circadian gene expression in mammalian tissue culture cells. *Cell*, **93**, 929–937.

Brown, S.A. & Schibler, U. (1999) The ins and outs of circadian timekeeping. *Curr. Opin. Genet. Dev.*, **9**, 588–594.

Carpenter, C.D., Kreps, J.A. & Simon, A.E. (1994) Genes encoding glycine-rich *Arabidopsis thaliana* proteins with RNA-binding motifs are influenced by cold treatment and an endogenous circadian rhythm. *Plant Physiol.*, **104**, 1015–1025.

Coupland, G. (2003) Shedding light on the circadian clock and the photoperiodic control of flowering. *Curr. Opin. Plant Biol.*, **6**, 13–19.

Daniel, X., Sugano, S. & Tobin, E.M. (2004) CK2 phosphorylation of CCA1 is necessary for its circadian oscillator function in Arabidopsis. *Proc. Natl. Acad. Sci. USA*, **101**, 3292–3297.

Devlin, P.F. (2002) Signs of the time: environmental input to the circadian clock. *J. Exp. Bot.*, **53**, 1535–1550.

Ditty, J.L., Williams, S.B. & Golden, S.S. (2003) A cyanobacterial circadian timing mechanism. *Annu. Rev. Genet.*, **37**, 513–543.

Dodd, A.N., Parkinson, K. & Webb, A.A.R. (2004) Independent circadian regulation of assimilation and stomatal conductance in the *ztl*-1 mutant of *Arabidopsis*. *New Phytologist*, **162**, 63–70.

Dowson-Day, M.J. & Millar, A.J. (1999) Circadian dysfunction causes aberrant hypocotyl elongation patterns in *Arabidopsis*. *Plant J.*, **17**, 63–71.

Doyle, M.R., Davis, S.J., Bastow, R.M., McWatters, H.G., Kozma-Bognar, L., Nagy, F., Millar, A.J. & Amasino, R. (2002) The *ELF4* gene controls circadian rhythms and flowering time in *Arabidopsis thaliana*. *Nature*, **419**, 74–75.

Dunlap, J.C. (1999) Molecular bases for circadian clocks. *Cell*, **96**, 271–290.

Edery, I. (1999) Role of posttranscriptional regulation in circadian clocks: lessons from *Drosophila*. *Chronobiol Int.*, **16**, 377–414.

Engelmann, W. & Johnsson, A. (1977) Attenuation of the petal movement rhythm in Kalanchoe with light pulses. *Physiol. Plant*, **43**, 68–76.

Engelmann, W., Simon, K. & Phen, C.J. (1992) Leaf movement rhythm in *Arabidopsis thaliana*. *Verlag der Zeitschrift für Naturforschung*, **47c**, 925–928.

Fankhauser, C. & Staiger, D. (2002) Photoreceptors in *Arabidopsis thaliana*: light perception, signal transduction and entrainment of the endogenous clock. *Planta*, **216**, 1–16.

Finlayson, S.A., Lee, I. J. & Morgan, P.W. (1998) Phytochrome B and the Regulation of Circadian Ethylene Production in Sorghum. *Plant Physiol.*, **116**, 17–25.

Fowler, S., Lee, K., Onouchi, H., Samach, A., Richardson, K., Morris, B., Coupland, G. & Putterill, J. (1999) GIGANTEA: a circadian clock-controlled gene that regulates photoperiodic flowering in *Arabidopsis* and encodes a protein with several possible membrane-spanning domains. *EMBO J.*, **18**, 4679–4688.

Gachon, F., Nagoshi, E., Brown, S.A., Ripperger, J. & Schibler, U. (2004) The mammalian circadian timing system: from gene expression to physiology. *Chromosoma.*

Giebultowicz, J.M. & Hege, D.M. (1997) Circadian clock in Malpighian tubules. *Nature*, **386**, 664.

Gorton, H.L., Williams, W.E., Binns, M.E., Gemmell, C.N., Leheny, E.A. & Shepherd, A.C. (1989) Circadian stomatal rhythms in epidermal peels from *Vicia faba*. *Plant Physiol.*, **90**, 1329–1334.

Grima, B., Chelot, E., Xia, R. & Rouyer, F. (2004) Morning and evening peaks of activity rely on different clock neurons of the *Drosophila* brain. *Nature*, **431**, 869–873.

Gutierrez, R.A., Ewing, R.M., Cherry, J.M. & Green, P.J. (2002) Identification of unstable transcripts in Arabidopsis by cDNA microarray analysis: rapid decay is associated with a group of touch- and specific clock-controlled genes. *PNAS*, **99**, 11513–11518.

Hall, A., Kozma-Bognar, L., Bastow, R.M., Nagy, F. & Millar, A.J. (2002) Distinct regulation of CAB and PHYB gene expression by similar circadian clocks. *Plant J.*, **32**, 529–537.

Hall, A., Bastow, R.M., Davis, S.J., Hanano, S., McWatters, H.G., Hibberd, V., Doyle, M.R., Sung, S., Halliday, K.J., Amasino, R.M. & Millar, A.J. (2003) The TIME FOR COFFEE gene maintains the amplitude and timing of *Arabidopsis* circadian clocks. *Plant Cell*, **15**, 2719–2729.

Hanano, S., Domagalska, M., Birkemeyer, C., Kopka, J. & Davis, S.J. (2004) Phytohormones maintain the circadian clock in *Arabidopsis thaliana*. *Abstract from the 15th International Conference on Arabidopsis Research, Berlin, Germany.*

Harmer, S.L., Hogenesch, J.B., Straume, M., Chang, H.S., Han, B., Zhu, T., Wang, X., Kreps, J.A. & Kay, S.A. (2000) Orchestrated transcription of key pathways in *Arabidopsis* by the circadian clock. *Science*, **290**, 2110–2113.

Heintzen, C., Nater, M., Apel, K. & Staiger, D. (1997) *At*GRP7, a nuclear RNA-binding protein as a component of a circadian-regulated negative feedback loop in *Arabidopsis thaliana*. *PNAS*, **94**, 8515–8520.

Heintzen, C., Melzer, S., Fischer, R., Kappeler, S., Apel, K. & Staiger, D. (1994) A light- and temperature-entrained circadian clock controls expression of transcripts encoding nuclear proteins with homology to RNA-binding proteins in meristematic tissue. *Plant J.*, **5**, 799–813.

Hennessey, T.L. & Field, C.B. (1992) Evidence of multiple circadian oscillators in bean plants. *J. Biol. Rhythms*, **7**, 105–113.

Jagota, A., de la Iglesia, H.O. & Schwartz, W.J. (2000) Morning and evening circadian oscillations in the suprachiasmatic nucleus in vitro. *Nat. Neurosci.*, **3**, 372–376.

Jin, X., Shearman, L.P., Weaver, D.R., Zylka, M.J., de Vries, G.J. & Reppert, S.M. (1999) A molecular mechanism regulating rhythmic output from the suprachiasmatic circadian clock. *Cell*, **96**, 57–68.

Jouve, L., Gaspar, T., Kevers, C., Greppin, H. & Degli Agosti, R. (1999) Involvement of indole-3-acetic acid in the circadian growth of the first internode of Arabidopsis. *Planta*, **209**, 136–142.

Kim, H.Y., Cote, G.G. & Crain, R.C. (1993) Potassium channels in *Samanea saman* protoplasts controlled by pPhytochrome and the biological clock. *Science*, **260**, 960–962.

King, R.W. (1975) Multiple circadian rhythms regulate photoperiodic flowering responses in *Chenopodium rubrum*. *Can. J. Bot.*, **53**, 2631–2638.

Kolar, C., Adam, E., Schäfer, E. & Nagy, F. (1995) Expression of tobacco genes for light-harvesting chlorophyll a/b binding proteins of photosystem II is controlled by two circadian oscillators in a developmentally regulated fashion. *Proc. Natl. Acad. Sci. USA*, **92**, 2174–2178.

Kolar, C., Fejes, E., Adam, E., Schafer, E., Kay, S. & Nagy, F. (1998) Transcription of Arabidopsis and wheat Cab genes in single tobacco transgenic seedlings exhibits independent rhythms in a developmentally regulated fashion. *Plant J.*, **13**, 563–569.

Kolar, J., Machackova, I., Eder, J., Prinsen, E., von Dongen, W., van Onckelen, H. & Illenerova, H. (1997) Melatonin: Occurence and daily rhythm in *Chenopodium rubrum*. *Phytochemistry*, **44**, 1407–1413.

Kramer, C., Loros, J.J., Dunlap, J.C. & Crosthwaite, S.K. (2003) Role for antisense RNA in regulating circadian clock function in *Neurospora crassa*. *Nature*, **421**, 948–952.

Kreps, J.A. & Simon, A.E. (1997) Environmental and genetic effects on circadian clock-regulated gene expression in *Arabidopsis*. *Plant Cell*, **9**, 297–304.

Kriegsfeld, L.J., Leak, R.K., Yackulic, C.B., LeSauter, J. & Silver, R. (2004) Organization of suprachiasmatic nucleus projections in Syrian hamsters (Mesocricetus auratus): an anterograde and retrograde analysis. *J. Comp. Neurol.*, **468**, 361–379.

Kuno, N., Moller, S.G., Shinomura, T., Xu, X., Chua, N.H. & Furuya, M. (2003) The novel MYB protein EARLY-PHYTOCHROME-RESPONSIVE1 is a componet of a slave circadian oscillator in *Arabidopsis*. *Plant Cell*, **15**, 2476–2488.

Le Minh, N., Damiola, F., Tronche, F., Schütz, G. & Schibler, U. (2001) Glucocorticoid hormones inhibit food-induced phase-shifting of peripheral circadian oscillators. *EMBO J.*, **20**, 7128–7136.

Lecharny, A., Tremolières, A. & Wagner, E. (1990) Correlation between the endogenous circadian rhythmicity in growth rate and fluctuations in oleic acid content in expanding stems of *Chenopodium rubrum* L. *Planta*, **182**, 211–215.

Liu, Y., Tsinoremas, N.F., Johnson, C.H., Lebedeva, N.V., Golden, S.S., Ishiura, M. & Kondo, T. (1995) Circadian orchestration of gene expression in cyanobacteria. *Genes Dev.*, **9**, 1469–1478.

Liu, Z., Taub, C.C. & McClung, C.R. (1996) Identification of an *Arabidopsis thaliana* ribulose-1,5-bisphosphate carboxylase/oxygenase activase (RCA) minimal promoter regulated by light and the circadian clock. *Plant Physiol.*, **112**, 43–51.

Lorkovic, Z.J. & Barta, A. (2002) Genome analysis: RNA recognition motif (RRM) and K homology (KH) domain RNA-binding proteins from the flowering plant *Arabidopsis thaliana*. *Nucleic Acids Res.*, **30**, 623–635.

Machackova, I., Chauvaux, N., Dewitte, W. & van Onckelen, H. (1997) Diurnal fluctuations in ethylene formation in *Chenopodium rubum*. *Plant Phys.*, **113**, 981–985.

Majercak, J., Chen, W.F. & Edery, I. (2004) Splicing of the period gene 3'-terminal intron is regulated by light, circadian clock factors, and phospholipase C. *Mol. Cell. Biol.*, **24**, 3359–3372.

Makino, S., Matsushika, A., Kojima, M., Oda, Y. & Mizuno, T. (2001) Light response of the circadian waves of the APRR1/TOC1 quintet: when does the quintet start singing rhythmically in *Arabidopsis*? *Plant Cell Physiol.*, **42**, 334–339.

Makino, S., Matsushika, A., Kojima, M., Yamashino, T. & Mizuno, T. (2002) The APRR1/TOC1 quintet implicated in circadian rhythms of *Arabidopsis thaliana*: I. Characterization with APRR1-overexpressing plants. *Plant Cell Physiol.*, **43**, 58–69.

Mas, P., Kim, W.Y., Somers, D.E. & Kay, S.A. (2003) Targeted degradation of TOC1 by ZTL modulates circadian function in *Arabidopsis thaliana*. *Nature*, **426**, 567–570.

Matsushika, A., Makino, S., Kojima, M., Yamashino, T. & Mizuno, T. (2002) The APRR1/TOC1 quintet implicated in circadian rhythms of *Arabidopsis thaliana*: II. Characterization with CCA1-overexpressing plants. *Plant Cell Physiol.*, **43**, 118–122.

Mayer, W.E. & Fischer, C. (1994) Protoplasts from *Phaseolus coccineus* L. pulvinar motor cells show circadian volume oscillations. *Chronobiol. Int.*, **11**, 156–164.

McClung, C.R. (2000) Circadian rhythms in plants: a millennial view. *Physiologia Plantarum*, **109**, 359–371.

McClung, C.R., Salome, P.A. & Michael, T.P. (2002) The *Arabidopsis* circadian system. In: *The Arabidopsis Book* (eds C.R. Sommerville & E.M. Meyerowitz), pp. 1–23. American Society of Plant Biologists, Rockville, MD.

McNeil, G.P., Zhang, X., Genova, G. & Jackson, F.R. (1998) A molecular rhythm mediating circadian clock output in *Drosophila*. *Neuron*, **20**, 297–303.

McWatters, H.G., Roden, L.C. & Staiger, D. (2001) Picking out parallels: plant circadian clocks in context. *Philos. Trans. R. Soc. Lond. B. Biol. Sci.*, **356**, 1735–1743.

Michael, T.P. & McClung, C.R. (2002) Phase-specific circadian clock regulatory elements in *Arabidopsis*. *Plant Physiol.*, **130**, 627–638.

Michael, T.P. & McClung, C.R. (2003) Enhancer trapping reveals widespread circadian clock transcriptional control in *Arabidopsis*. *Plant Physiol.*, **132**, 629–639.

Michael, T.P., Salome, P.A. & McClung, C.R. (2003) Two Arabidopsis circadian oscillators can be distinguished by differential temperature sensitivity. *Proc. Natl. Acad. Sci. USA*, **100**, 6878–6883.

Millar, A.J. (1998) The cellular organization of circadian rhythms in plants: not one but many clocks. In: *Biological Rhythms and Photoperiodism in Plants* (eds P.J. Lumsden & A.J. Millar), pp. 51–68. BIOS Scientific Publishers Ltd, Oxford.

Millar, A.J. (2004) Input signals to the plant circadian clock. *J. Exp. Bot.*, **55**, 277–283.

Millar, A. & Kay, S. (1991) Circadian control of *cab* gene transcription and mRNA accumulation in *Arabidopsis*. *Plant Cell*, **3**, 541–550.

Millar, A., Carré, I.A., Strayer, C.S., Chua, N.H. & Kay, S. (1995) Circadian clock mutants in *Arabidopsis* identified by luciferase imaging. *Science*, **267**, 1161–1163.

Mittag, M., Lee, D.H. & Hastings, J.W. (1994) Circadian expression of the luciferin-binding protein correlates with the binding of a protein to the 3' untranslated region of its mRNA. *Proc. Natl. Acad. Sci. USA*, **91**, 5257–5261.

Mizoguchi, T., Wheatley, K., Hanzawa, Y., Wright, L., Mizoguchi, M., Song, H.R., Carré, I.A. & Coupland, G. (2002) *LHY* and *CCA1* are partially redundant genes required to maintain circadian rhythms in *Arabidopsis*. *Dev. Cell.*, **2**, 629–641.

Morikawa, K., Ito, S., Tsunoyama, Y., Nakahira, Y., Shiina, T. & Toyoshima, Y. (1999) Circadian-regulated expression of a nuclear-encoded plastid sigma factor gene (sigA) in wheat seedlings. *FEBS Lett.*, **451**, 275–278.

Morse, D., Milos, P.M., Roux, E. & Hastings, J.W. (1989) Circadian regulation of bioluminescence in Gonyaulax involves translational control. *Proc. Natl. Acad. Sci. USA*, **86**, 172–176.

Murakami, M., Yamashino, T. & Mizuno, T. (2004) Characterization of circadian-associated APRR3 pseudo-response regulator belonging to the APRR1/TOC1 quintet in *Arabidopsis thaliana*. *Plant Cell Physiol.*, **45**, 645–650.

Murakami-Kojima, M., Nakamichi, N., Yamashino, T. & Mizuno, T. (2002) The APRR3 component of the clock-associated APRR1/TOC1 quintet is phosphorylated by a novel protein kinase belonging to the WNK family, the gene for which is also transcribed rhythmically in *Arabidopsis thaliana*. *Plant Cell Physiol.*, **43**, 675–683.

Nagoshi, E., Saini, C., Bauer, C., Laroche, T., Naef, F. & Schibler, U. (2004) Circadian gene expression in individual fibroblasts; cell-autonomous and self-sustained oscillators pass time to daughter cells. *Cell*, **119**, 693–705.

Nagy, F., Kay, S.A. & Chua, N.-H. (1988) A circadian clock regulates transcription of the wheat *Cab-1* gene. *Genes Dev.*, **2**, 376–382.

Nakamichi, N., Matsushika, A., Yamashino, T. & Mizuno, T. (2003) Cell autonomous circadian waves of the APRR1/TOC1 quintet in an established cell line of *Arabidopsis thaliana*. *Plant Cell*, **44**, 360–365.

Nakamichi, N., Ito, S., Oyama, T., Yamashino, T., Kondo, T. & Mizuno, T. (2004) Characterization of plant circadian rhythms by employing *Arabidopsis* cultured cells with bioluminescence reporters. *Plant Cell Physiol.*, **45**, 57–67.

Neuhaus, G., Bowler, C., Kern, R. & Chua, N.H. (1993) Calcium/Calmodulin-dependent and -idependent phytochrome signal transduction pathways. *Cell.*, **73**, 937–952.

Onai, K., Okamoto, K., Nishimoto, H., Morioka, C., Hirano, M., Kami-Ike, N. & Ishiura, M. (2004) Large-scale screening of Arabidopsis circadian clock mutants by a high-throughput real-time bioluminescence monitoring system. *Plant J.*, **40**, 1–11.

Paasch, K., Lein, C., Nessiem, M. & Neumann, K.H. (1997) Changes in the concentration of some phytohormones in cultured root explants and in intact carrot plants (*Daucus carota* L.) during the day. *J. Appl. Bot.*, **71**, 85–89.

Park, D.H., Somers, D.E., Kim, Y.S., Choy, Y.H., Lim, H.K., Soh, M.S., Kim, H.J., Kay, S.A. & Nam, H.G. (1999) Control of circadian rhythms and photoperiodic flowering by the *Arabidopsis* GIGANTEA gene. *Science*, **285**, 1579–1582.

Pilgrim, M.L., Caspar, T., Quail, P.H. & McClung, C.R. (1993) Circadian and light-regulated expression of nitrate reductase in *Arabidopsis*. *Plant Mol. Biol.*, **23**, 349–364.

Pittendrigh, C.S. & Daan, S. (1976) A functional analysis of circadian pacemakers in nocturnal rodents. V A clock for all seasons. *J. Comp. Physiol.*, **106**, 333–355.

Plautz, J.D., Kaneko, M., Hall, J.C. & Kay, S.F. (1997) Independent photoreceptive circadian clocks throughout *Drosophila*. *Science*, **278**, 1632–1635.

Reed, J.W., Nagpal, P., Bastow, R.M., Solomon, K.S., Dowson-Day, M.J., Elumalai, R.P. & Millar, A.J. (2000) Independent action of ELF3 and phyB to control hypocotyl elongation and flowering time. *Plant Physiol.*, **122**, 1149–1160.

Ripperger, J.A., Shearman, L.P., Reppert, S.M. & Schibler, U. (2000) CLOCK, an essential pacemaker component, controls expression of the circadian transcription factor DBP. *Genes Dev.*, **14**, 679–689.

Robinson, B.G., Frim, D.M., Schwartz, W.J. & Majzoub, J.A. (1988) Vasopressin mRNA in the suprachiasmatic nuclei: daily regulation of polyadenylate tail length. *Science*, **241**, 342–344.

Roenneberg, T. & Morse, D. (1993) Two circadian oscillators in one cell. *Nature*, **362**, 362–364.

Rudolf, F., Wehrle, F. & Staiger, D. (2004) Slave to the rhythm. *The Biochemist*, **26**, 11–13.

Sai, J. & Johnson, C.H. (1999) Different circadian oscillators control Ca^{2+} fluxes and lhcb gene expression. *Proc. Natl. Acad. Sci. USA*, **96**, 11659–11663.

Sato, E., Nakamichi, N., Yamashino, T. & Mizuno, T. (2002) Abberant expression of the *Arabidopsis* circadian-regulated *APRR5* gene belonging to the APRR1/TOC1 quintet results in early flowering and hypersenitiveness to light in early photomorphogenesis. *Plant Cell Physiol.*, **43**, 1374–1385.

Satter, R.L. & Galston, A.W. (1981) Mechanisms of Control of Leaf Movements. *Annu. Rev. Plant Physiol.*, **32**, 83–110.

Schaffer, R., Ramsay, N., Samach A, C., S., Putterill, J., Carré, I.A. & Coupland, G. (1998) The late elongated hypocotyl mutation of *Arabidopsis* disrupts circadian rhythms and the photoperiodic control of flowering. *Cell.*, **93**, 1219–1229.

Schaffer, R., Landgraf, J., Accerbi, M., Simon, V., Larson, M. & Wisman, E. (2001) Microarray analysis of diurnal and circadian-regulated genes in *Arabidopsis*. *Plant Cell*, **13**, 113–123.

So, W. & Rosbash, M. (1997) Post-transcriptional regulation contributes to Drosophila clock gene mRNA cycling. *EMBO J.*, **16**, 7146–7155.

Somers, D.E., Devlin, P.F. & Kay, S.A. (1998a) Phytochromes and cryptochromes in the entrainment of the *Arabidopsis* circadian clock. *Science*, **282**, 1488–1490.

Somers, D.E., Webb, A.A., Pearson, M. & Kay, S.A. (1998b) The short-period mutant, toc1-1, alters circadian clock regulation of multiple outputs throughout development in *Arabidopsis thaliana*. *Development*, **125**, 485–494.

Staiger, D. (2002) Circadian rhythms in Arabidopsis: time for nuclear proteins. *Planta*, **214**, 334–344.

Staiger, D. & Apel, K. (1999) Circadian clock-regulated expression of an RNA-binding protein in *Arabidopsis*: characterization of a minimal promoter element. *Mol. Gen. Genet*, **261**, 811–819.

Staiger, D. & Heintzen, C. (1999) The circadian system of *Arabidopsis thaliana*: forward and reverse genetic approaches. *Chronobiol. Int.*, **16**, 1–16.

Staiger, D., Apel, K. & Trepp, G. (1999) The Atger3 promoter confers circadian clock-regulated transcription with peak expression at the beginning of the night. *Plant Mol. Biol.*, **40**, 873–882.

Staiger, D., Zecca, L., Kirk, D.A., Apel, K. & Eckstein, L. (2003a) The circadian clock regulated RNA-binding protein AtGRP7 autoregulates its expression by influencing alternative splicing of its own pre-mRNA. *Plant J.*, **33**, 361–371.

Staiger, D., Allenbach, L., Salathia, N., Fiechter, V., Davis, S.J., Millar, A.J., Chory, J. & Fankhauser, C. (2003b) The *Arabidopsis* SRR1 gene mediates phyB signaling and is required for normal circadian clock function. *Genes Dev.*, **17**, 256–268.

Stokkan, K.A., Yamazaki, S., Tei, H., Sakaki, Y. & Menaker, M. (2001) Entrainment of the circadian clock in the liver by feeding. *Science*, **291**, 490–493.

Stoleru, D., Peng, Y., Agosto, J. & Rosbash, M. (2004) Coupled oscillators control morning and evening locomotor behaviour of *Drosophila*. *Nature*, **431**, 862–868.

Strayer, C., Oyama, T., Schultz, T.F., Raman, R., Somers, D.E., Mas, P., Panda, S., Kreps, J.A. & Kay, S.A. (2000) Cloning of the *Arabidopsis* clock gene TOC1, an autoregulatory response regulator homolog. *Science*, **289**, 768–771.

Sugano, S., Andronis, C., Ong, M.S., Green, R.M. & Tobin, E.M. (1999) The protein kinase CK2 is involved in regulation of circadian rhythms in *Arabidopsis*. *PNAS*, **96**, 12362–12366.

Thain, S.C., Hall, A. & Millar, A.J. (2000) Functional independence of circadian clocks that regulate plant gene expression. *Curr. Biol.*, **10**, 951–956.

Thain, S.C., Murtas, G., Lynn, J.R., McGrath, R.B. & Millar, A.J. (2002) The circadian clock that controls gene expression in *Arabidopsis* is tissue specific. *Plant Physiol.*, **130**, 102–110.

Thain, S.C., Vandenbussche, F., Laarhoven, L.J., Dowson-Day, M.J., Wang, Z.Y., Tobin, E.M., Harren, F.J., Millar, A.J. & Van Der Straeten, D. (2004) Circadian rhythms of ethylene emission in *Arabidopsis*. *Plant Physiol.*, **136(3)**, 3751–3761.

Tsinoremas, N.F., Ishiura, M., Kondo, T., Andersson, C.R., Tanaka, K., Takahashi, H., Johnson, C.H. & Golden, S.S. (1996) A sigma factor that modifies the circadian expression of a subset of genes in cyanobacteria. *EMBO J.*, **15**, 2488–2495.

van Nocker, S. & Vierstra, R.D. (1993) Two cDNAs from *Arabidopsis thaliana* encode putative RNA binding proteins containing glycine-rich domains. *Plant Mol. Biol.*, **21**, 695–699.

Viswanathan, N. & Subbaraj, R. (1983) Action of gibberellic acid in effecting phase shifts of the circadian leaf-movement rhythm of a cotton plant, *Gossypium hirsutum*. *Can. J. Bot.*, **61**, 2527–2529.

Waltenberger, H., Schneid, C., Grosch, J.O., Bareiss, A. & Mittag, M. (2001) Identification of target mRNAs for the clock-controlled RNA-binding protein Chlamy 1 from *Chlamydomonas reinhardtii*. *Mol. Genet. Genomics*, **265**, 180–188.

Wang, Z.Y. & Tobin, E.M. (1998) Constitutive expression of the CIRCADIAN CLOCK ASSOCI-ATED1 (CCA1) gene disrupts circadian rhythms and suppresses its own expression. *Cell*, **93**, 1207–1217.

Wang, Z.Y., Kenigsbuch, D. & Tobin, E.M. (1997) A *Myb*-related transcription factor is involved in the phytochrome regulation of an *Arabidopsis Lhcb* gene. *Plant Cell*, **9**, 491–499.

Welsh, D.K., Logothetis, D.E., Meister, M. & Reppert, S.M. (1995) Individual neurons dissociated from rat suprachiasmatic nucleus express independently phased circadian firing rhythms. *Neuron*, **14**, 697–706.

Whitmore, D., Foulkes, N.S., Strahle, U. & Sassone-Corsi, P. (1998) Zebrafish *Clock* rhythmic expression reveals independent peripheral circadian oscillators. *Nat. Neurosci*, **1**, 701–707.

Yamaguchi, S., Isejima, H., Matsuo, T., Okura, R., Yagita, K., Kobayashi, M. & Okamura, H. (2003) Synchronization of cellular clocks in the suprachiasmatic nucleus. *Science*, **302**, 1408–1412.

Yamashino, T., Matsushika, A., Fujimori, T., Sato, S., Kato, T., Tabata, S. & Mizuno, T. (2003) A link between circadian-controlled bHLH factors and the APRR1/TOC1 quintet in *Arabidopsis thaliana*. *Plant Cell Physiol.*, **44**, 619–629.

Yamazaki, S., Numano, R., Abe, M., Hida, A., Takahashi, R., Ueda, M., Block, G.D., Sakai, Y., Menaker, M. & Tei, H. (2000) Resetting central and peripheral circadian oscillators in transgenic rats. *Science*, **288**, 682–685.

Yoo, S.H., Yamazaki, S., Lowrey, P.L., Shimomura, K., Ko, C.H., Buhr, E.D., Siepka, S.M., Hong, H.K., Oh, W.J., Yoo, O.J., Menaker, M. & Takahashi, J.S. (2004) PERIOD2::LUCIFERASE real-time reporting of circadian dynamics reveals persistent circadian oscillations in mouse peripheral tissues. *Proc. Natl. Acad. Sci. USA*, **101**, 5339–5346.

Young, M.W. & Kay, S.A. (2001) Time zones: a comparative genetics of circadian clocks. *Nat. Rev. Genet.*, **2**, 702–715.

Zhao, B., Schneid, C., Iliev, D., Schmidt, E.M., Wagner, V., Wollnik, F. & Mittag, M. (2004) The circadian RNA-binding protein CHLAMY1 represents a novel type heteromer of RNA recognition motif and lysine homology domain-containing subunits. *Eukaryot. Cell*, **3**, 815–825.

4   Entrainment of the circadian clock

David E. Somers

4.1   Introduction

To be useful as a time-keeping mechanism, the endogenous circadian clock must connect with the external environment. While we often assess the free-running period of a mutant or a plant under constant conditions, plants do not typically live in a non-cyclic environment. The clock must confer advantages to the plant while growing in the normal world of light/dark or temperature cycles. Since the free-running period (FRP) is rarely exactly 24 h, each day the environment entrains the endogenous oscillator to fit the 24 h rhythm of the spinning earth. This requires shifting the circadian system forward or backward in time to get a match between the environment and the oscillator. A successfully entrained clock will not only oscillate with a 24 h rhythm but, more importantly, all the processes controlled by that clock will establish a specific and precise phase relationship with the light/dark (or temperature) cycles that act as the entraining stimuli. When stably entrained, the time of peak expression for a given process will repeatedly occur at a specific time, relative to a particular phase of the environment (e.g. sunrise or sunset). It is in the maintenance of a stable phase relationship between the external environment and the internal, clock-controlled processes that the true utility of the circadian clock lies.

The process of entrainment, then, requires that the plant perceives the environment and adjusts the phase of the oscillator appropriately each day. In principle, this adjustment process can occur via two mechanisms. The system could respond to environmental cues at particular, discrete times of the day, or the system could be continually responsive and adjust the phase of the oscillator throughout the day. The former is called discrete or non-parametric entrainment, and the latter is referred to as continuous or parametric entrainment (Johnson *et al.*, 2003). Work by Colin Pittendrigh, largely through the study of *Drosophila* eclosion rhythms, lends strong support for the discrete model of entrainment (Pittendrigh, 1981). Also supporting this model is the observation that the phase resetting ability of discrete light pulses varies throughout the circadian cycle. At some circadian times a light pulse will advance the phase of the clock, while at other times a phase delay will result. A systematic application of light pulses along the circadian cycle will generate one of the characteristic descriptors of the circadian clock: a phase response curve (Fig. 4.1).

In contrast to the discrete entrainment model, Juergen Aschoff observed that the FRP of many organisms maintained in continuous light varies depending on the light intensity (Aschoff's rule; Aschoff, 1979). This led to the notion of parametric entrainment, whereby the clock is continually phase-adjusted throughout the photoperiod, resulting in a net change in free-running period.

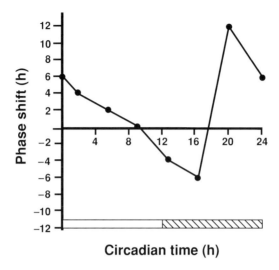

Figure 4.1 Red light phase-response curve for Arabidopsis. Plants grown in 12 h white light/12 h dark cycles were transferred to a constant dim red light background. Different sets of plants were given one 3-h high intensity light pulse (100X background) at different times during the first circadian cycle and returned to dim red light. Stable phase shifts in peak *CAB2::luciferase* expression (advances are positive; delays are negative) were calculated relative to unpulsed controls. Phase shifts are plotted relative to the circadian time (CT; the endogenous period/24 h) of the pulse. Open bar is subjective day; hatched bar is subjective night. Modified from Devlin and Kay (2001).

The question of which of these two models (or both) is the most accurate in describing the process of entrainment is not yet addressable at the molecular level. Many systems, as in plants, are still in search of all the constituent components of the oscillator, and of the photic input pathway. This review will summarize what we currently know about the components involved in the entrainment of the plant circadian oscillator.

4.2   Light perception and entrainment

4.2.1   Phytochromes

Light is the primary and best-understood entraining stimulus for plants. The starting point has been the identification of the photoreceptors involved in photoentrainment. The well-characterized red light photoreceptors (the phytochromes) were the likely candidates to consider. The phytochromes rely on a covalently attached linear tetrapyrolle (phytochromobilin) as the light-absorbing moiety, and are activated by red light. Characteristic of the family, many phytochrome-mediated responses are reversed by subsequent far-red light illumination. There are two classes of phytochrome molecules; light-labile and light-stable forms. Phytochrome A (phyA) encodes the light-labile phytochrome, which is degraded rapidly in the presence of light. Phytochromes B, C, D and E in *Arabidopsis* are the light stable forms. Most of the phytochromes in the plant are phyA and phyB, and mutant analyses confirm that phys C, D and E are largely additive to the effects of phyB (Monte *et al.*, 2003). *In vitro* assays show that phytochromes can autophosphorylate in response to red light, and substrates have been identified that can, in turn, be phosphorylated by this activated form of phytochrome. Although results such as these strongly suggest that phytochrome acts as a light-activated serine-threonine protein kinase, unequivocal *in vivo* evidence for this conclusion is still lacking (Quail, 2002).

Null mutations in the phytochrome genes were used to demonstrate the role of each of the five members in light entrainment of the clock. Fluence rate response curves of wild-type *Arabidopsis* in red and blue light follow Aschoff's Rule in showing the expected dependency of free-running period on irradiance for diurnal organisms: period shortens with increasing light intensity (Somers *et al.*, 1998b, 2000). In all cases tested, loss of one or more phytochromes causes a reduced sensitivity to red light, as indicated by a longer period, relative to wild-type. Overexpression of phyB shortens period in red light. However, phyA and phyB appear to work over different fluence rates in mediating red light signaling to the clock. At high red light fluence rates (more than 5 μmol m^{-2} s^{-1}) period is lengthened by 1.5–2 h in *phyB* mutants, while loss of phyA has no effect. Conversely, at fluence rates below 5 μmol m^{-2} s^{-1} *phyA* mutants cause period lengthening, and the absence of phyB has no effect (Fig. 4.2; Somers *et al.*, 1998a). The *phyAphyB* double mutant is additive, with a free-running period consistently 2 h longer than wild-type over a 200-fold range in red light intensity (Devlin & Kay, 2000). Additional removal of other light-stable

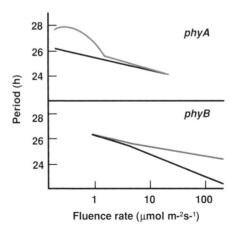

Figure 4.2 Non-overlapping fluence-rate range effects of the *phyA* and *phyB* mutants on free-running period in red light. Different sets of plants grown in 12 h white light/12 h dark cycles were transferred to a range of different constant red light intensities for 4 days and the free-running period of *CAB2::luciferase* expression was determined. Black line: WT; gray line: *phyA* or *phyB* mutant. Modified from Somers *et al.* (1998).

phytochromes (e.g. phyD or phyE) to create triple mutants causes a slight further period lengthening. Taken together, the phy mutant analyses clearly demonstrate that phyA and phyB are the primary phys responsible for mediating parametric red light entrainment of the circadian clock.

4.2.2 Cryptochromes

The cryptochromes are flavin-binding proteins. Two different chromophores, a light harvesting folate or deazaflavin and a catalytic flavin adenine dinucleotide (FMN), reside within the amino terminal portion of the protein (Lin, 2002; Cashmore, 2003; Lin & Shalitin, 2003). The cryptochromes show strong similarity to DNA photolyases, but do not possess photolyase activity. Both members of the cryptochrome family in *Arabidopsis* (*cry1* and *cry2*) autophosphorylate in response to blue light, and may act as blue-light activated protein kinases (Shalitin *et al.*, 2002, 2003; Bouly *et al.*, 2003).

Although the two cryptochrome blue light receptors (cry1 and cry2) do not absorb in the red, the *cry1 cry2* double mutant shows significantly longer periods than wild-type at red light fluence rates less than 20 μmol m^{-2} s^{-1}. Additionally, at low-fluence white light, the period in the *phyA cry1* double mutant is not an additive effect of both mutations, but is lengthened similar to that of the *cry1* single mutant alone (Devlin & Kay, 2000). Since free-running period in the *phyA* mutant is longer than WT at low-fluence white light, the epistasis of *cry1* to *phyA*, together with the red light effects in the *cry1 cry2* double mutants, suggest that cry1 may act as a downstream red light signaling intermediate in phyA signaling to the clock (Devlin & Kay, 2000). *In vitro* assays show a phyA-mediated red light dependent phosphorylation of cry1 and 2, consistent with the above idea (Ahmad *et al.*, 1998). Additional tests of the *phyA cry1* double mutant in red light and blue light alone are necessary to confirm this hypothesis.

If a potential role for the cryptochromes in red light signaling to the clock is unexpected, it is not surprising to find effects of the *cry1* and *cry2* mutants on period in blue light. Absence of cry1 causes a 2–3 h period lengthening at high and low fluence rates, with little effect at intermediate fluence rates (Fig. 4.3). Overexpression of cry1 causes period shortening. The *cry2* mutant alone has a minor effect on period in blue light (Somers *et al.*, 1998a). The *cry1 cry2* double mutant, however, causes a consistently strong 3–4 h lengthening of period over a 50-fold change in fluence rate (Fig. 4.3) (Devlin & Kay, 2000). This suggests some degree of redundancy in the action of the two photoreceptors. Additionally, unlike the loss of rhythmicity observed in mammalian systems in the complete absence of cryptochromes (van der Horst *et al.*, 1999), the plant circadian system continues to function in blue light (and darkness) without cryptochromes. PhyA mediates low fluence blue light signaling to the clock (Somers *et al.*, 1998a), but this may be dependent on the presence of cry1. These results can be coupled with the observed continued cycling of the *phyA phyB* double mutant in red light, and similar robust cycling of leaf movement in the *phyA phyB cry1 cry2* quadruple mutant in white light (Yanovsky *et al.*, 2000),

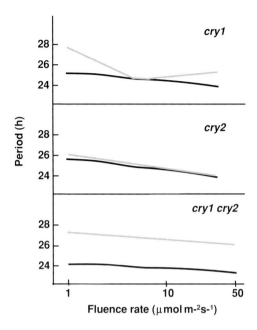

Figure 4.3 Partial redundancy of the *cry1* and *cry2* photoreceptors in mediating blue light control of free-running period. Different sets of plants grown in 12 h white light/12 h dark cycles were transferred to a range of different constant blue light intensities and the free-running period of *CAB2::luciferase* expression was determined. Black line: WT; gray line: *cry1, cry2 or cry1 cry2* mutant. Modified from Devlin and Kay (2000).

to suggest that there are multiple phototransduction pathways to entrain the plant circadian clock.

Most tests have been designed to determine the role of a candidate gene in the clock by observing the free-running period in constant light conditions. A shortening or lengthening of period implies some alteration either in the phototransduction pathway to the central oscillator, or in some component that impinges on the activity of the oscillator. When tested over a range of fluence rates, this assay can provide information on the state of the parametric entrainment system. An alteration in the slope, relative to wild-type, suggests a change in the entrainment pathway or in its robustness. Few experiments have been designed to test the effects of a mutant allele on the discrete, or non-parametric photoentrainment properties of the clock. One such test examined the effects of the loss of phyA on the ability to entrain to 10 h/10 h blue/dark cycles. There was a clear reduction in the ability to re-entrain at low blue light fluence rates (less than 2 μmol m^{-2} s^{-1}), often taking 2–3 cycles to match the wild-type, which was able to re-entrain immediately. Re-entrainment at high blue light fluence rates was the same in *phyA* mutant and WT (sp). Interestingly, these results match the fluence rates over which a phyA deficiency causes a lengthening of period under continuous blue light. The similarity of the fluence rate range over which phyA affects entrainment under these two conditions supports the notion that light signaling to the clock under parametric and non-parametric entrainment may be similar, at least via some photoreceptors (Somers *et al.*, 1998a).

A contrary view has developed from an observed change in the *phase* of the free-running period in the *phyB* mutant after release into constant white light. Loss of phyB results in a 2–3 h phase advance in the expression of *CAB2::luciferase*, in leaf movement rhythms and in the cycling of CO_2 uptake (Salome *et al.*, 2002). There is no effect on free-running period in white light. Interestingly, when entrained to temperature cycles in constant light, and released into constant temperature and light, the phase of WT and *phyB* mutants is identical. These results suggest that under parametric entrainment (constant white light), loss of phyB is inconsequential, while the discrete (non-parametric) entrainment of light/dark cycles does require functional phyB to attain the proper phase angle. These data also demonstrate that at least one difference between the light and temperature entrainment pathways is the dispensability of phyB to temperature entrainment.

4.2.3   A novel family of blue light photoreceptors

Members of the *ZEITLUPE* family of proteins was first identified by mutant screens, and by partial sequence homology to the phototropin class of blue light photoreceptors (Millar *et al.*, 1995; Kiyosue & Wada, 2000; Nelson *et al.*, 2000; Somers *et al.*, 2000). The three family members (ZEITLUPE [LKP1, ADO1], LKP2 [ADO2] and FKF1 [ADO3]) share strong amino acid sequence similarity (70–80% identity) throughout the length of the proteins. They possess three distinct domains that have been well characterized in other proteins, but are found together uniquely in this family: the LOV (*L*ight, *O*xygen, *V*oltage) domain, the F-box domain, and kelch motifs.

LOV/PAS domains were originally identified in proteins that function as sensors of light or oxygen and in members of the EAG family of voltage-gated potassium channel proteins. LOV domains are a subset of the larger group of PAS (Per Arnt Sim, named after the proteins first found to contain these motifs) domains found in a diverse variety of organisms and proteins including PERIOD, a key component of the *Drosophila* and mammalian clock (Huala *et al.*, 1997; Taylor & Zhulin, 1999). These domains can act as sites for protein–protein interactions and can also mediate ligand binding speices (Ballario *et al.*, 1998). The crystal structures of several LOV/PAS domains have been solved and they have been found to comprise a β-sheet core with α-helices on either side forming a hydrophobic pocket allowing binding of a ligand (Gong *et al.*, 1998; Pellequer *et al.*, 1999).

The LOV domain, near the amino terminus of ZTL, has been shown to bind flavin in the two other divergent types of protein: the phototropins (Chapter 5) and WHITE COLLAR-1 (WC-1) (Crosson & Moffat, 2001; Crosson *et al.*, 2003). The phototropins mediate blue-light dependent photomovement in plants, and autophosphorylate in response to blue light. Blue light absorbance is conferred by the flavin mononucleotide (FMN) bound to each of the two LOV domains (Christie *et al.*, 2002). The *phot1* mutant has no detectable effect on free-running period in blue light (Somers & Kay, unpublished), but the *phot2* and the *phot1 phot2* double mutants have not been tested.

WC-1 is required for all blue light responses in *Neurospora crassa*, as well as for circadian clock function (Liu, 2003). Flavin adenine dinucleotide (FAD) binds the WC-1 LOV domain and confers blue-light dependent DNA binding (Froehlich *et al.*, 2002; He *et al.*, 2002). The LOV domains of FKF1 and ZTL have been shown to bind FMN *in vitro*, and can even partially rescue WC-1 activity when substituted for the endogenous LOV domain (Cheng *et al.*, 2003; Imaizumi *et al.*, 2003). These results strongly suggest that the ZTL family may define a new class of blue-light photoreceptor. One striking difference, however, is the absence of a dark recovery of the photoactivated FKF1 LOV domain. *In vitro*, the phototropin LOV domains can undergo a light-activated photocycle in which the light-altered absorption spectra returns to the inactivated state after some period in the dark (Salomon *et al.*, 2000). The absence of this cycle in the isolated FKF1 LOV domain suggests that light absorption may be a one-way process and the molecule is permanently activated (or inactivated) when illuminated, to be later removed from the system in some other way. Alternatively, a partner molecule may be required to assist in completing the return to the ground state, or the *in vitro* results relying on just a portion of the full molecule may be inaccurate. Development of a light-dependent *in vitro* assay will be needed to fully determine whether this is a new blue light photoreceptor family.

The two carboxy-terminal motifs (F-box and kelch domains) of the ZTL family of proteins act together as a key component of a Skp1-F-box-Cullin (SCF) E3 ubiquitin ligase. This class of ubiquitin ligases relies on the F-box protein to link a target protein to the larger portion of the E3 complex to complete the ubiquitination process that sends the targeted polypeptide to the 26S proteasome for proteolysis (Cardozo & Pagano, 2004; Vierstra, 2003). The 45-amino acid F-box domain physically interacts with the Skp1-like protein (in *Arabidopsis*: ASK [*Arabidopsis* Skp1-like]) of the complex, while the carboxy-terminal region of the F-box protein associates with the proteolytic target. Evidence that ZTL participates in SCF complex formation *in vivo* comes from co-immunoprecipitation experiments, which demonstrates the necessity of the F-box domain, and the dispensability of the LOV domain, in associating with ASK/Cullin complex (Han *et al.*, 2004). The six kelch motifs likely fold into a β-propeller and act to facilitate the interaction with the proteolytic target (Adams *et al.*, 2000; Somers, 2001). The same functions for the F-box and kelch domains are presumed for FKF1 and LKP2.

Loss-of-function and null mutations in *ZTL* lengthen free-running period by 3 h in white, red and blue light at high fluence rates, but cause a much more severe effect (8–9 h lengthening) at low fluence rates (Somers *et al.*, 2000). Conversely, overexpression of ZTL shortens period and flattens the fluence rate response curve to nearly horizontal at high levels of overexpression (Somers *et al.*, 2004). This change in slope strongly indicates that the phototransduction pathway has been altered. Interestingly, a single point mutation in the ZTL LOV domain eliminates the fluence rate-dependence caused by the null allele, and period is consistently 3–4 h longer over all fluence rates (Kevei *et al.*, 2005). Taken together with the likely FMN-binding of the LOV domain, these results suggest that ZTL is part of the photoentrainment pathway under continuous light.

LKP2 is the family member most closely related to ZTL, and overexpression also results in arrhythmicity (Schultz *et al.*, 2001). However, the *lkp2* mutant has no apparent circadian phenotype, and the period of the *ztl lkp2* double mutant is identical to the *ztl* single mutant (D. Somers, unpublished data). It may mediate other, less obvious processes, or enhance ZTL activity in a subtle way.

In contrast, neither the overexpression of FKF1 nor the *fkf1* mutant shows any defect in circadian period (Imaizumi *et al.*, 2003). FKF1 appears to be dedicated to a role in the control of photoperiodic timing, downstream of the circadian clock (Imaizumi *et al.*, 2003).

4.3   Signal transduction and interaction

4.3.1   ZTL and phototransduction: phy and cry photoreceptors

Mutations in the phytochromes, cryptochromes and ZTL cause period lengthening when absent. Additionally, yeast two-hybrid tests show that phyB-ZTL and cry1-ZTL interactions are plausible *in planta* (Jarillo *et al.*, 2001). Therefore, what is the relationship between the known photoreceptors (phys and crys), and the putative new photoreceptor, ZTL? Period lengthening under red light in *ztl* null mutants cannot be explained by direct red light absorption by the LOV domain. However, red-light-activated Pfr phytochrome could act as an intermediary to convey red light irradiance information to the clock via a phy-ZTL interaction. A direct signaling pathway of this kind would mean that *ztl* mutations should be epistatic to phyB mutations in red light. *ztl phyB* double mutants were tested under red light, and period lengthening was greatly enhanced (31–33 h) over the *phyB* (25.5 h) and *ztl-1* (28.1 h) single mutants (D. Somers, unpublished data). These results suggest that phyB and ZTL act in parallel, or additively, in the mediation of light signaling to the clock, although it remains possible that phyB acts in part through ZTL as well. It is also possible that other light-stable phys act through ZTL (Fig. 4.4).

Similar experiments testing the *cry2 ztl* double mutants in red light show a surprising shortening of period, relative to the *ztl* single mutant. The *cry2* mutant has no effect on period under red light, but causes a 2 h period shortening (26 h vs. 28 h) when combined with *ztl-1* (D. Somers, unpublished). Cry2 and phyB interact *in vivo* (Mas *et al.*, 2000), and genetic evidence points to cry2 as a negative regulator of phyB activity during floral induction (Mockler *et al.*, 1999). Hence, in the absence of cry2, red light signaling via phyB may be enhanced, partially counteracting the long period effect of the *ztl* mutation. In this way, cry2 may act as a negative regulator of phyB-mediated phototransduction to the clock (Fig. 4.4).

4.3.2   ZTL and phototransduction: ELF3

ELF3 plays a key role in circadian phototransduction, though its molecular function is still uncertain (Hicks *et al.*, 2001; Liu *et al.*, 2001). Under continuous

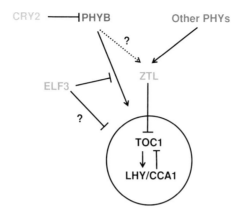

Figure 4.4 Working model of a portion of the red light phototransduction pathway to the Arabidopsis circadian oscillator. The above relationship among multiple photoreceptors and signaling intermediates is proposed based on known genetic and molecular interactions. See text for details.

light, *elf3* mutants become arrhythmic very quickly, as measured by promoter-luciferase reporter assays and by leaf movement (Hicks *et al.*, 1996; Covington *et al.*, 2001). Overexpression of ELF3 (ELF3 OX) causes a light dependent lengthening of period (26.5 h) under continuous light (Covington *et al.*, 2001). In darkness, however, rhythmicity in the *elf3* mutant is sustained and the overexpressor has no effect on period. These results highlight the light-dependent nature of ELF3 function that is also evident in the aberrantly long hypocotyls seen in continuous red and blue light, but not in darkness, where *elf3* plants appear WT (Zagotta *et al.*, 1996). The molecular basis of the arrhythmicity and long hypocotyl phenotypes of the *elf3* mutants is still unclear, but the extremely early flowering is linked to changes in *CONSTANS* (*CO*) and *FLOWERING TIME* (*FT*) expression. The mRNA levels of both genes are strongly upregulated in the absence of ELF3, and strongly repressed in ELF3 OX (Suarez-Lopez *et al.*, 2001) (Kim *et al.*, 2005). These results fit well with the current model of photoperiodic flowering, whereby coincidence of high CO levels during the photoperiod results in increased *FT* transcription, leading to the floral transition (Yanovsky & Kay, 2002; Valverde *et al.*, 2004). Therefore, one function of ELF3 is to negatively regulate CO levels.

elf3 mutations increase light sensitivity to a number of photodependent processes, leading to the notion that it normally de-sensitizes phototransduction systems to light in general, and to the circadian clock in particular. The acute response of *CAB2* expression to red light pulses is accentuated in *elf3* mutants and diminished in the ELF3 OX (Anderson *et al.*, 1997; Covington *et al.*, 2001). Most revealing, however, is the effect of ELF3 dosage on red light and blue light PRCs. The *elf3* mutant increases the amplitude of both PRCs, particularly during subjective night. Conversely, the ELF3 OX mutes the PRC amplitude, particularly in red light. The amplitude of a PRC can often be adjusted by the intensity of the light delivered during the protocol: higher intensity light pulses often result in greater phase shifts, with a strong discontinuity between advances and delays (Type 0 PRC). Type 1 PRCs display small phase-shifts, show a continuous transition between advances and delays, and can often be generated by weak light pulses. Hence, the *elf3*/ELF3 OX results can be interpreted as a modulation of light input to the clock, dependent on ELF3 dosage. Absence of ELF3 causes hypersensitivity of the entrainment pathway to light, particularly during subjective night, and high levels of ELF3 diminish phototransduction to the oscillator (Covington *et al.*, 2001).

A 'dark release' assay was devised to ascertain when during the circadian cycle ELF3 might act (McWatters *et al.*, 2000). The timing of the first peak of *CAB::luciferase* activity after LD entrainment and release into the dark was used to infer whether the circadian system continues to operate during a two-day period in constant light (LL), and is simply masked in the *elf3* mutant, or whether the clock has in fact stopped. Absence of *elf3* appeared to only affect the clock after 10 h in LL, after which peak *CAB::luc* expression was determined solely by the light-to-dark transition. Additionally, the normally 'clock-gated' acute response to light is largely absent in *elf3* mutants, with the system highly sensitive to light pulses at all circadian times (McWatters *et al.*, 2000). The authors come to a similar conclusion as that derived from the PRC experiments: ELF3 acts to modulate light input to the circadian system, and it is acting near the subjective dusk and into the subjective night.

Double mutant combinations between the *elf3*-1 mutant, the ELF3 OX and ZTL were created to test the roles of both proteins relative to each other. ZTL protein levels are post-transcriptionally regulated and cycle robustly under LD, with peak expression near ZT 13 (Kim *et al.*, 2003b; Somers *et al.*, 2004). ELF3 protein levels are rhythmic and peak around the same time (Liu *et al.*, 2001), so a direct physical or genetic interaction appeared possible. However, *ztl* mutant/ELF3 OX individuals have a strongly reduced rhythmic amplitude (*CAB::luc*) and longer period (30 h) than either line alone, suggesting an additive interaction between the two. ZTL OX or *ztl* mutant combinations with *elf3-1* result in *elf3*-like arrhythmicity in LL. However, when the timing of the first peak of *CAB::luc* expression is examined after entrained plants are released into constant dark (DD), it is apparent that the effect of ELF3 and ZTL on the circadian system is largely additive (Kim *et al.*, 2005) (Fig. 4.4).

As an additional complication to this network, ELF3 and phyB can interact in two-hybrid assays and *in vitro*. ELF3 interacts specifically with the C-terminus of phyB, linking ELF3 directly to phytochrome signal transduction (Liu *et al.*, 2001). The phenotypes of the *elf3-1 phyB* double mutant and the *phyB* ELF3 OX plants show *phyB* is largely epistatic to the ELF3 phenotype, although circadian phenotypes were not assessed in these tests (Reed *et al.*, 2000; Liu *et al.*, 2001). Additionally, ELF3 overexpression is very effective in lengthening period under constant blue light, demonstrating that blue light signaling to the clock is also modulated by ELF3 (Covington *et al.*, 2001).

4.3.3 Other potential photoentrainment intermediates

Like *ELF3*, *ELF4* was first identified as an early flowering mutant. Similar to the *elf3-1* mutant, *elf4-1* has very long hypocotyls under short days and disrupted rhythmicity under constant light. The lesion is more extensive than in *elf3* mutants, as rhythmicity in darkness is also affected. ELF4 mutations appear to cause a loss in the precision of the circadian clock, with some individuals cycling with unusually short periods, and others showing aberrantly long periods (Doyle *et al.*, 2002). The sequence of the predicted ELF4 polypeptide shows no clear relationship to other known proteins. Red light dependent expression of ELF4 is diminished in *phyB* mutants, which may be part of the explanation for the long hypocotyl phenotype (Tepperman *et al.*, 2004). It is this and the similarity to the *elf3* mutant phenotype that suggests a role for ELF4 in the light input pathway, but further work is necessary to confirm this.

The *tic* mutation shares with *elf4-1* a loss in the precision of circadian cycling. This defect is not as severe as in *elf4-1*, and most of the plants show short period cycling of clock-regulated gene expression in the light and dark (Hall *et al.*, 2003). But not all phenotypes are affected in the same way: leaf movement rhythms are close to wild-type period, or slightly longer. The *tic* mutation also affects circadian gating of light input, causing an enhanced level of sensitivity to light pulses, similar to the effects of *elf3* mutations. *TIC* has not been cloned, but indirect 'release' assays suggest that it may normally act late at night or near dawn. *tic1 elf3-1* plants are severely arrhythmic under constant conditions, while under light/dark cycles the plants show no sign of anticipation of dawn. These results are consistent with the two genes acting at different points in the circadian cycle (ELF3 near dusk (McWatters *et al.*, 2000) and TIC near dawn), with the loss of both resulting in completely dysfunctional circadian system (Hall *et al.*, 2003).

The SRR1 protein is another example of the close linkage between light signaling and the circadian system. Originally identified as a mutant impaired in phyB signaling (*srr1; sensitivity to red light reduced 1*), close examination of the phenotype uncovered a wide-ranging effect on circadian cycling (Staiger *et al.*, 2003). All clock-regulated genes that were tested (*CAB, CCR2, CCA1, CAT3, TOC1, GI*, and *FKF1*) showed 2–3 h shorter periods, more rapid dampening, and reduced amplitude of the oscillation in the *srr1* mutant background in LL. Leaf movement rhythms were also shortened. When entrained *srr1* mutants were released into DD for up to 3 days, the period was similarly shorter, indicating a role beyond only phyB-mediated phototransduction. Like ELF3 and ELF4, SRR1 has no obvious protein motifs that place it into a class of previously characterized domains. However, it shares strong sequence similarity with putative metazoan homologs. The function of the mammalian and *Drosophila* genes is equally unknown, but the sequence similarities with SRR1 suggest the possibility of a conserved function across eukaryotic circadian systems (Staiger *et al.*, 2003).

The *Arabidopsis* five-member *PRR* gene family (pseudoresponse regulators) is so-called because of close similarity to the response regulator component of the bacterial two-component phosphorelay signaling system (see Chapters 1 and 2 and Mizuno (2004)). All five members have a lesser or greater effect on free-running period, but the molecular role of each is still unclear. Interestingly, some of the single gene knockout mutants show no circadian defect in constant darkness, although short and long free-running periods occur in constant light. The *prr9*, *prr7* and *prr5* mutants cause period lengthening (*prr9* and 7) or period shortening (*prr5*) in red and blue light but are WT in DD. These results suggest roles for these genes in the photoentrainment pathway, rather than in the central oscillator itself (Eriksson *et al.*, 2003; Farre *et al.*, 2005). However, the *prr9 prr7* double mutant causes a strongly dampened and distinctly longer free-running period in DD, contrasting the clear absence of a DD phenotype in the two single mutants. A similar unexpected synergism in the same double mutant is also apparent when tested under thermal entrainment conditions (see below).

Another *PRR* double mutant combination has recently been tested. The *prr5 prr7* double knockout results in very low amplitude and short period cycling of message levels in LL, which rapidly damps to arrhythmicity (Nakamichi *et al.*, 2005). This holds for all genes tested, including *TOC1*, *CCA1*, *LHY* and *GI*. In DD, a similar rapid damping of rhythmicity occurs, indicating that these genes are also required for sustained cycling in the absence of light input.

Clearly, the line between phototransduction pathway and oscillator becomes blurred as increasing numbers of components are removed from the system. Molecular or biochemical pairings of the members of these photoreceptor and signaling families will greatly help sort out their roles.

4.4   Coupling to the central oscillator

The plethora of photoreceptors and their various interaction and interactors need, eventually, to connect to core elements of the circadian oscillator. The precise nature of all the core clock components is still lacking, but of the three proteins that are thought to comprise some of the essential elements (CCA1/LHY/TOC1), one has been molecularly linked to the phototransduction pathway.

TOC1/PRR1 is the best characterized of the *PRR* gene family (Chapter 1 and 2 and (Mizuno, 2004)). *toc1* mutants shorten free-running period and *TOC1* over-expressors lengthen period, and these effects are evident in both the light and the dark (Somers *et al.*, 1998b; Strayer *et al.*, 2000; Mas *et al.*, 2003a). These phenotypes are reciprocal to the ZTL loss- and gain-of-function lines, and led to the discovery of a genetic and molecular interaction between the two. *TOC1* is clock-regulated, with peak mRNA and protein levels during the subjective night, in phase very much with ZTL. In *ztl* mutants, TOC1 levels are damped high, although *TOC1* mRNA continues to cycle, suggesting that ZTL acts post-transcriptionally to regulate TOC1. In the wild-type, extended darkness causes TOC1 levels to diminish, and conversely, protein levels damp high in LL. The stabilization of TOC1 levels in DD in the *ztl* mutant background has led to the suggestion that ZTL, in its role as an F-box protein, normally leads to degradation of TOC1 in the dark (Mas *et al.*, 2003b). Co-immunoprecipitation of TOC1 and ZTL from plant extracts supports this notion, as does the loss of this interaction in the *ztl-1* background. The *ztl-1* allele does not affect ztl levels, nor does it alter the normal cycling of the protein (Somers *et al.*, 2004). However, the point mutation in the third kelch domain appears sufficient to abrogate the TOC1/ZTL interaction, eliminating ZTL from the TOC1 immunoprecipitates. The consequent accumulation of TOC1 leads to the long period phenotype, which the mild TOC1 OX phenocopies (Mas *et al.*, 2003a).

These results now provide a testable model of photoentrainment. Can short pulses of light (or dark) reset the oscillator via a consequent rapid change in TOC1 levels? Is this the primary resetting gateway to the oscillator, or are there multiple, parallel photoentrainment paths, affecting other oscillator components (e.g. CCA1/LHY) as well? Will phase resetting be compromised in the *ztl* mutant background due to the inability to effect sufficiently large changes in TOC1 levels? If so, the PRC of *ztl* mutants should be markedly different from WT. As noted earlier, the LOV domain of the ZTL family may not revert to the ground state after light absorption. This may mean that ZTL is only active in the dark, where it can effect TOC1 degradation, and becomes *light-inactivated* during the photoperiod.

Until recently, a very different model of light-dependent phase resetting had been considered. phyB is nuclear localized after red light stimulation, suggesting a very short signal transduction pathway between light perception and gene transcription. PIF3, a bHLH transcription factor, interacts with Pfr_{phyB} and can also bind the G-box, a DNA sequence found in many light-regulated promoters, including the clock-associated genes *CCA1* and *LHY* (Martinez-Garcia *et al.*, 2000). *In vitro* evidence for PIF3 binding to phyB and DNA suggested the possibility of a very short phototransduction pathway toward the activation of clock-related genes. However, a recent revision in the phenotype of the *pif3* mutant, from one of *hypo*sensitivity to one of *hyper*sensitivity to red light, has caused a re-thinking of this model (Kim *et al.*, 2003a; Bauer *et al.*, 2004; Monte *et al.*, 2004; Park *et al.*, 2004). Now PIF3 appears to be a negative regulator of phyB signaling, during certain times and for certain processes in development. Rapid, red-light-induced proteolysis of PIF3 is observed *in vivo*, though the protein can re-accumulate in the dark (Monte *et al.*, 2004). However, *pif3* mutants have no circadian period defects, effectively eliminating this protein from inclusion in the circadian cycle (Monte *et al.*, 2004; Oda *et al.*, 2004).

4.5   Circadian gating of photic input

Implicit in the shape of the PRC is the notion that the clock regulates its own responsiveness to light. At certain times during the circadian cycle a light pulse can reset the oscillator; at other times the same input has no effect. Additionally, the extent of certain clock-controlled outputs, such as gene expression, is 'gated' by the clock over the course of the circadian day. For example, the magnitude of the rapid upregulation of CAB expression in response to light (the 'acute' response) varies depending on the circadian phase, with very little expression during the subjective night (Millar & Kay, 1996). Other clock-gated outputs include hypocotyl expansion, and the shade avoidance response (Dowson-Day & Millar, 1999; Salter *et al.*, 2003)

In principle, gating of light input by the clock could occur at two separate points. The primary light signaling pathway that leads to the alteration in clock components that effect resetting could itself be under clock control. Alternatively, one or more of the output pathways leading from oscillator activity could act to gate additional light input to affect other outputs. In this instance, the gating output would in effect be a sub-oscillator, or slave, of the central oscillator that modulates the effect of a photic signal downstream of the clock. The aforementioned acute response, which depends on brief light pulses to be observable, is likely an example of the latter, since resetting of the central oscillator by short light pulses is not effective in *Arabidopsis*.

Reports of the cycling of photoreceptor-promoter-reporter genes and mRNA levels under LD or LL have led to models incorporating gating of light input to the clock through phase-dependent changes in photoreceptor abundance (Bognar *et al.*, 1999; Hall *et al.*, 2001, 2002; Toth *et al.*, 2001). However, the critical light-sensing components are the phy and cry proteins, and a recent study characterized all five phytochrome protein expression patterns under diurnal and circadian light cycles. Under light/dark cycles some cycling of phyA, B and C is observed, but it is out of phase with previously reported cycling of PHY mRNA levels. In addition, the peaks of the oscillations in the phyA, B, and C apoproteins occur during the dark phase, when the mRNA levels are reported to be low (Toth *et al.*, 2001). The authors conclude that, although there are often robust responses of *PHY* promoter activities to light cycles, oscillations in *PHY* mRNAs are not rapidly converted to corresponding oscillations in protein levels. Under LL phy protein cycling was of extremely low amplitude, suggesting that the diurnal cycling of phyA, B, and C may depend more on their stabilities in light than on their transcriptional or post-transcription mRNA regulation (Sharrock & Clack, 2002).

An additional factor, however, lies in the question of whether 'freshly-made' phytochromes are qualitatively different from the pre-existing 'bulk' phytochrome (Bognar *et al.*, 1999). While this may be difficult to discern, the qualitative difference between nuclear-localized and cytoplasmic phytochrome may be more relevant. Light-dependent nuclear import of phytochromes (Kircher *et al.*, 2002), coupled with the nuclear localization of candidate clock components (CCA1/ LHY/ PRRs) may provide a robust way of gating light input to the clock. It will be important to determine if circadian regulation of nuclear import occurs under LL or DD, which would be a clock-regulated way of changing the local (nuclear) concentration of the photoreceptors, and consequently photic input.

As noted in an earlier section, ELF3 is a prime candidate as a gating agent of light input to the clock, by virtue of the specific phase of maximum expression (subjective dusk) that coincides with the apparent time of maximum action (attenuation of light input). ELF3 currently stands as the best example of the 'zeitnehmer' concept, which holds that clock-regulated control of light input helps stabilize the system as a whole (Roenneberg & Merrow, 1998, 2003). Manipulation of ELF3 expression to other phases of expression (e.g. dawn) would be an interesting way to test this concept experimentally.

4.6 Thermal entrainment

Temperature is the other major environmental factor that can reset circadian clocks (Rensing & Ruoff, 2002). A temperature variation of only 4°C (12 h 24°C/12 h 20°C) is sufficient to stably entrain *Arabidopsis* under constant light (Somers *et al.*, 1998b; Michael & McClung, 2002). Based on the expression peaks of a *CAB2::luciferase* reporter, the warm temperature (24°C) corresponds to the light period of a standard light/dark entrainment protocol, and the cold temperature (20°C) corresponds to the dark period. Recently, stronger temperature cycles (22° C/12°C) have been used to establish temperature phase response curves (tPRCs) in *Arabidopsis*. Using three different promoter::luciferase fusions (*TOC1::LUC, CAB2::LUC* and *CAT3::LUC*), the authors demonstrate two significantly different classes of tPRCs. Both the amplitudes and the times of the peaks and troughs of phase advances and delays are altered, with the amplitude of the *CAT3::LUC* tPRC markedly more robust than the identical tPRCs derived from the other two reporters (Michael *et al.*, 2003). These data support the notion of two distinct circadian oscillators in *Arabidopsis*. However, it is not clear whether they are equivalent and parallel oscillators, whether they are interactively linked, or whether one may be a subservient 'slave' to a primary central oscillator. A parallel study using the same reporters to generate a light PRC has not been conducted, which would be a useful test of how similar the temperature and light entrainment mechanisms are.

At the molecular level, very recent work has begun to examine the effects of specific single and double mutant knockouts on temperature entrainment. Unexpectedly, the *prr7prr9* double mutant is extremely unresponsive to normally effective temperature cycles, although the single mutants alone show no such defect (Salome & McClung, 2005). Most striking is the effect of the double mutation on the tPRC. Over much of the circadian day the *prr7prr9* double mutant is largely unresponsive to temperature pulses. The effect is particularly strong during the late subjective night when WT plants show strong phase delays in response to temperature pulses (Michael *et al.*, 2003; Salome & McClung, 2005), while the double mutant phase advances or delays 1 h or less. This inability to entrain to temperature cycles is not mirrored under light/dark entrainment protocols, although free-running period is very long (see above). Thus, the double mutant demonstrates the importance of these genes in both the light (parametric) and temperature (parametric and nonparametric) entrainment pathways. A light PRC for this double mutant would help determine how light signaling in this background has changed.

Surprisingly, very similar results are seen in the *prr5prr7* double mutant. As noted earlier, the absence of both genes strongly shortens free-running period with a rapid transition to arrhythmicity for most genes tested. Under light/dark cycles the amplitude of mRNA cycling is strongly reduced in these genes, but under thermal cycles the dampening effect is more severe. This applies to a wide range of clock-controlled genes (*CCA1, LHY, TOC1, PRR3, GI* and *CAB2*), all of which damp to high levels of expression (Nakamichi *et al.*, 2005). Those transcripts that do show cycling are phase advanced, as in light/dark cycles, and they do not coincide with the temperature transitions. This indicates that the clock is still entraining to thermal cycles, and that PRR5 and PRR7 do not act exclusively in the photoentrainment pathway.

The rate of every biochemical reaction is subject to temperature effects, making it exceedingly difficult to determine what might be the temperature sensor that initiates a 'temperature input pathway'. While there may be no one molecule that serves as the initiation point of temperature entrainment, screens for mutants that cannot respond to temperature entraining cycles might identify components that are particularly sensitive to temperature fluctuations.

Despite the aforementioned experimental advances, we are lacking even the slightest hint of a molecular understanding of temperature entrainment in plants. With only a few molecular players in hand, *Neurospora* researchers have devised a working model of temperature entrainment that relates changes in absolute levels of *FREQUENCY (FRQ)* protein to temperature resetting (Liu *et al.*, 1998; Liu, 2003). Given the increasing number of clock-associated components that have been identified in *Arabidopsis*, the time may be ripe for similar experiments to be undertaken in the plant field.

4.7   Conclusion

The mechanism of entrainment remains an elusive goal. With an array of at least seven to ten photoreceptors dedicated to relatively narrow and specific wavebands, plants have much more flexibility, and therefore more potential complexity, in how to make the connection between the photic environment and the oscillator. Evidence for photoreceptor-dependent modulation of the activity of other photoreceptors (e.g. cry/phy and cry/phy/ZTL family interactions) furthers the emerging notion of a web of photic interactions. Beyond that first step of light perception, it is also clear that a number of modifiers/modulators of phototransduction (e.g. ELF3, ELF4, TIC) play important roles in maintaining the appropriate level of signal to the oscillator.

While the complexity may appear overwhelming, one message coming from all this may be the likelihood of multiple photoentrainment pathways. The notion of all phototransduction converging on a single component of the oscillator to effect a phase shift may be attractive, but likely to be naïve and wrong. Given the gene-family nature of the current candidate and emerging 'core' clock components (e.g. CCA1/LHY; the PRRs; the PIF/PIL bHLH transcription factors; and the DOF transcription factors), it appears more likely that many of these, under some circumstances or another, may be targets of light-induced change that can effect stable re-entrainment. The challenge ahead lies in choosing well the ways that will best allow one to tease out that first strand of the entrainment pathway web. Even with an incomplete suite of players, efforts in this direction will lead to new insights into the circadian system.

References

Adams, J., Kelso, R. & Cooley, L. (2000) The kelch repeat superfamily of proteins: propellers of cell function. *Trends Cell Biol.*, **10**, 17–24.

Ahmad, M., Jarillo, J.A., Smirnova, O. & Cashmore, A.R. (1998) The cry11 blue light photoreceptor of *Arabidopsis* interacts with phytochrome A in vitro. *Molecular Cell*, **1**, 939–948.

Anderson, S.L., Somers, D.E., Millar, A.J., Hanson, K., Chory, J. & Kay, S.A. (1997) Attenuation of phytochrome A and B signaling pathways by the *Arabidopsis* circadian clock. *Plant Cell*, **9**, 1727–1743.

Aschoff, J. (1979) Circadian rhythms: influences of internal and external factors on the period measured in constant conditions. *Z. Tierpsychol.*, **49**, 225–249.

Ballario, P., Talora, C., Galli, D., Linden, H. & Macino, G. (1998) Roles in dimerization and blue light photoresponse of the PAS and LOV domains of Neurospora crassa white collar proteins. *Mol. Microbiol.*, **29**, 719–729.

Bauer, D., Viczian, A., Kircher, S., Nobis, T., Nitschke, R., Kunkel, T., Panigrahi, K.C., Adam, E., Fejes, E., Schafer, E. & Nagy, F. (2004) Constitutive photomorphogenesis 1 and multiple photoreceptors control degradation of phytochrome interacting factor 3, a transcription factor required for light signaling in Arabidopsis. *Plant Cell*, **16**, 1433–1445.

Bognar, L.K., Hall, A., Adam, E., Thain, S.C., Nagy, F. & Millar, A.J. (1999) The circadian clock controls the expression pattern of the circadian input photoreceptor, phytochrome B. *Proc. Natl. Acad. Sci. USA*, **96**, 14652–14657.

Bouly, J.P., Giovani, B., Djamei, A., Mueller, M., Zeugner, A., Dudkin, E.A., Batschauer, A. & Ahmad, M. (2003) Novel ATP-binding and autophosphorylation activity associated with *Arabidopsis* and human cryptochrome-1. *Eur. J. Biochem.*, **270**, 2921–2928.

Cardozo, T. & Pagano, M. (2004) The SCF ubiquitin ligase: insights into a molecular machine. *Nat. Rev. Mol. Cell Biol.*, **5**, 739–751.

Cashmore, A.R. (2003) Cryptochromes: enabling plants and animals to determine circadian time. *Cell*, **114**, 537–543.

Cheng, P., He, Q., Yang, Y., Wang, L. & Liu, Y. (2003) Functional conservation of light, oxygen, or voltage domains in light sensing. *Proc. Natl. Acad. Sci. USA*, **100**, 5938–5943.

Christie, J.M., Swartz, T.E., Bogomolni, R.A. & Briggs, W.R. (2002) Phototropin LOV domains exhibit distinct roles in regulating photoreceptor function. *Plant J.*, **32**, 205–219.

Covington, M.F., Panda, S., Liu, X.L., Strayer, C.A., Wagner, D.R. & Kay, S.A. (2001) Elf3 modulates resetting of the circadian clock in *Arabidopsis*. *Plant Cell*, **13**, 1305–1316.

Crosson, S. & Moffat, K. (2001) Structure of a flavin-binding plant photoreceptor domain: Insights into light-mediated signal transduction. *Proc. Natl. Acad. Sci. USA*, **98**, 2995–3000.

Crosson, S., Rajagopal, S. & Moffat, K. (2003) The LOV domain family: photoresponsive signaling modules coupled to diverse output domains. *Biochemistry*, **42**, 2–10.

Devlin, P.F. & Kay, S.A. (2000) Cryptochromes are required for phytochrome signaling to the circadian clock but not for rhythmicity. *Plant Cell*, **12**, 2499–2510.

Dowson-Day, M.J. & Millar, A.J. (1999) Circadian dysfunction causes aberrant hypocotyl elongation patterns in *Arabidopsis*. *Plant J.*, **17**, 63–71.

Doyle, M.R., Davis, S.J., Bastow, R.M., McWatters, H.G., Kozma-Bognar, L., Nagy, F., Millar, A.J. & Amasino, R.M. (2002) The ELF4 gene controls circadian rhythms and flowering time in *Arabidopsis thaliana*. *Nature*, **419**, 74–77.

Eriksson, M.E., Hanano, S., Southern, M.M., Hall, A. & Millar, A.J. (2003) Response regulator homologues have complementary, light-dependent functions in the Arabidopsis circadian clock. *Planta*, **218**, 159–162.

Farre, E.M., Harmer, S.L., Harmon, F.G., Yanovsky, M.J. & Kay, S.A. (2005) Overlapping and distinct roles of PRR7 and PRR9 in the *Arabidopsis* circadian clock. *Curr. Biol.*, **15**, 47–54.

Froehlich, A.C., Liu, Y., Loros, J.J. & Dunlap, J.C. (2002) White Collar-1, a circadian blue light photoreceptor, binding to the frequency promoter. *Science*, **297**, 815–819.

Gong, W., Hao, B., Mansy, S.S., Gonzalez, G., Gilles-Gonzalez, M.A. & Chan, M.K. (1998) Structure of a biological oxygen sensor: a new mechanism for heme-driven signal transduction. *Proc. Natl. Acad. Sci. USA*, **95**, 15177–15182.

Hall, A., Bastow, R.M., Davis, S.J., Hanano, S., McWatters, H.G., Hibberd, V., Doyle, M.R., Sung, S., Halliday, K.J., Amasino, R.M. & Millar, A.J. (2003) The TIME FOR COFFEE gene maintains the amplitude and timing of *Arabidopsis* circadian clocks. *Plant Cell*, **15**, 2719–2729.

Hall, A., Kozma-Bognar, L., Bastow, R.M., Nagy, F. & Millar, A.J. (2002) Distinct regulation of CAB and PHYB gene expression by similar circadian clocks. *Plant J.*, **32**, 529–537.

Hall, A., Kozma-Bognar, L., Toth, R., Nagy, F. & Millar, A.J. (2001) Conditional circadian regulation of PHYTOCHROME A gene expression. *Plant Physiol.*, **127**, 1808–1818.

Han, L., Mason, M., Risseeuw, E.P., Crosby, W.L. & Somers, D.E. (2004) Formation of an SCF complex is required for proper regulation of circadian timing. *Plant J.*, **40**, 291–301.

He, Q., Cheng, P., Yang, Y., Wang, L., Gardner, K.H. & Liu, Y. (2002) White Collar-1, a DNA binding transcription factor and a light sensor. *Science*, **297**, 840–843.

Hicks, K.A., Albertson, T.M. & Wagner, D.R. (2001) Early flowering3 encodes a novel protein that regulates circadian clock function and flowering in *Arabidopsis*. *Plant Cell*, **13**, 1281–1292.

Hicks, K.A., Millar, A.J., Carré, I.A., Somers, D.E., Straume, M., Meeks-Wagner, R. & Kay, S.A. (1996) Conditional circadian dysfunction of the Arabidopsis early-flowering 3 mutant. *Science*, **274**, 790–792.

Huala, E., Oeller, P.W., Liscum, E., Han, I.S., Larsen, E. & Briggs, W.R. (1997) *Arabidopsis* NPH1: A protein kinase with a putative redox-sensing domain. *Science*, **278**, 2120–2123.

Imaizumi, T., Tran, H.G., Swartz, T.E., Briggs, W.R. & Kay, S.A. (2003) FKF1 is essential for photoperiodic-specific light signaling in *Arabidopsis*. *Nature*, **426**, 302–306.

Jarillo, J.A., Capel, J., Tang, R.H., Yang, H.Q., Alonso, J.M., Ecker, J.R. & Cashmore, A.R. (2001) An *Arabidopsis* circadian clock component interacts with both cry1 and phyB. *Nature*, **410**, 487–490.

Johnson, C.H., Elliott, J.A. & Foster, R. (2003) Entrainment of circadian programs. *Chronobiol. Int.*, **20**, 741–774.

Kevei, E., Gyula, P., Hall, A., Kozma-Bognar, L., Kim, W.Y., Eriksson, M.E., Tóth, R., Hanano, S., Fehér, B., Southern, M.M., Bastow, R.M., Hibberd, V., Davis, S.J., Somers, D.E., Nagy, F. & Millar, A.J. (2005) Functional analysis of the circadian clock gene *ZEITLUPE* by forward genetics. Submitted.

Kim, J., Yi, H., Choi, G., Shin, B., Song, P.S. & Choi, G. (2003a) Functional characterization of phytochrome interacting factor 3 in phytochrome-mediated light signal transduction. *Plant Cell*, **15**, 2399–2407.

Kim, W.Y., Geng, R. & Somers, D.E. (2003b) Circadian phase-specific degradation of the F-box protein ZTL is mediated by the proteasome. *Proc. Natl. Acad. Sci. USA*, **100**, 4933–4938.

Kim, W.Y., Hicks, K.A. & Somers, D. E. (2005) Independent roles for *ELF3* and *ZTL* in the control of circadian timing, hypocotyl length and flowering time. *Plant Physiology* (in revision).

Kircher, S., Gil, P., Kozma-Bognar, L., Fejes, E., Speth, V., Husselstein-Muller, T., Bauer, D., Adam, E., Schafer, E. & Nagy, F. (2002) Nucleocytoplasmic partitioning of the plant photoreceptors phytochrome A, B, C, D, and E is regulated differentially by light and exhibits a diurnal rhythm. *Plant Cell*, **14**, 1541–1555.

Kiyosue, T. & Wada, M. (2000) LKP1 (LOV kelch protein 1): a factor involved in the regulation of flowering time in *Arabidopsis*. *Plant J.*, **23**, 807–815.

Lin, C. (2002) Blue light receptors and signal transduction. *Plant Cell*, **14**, S207–S225.

Lin, C. & Shalitin, D. (2003) Cryptochrome structure and signal transduction. *Annu. Rev. Plant Biol.*, **54**, 469–496.

Liu, X.L., Covington, M.F., Fankhauser, C., Chory, J. & Wagner, D.R. (2001) ELF3 encodes a circadian clock-regulated nuclear protein that functions in an *Arabidopsis* PHYB signal transduction pathway. *Plant Cell*, **13**, 1293–1304.

Liu, Y. (2003) Molecular mechanisms of entrainment in the *Neurospora* circadian clock. *J. Biol. Rhythms*, **18**, 195–205.

Liu, Y., Merrow, M., Loros, J.J. & Dunlap, J.C. (1998) How temperature changes reset a circadian oscillator. *Science*, **281**, 825–829.

Martinez-Garcia, J.F., Huq, E. & Quail, P.H. (2000) Direct targeting of light signals to a promoter element-bound transcription factor. *Science*, **288**, 859–863.

Mas, P., Alabadi, D., Yanovsky, M.J., Oyama, T. & Kay, S.A. (2003a) Dual role of TOC1 in the control of circadian and photomorphogenic responses in *Arabidopsis*. *Plant Cell*, **15**, 223–236.

Mas, P., Devlin, P.F., Panda, S. & Kay, S.A. (2000) Functional interaction of phytochrome B and cryptochrome 2. *Nature*, **408**, 207–211.

Mas, P., Kim, W.Y., Somers, D.E. & Kay, S.A. (2003b) Targeted degradation of TOC1 by ZTL modulates circadian function in *Arabidopsis thaliana*. *Nature*, **426**, 567–570.

McWatters, H.G., Bastow, R.M., Hall, A. & Millar, A.J. (2000) The ELF3 zeitnehmer regulates light signaling to the circadian clock. *Nature*, **408**, 716–720.

Michael, T.P. & McClung, C.R. (2002) Phase-specific circadian clock regulatory elements in *Arabidopsis*. *Plant Physiol. JID-0401224*, **130**, 627–638.

Michael, T.P., Salome, P.A. & McClung, C.R. (2003) Two *Arabidopsis* circadian oscillators can be distinguished by differential temperature sensitivity. *Proc. Natl. Acad. Sci. USA*, **100**, 6878–6883.

Millar, A.J., Carré, I.A., Strayer, C.A., Chua, N. H. & Kay, S.A. (1995) Circadian clock mutants in *Arabidopsis* identified by luciferase imaging. *Science*, **267**, 1161–1163.

Millar, A.J. & Kay, S.A. (1996) Integration of circadian and phototransduction pathways in the network controlling *CAB* gene transcription in *Arabidopsis*. *Proc. Natl. Acad. Sci. USA*, **93**, 15491–15496.

Mizuno, T. (2004) Plant response regulators implicated in signal transduction and circadian rhythm. *Curr. Opin. Plant Biol.*, **7**, 499–505.

Mockler, T.C., Guo, H., Yang, H., Duong, H. & Lin, C. (1999) Antagonistic actions of Arabidopsis cryptochromes and phytochrome B in the regulation of floral induction. *Development*, **126**, 2073–2082.

Monte, E., Alonso, J.M., Ecker, J.R., Zhang, Y., Li, X., Young, J., Austin-Phillips, S. & Quail, P.H. (2003) Isolation and characterization of phyC mutants in *Arabidopsis* reveals complex crosstalk between phytochrome signaling pathways. *Plant Cell*, **15**, 1962–1980.

Monte, E., Tepperman, J.M., Al Sady, B., Kaczorowski, K.A., Alonso, J.M., Ecker, J.R., Li, X., Zhang, Y. & Quail, P.H. (2004) The phytochrome-interacting transcription factor, PIF3, acts early, selectively, and positively in light-induced chloroplast development. *Proc. Natl. Acad. Sci. USA*, **101**, 16091–16098.

Nakamichi, N., Kita, M., Ito, S., Sato, E., Yamashino, T. & Mizuno, T. (2005) The Arabidopsis pseudo-response regulators, PRR5 and PRR7, coordinately play essential roles for circadian clock function. *Plant Cell Physiol.*

Nelson, D.C., Lasswell, J., Rogg, L.E., Cohen, M.A. & Bartel, B. (2000) FKF1, a clock-controlled gene that regulates the transition to flowering in Arabidopsis. *Cell*, **101**, 331–340.

Oda, A., Fujiwara, S., Kamada, H., Coupland, G. & Mizoguchi, T. (2004) Antisense suppression of the *Arabidopsis* PIF3 gene does not affect circadian rhythms but causes early flowering and increases FT expression. *FEBS Lett.*, **557**, 259–264.

Park, E., Kim, J., Lee, Y., Shin, J., Oh, E., Chung, W.I., Liu, J.R. & Choi, G. (2004) Degradation of phytochrome interacting factor 3 in phytochrome-mediated light signaling. *Plant Cell Physiol.*, **45**, 968–975.

Pellequer, J.L., Brudler, R. & Getzoff, E.D. (1999) Biological sensors: More than one way to sense oxygen. *Curr. Biol.*, **9**, R416–R418.

Pittendrigh, C.S. (1981) Circadian systems: entrainment. In: *Handbook of Behavioral Neurobiology 4: Biological Rhythms* (ed J. Aschoff). New York: Plenum Press.

Quail, P.H. (2002) Phytochrome photosensory signaling networks. *Nat Rev Mol Cell Biol JID - 100962782*, **3**, 85–93.

Reed, J.W., Nagpal, P., Bastow, R.M., Solomon, K.S., Dowson-Day, M.J., Elumalai, R.P. & Millar, A.J. (2000) Independent action of ELF3 and phyB to control hypocotyl elongation and flowering time. *Plant Physiol.*, **122**, 1149–1160.

Rensing, L. & Ruoff, P. (2002) Temperature effect on entrainment, phase shifting, and amplitude of circadian clocks and its molecular bases. *Chronobiol. Int.*, **19**, 807–864.

Roenneberg, T. & Merrow, M. (1998) Molecular circadian oscillators: An alternative hypothesis. *Journal of Biol. Rhythms*, **13**, 167–179.

Roenneberg, T. & Merrow, M. (2003) The network of time: understanding the molecular circadian system. *Curr. Biol.*, **13**, R198–R207.

Salome, P.A. and& McClung, C.R. (2005) Pseudo-response regulator 7 and 9 are partially redundant genes essential for the temperature responsiveness of the *Arabidopsis* circadian clock. *Plant Cell.*

Salome, P.A., Michael, T.P., Kearns, E.V., Fett-Neto, A.G., Sharrock, R.A. & McClung, C.R. (2002) The out of phase 1 mutant defines a role for PHYB in circadian phase control in *Arabidopsis*. *Plant Physiol.*, **129**, 1674–1685.

Salomon, M., Christie, J.M., Knieb, E., Lempert, U. & Briggs, W.R. (2000) Photochemical and mutational analysis of the FMN-binding domains of the plant blue light receptor, phototropin. *Biochemistry*, **39**, 9401–9410.

Salter, M.G., Franklin, K.A. & Whitelam, G.C. (2003) Gating of the rapid shade-avoidance response by the circadian clock in plants. *Nature*, **426**, 680–683.

Schultz, T.F., Kiyosue, T., Yanovsky, M., Wada, M. & Kay, S.A. (2001) A role for LKP2 in the circadian clock of *Arabidopsis*. *Plant Cell*, **13**, 2659–2670.

Shalitin, D., Yang, H., Mockler, T.C., Maymon, M., Guo, H., Whitelam, G.C. & Lin, C. (2002) Regulation of *Arabidopsis* cryptochrome 2 by blue-light-dependent phosphorylation. *Nature JID-0410462*, **417**, 763–767.

Shalitin, D., Yu, X., Maymon, M., Mockler, T. & Lin, C. (2003) Blue light-dependent in vivo and in vitro phosphorylation of *Arabidopsis* cryptochrome 1. *Plant Cell*, **15**, 2421–2429.

Sharrock, R.A. & Clack, T. (2002) Patterns of expression and normalized levels of the five *Arabidopsis* phytochromes. *Plant Physiol.*, **130**, 442–456.

Somers, D.E. (2001) Clock-associated genes in *Arabidopsis*: a family affair. *Philos. Trans. R. Soc. Lond. B. Biol. Sci.*, **356**, 1745–1753.

Somers, D.E., Devlin, P.F., Kay, S.A. (1998a) Phytochromes and cryptochromes in the entrainment of the *Arabidopsis* circadian clock. *Science*, **282**, 1488–1490.

Somers, D.E., Kim, W.Y. & Geng, R. (2004) The F-Box protein ZEITLUPE confers dosage-dependent control on the circadian clock, photomorphogenesis, and flowering time. *Plant Cell*, **16**, 769–782.

Somers, D.E., Schultz, T.F., Milnamow, M. & Kay, S.A. (2000) *ZEITLUPE* encodes a novel clock-associated PAS protein from *Arabidopsis*. *Cell*, **101**, 319–329.

Somers, D.E., Webb, A.A.R., Pearson, M. & Kay, S. (1998b) The short-period mutant, *toc1-1*, alters circadian clock regulation of multiple outputs throughout development in *Arabidopsis thaliana*. *Development*, **125**, 485–494.

Staiger, D., Allenbach, L., Salathia, N., Fiechter, V., Davis, S.J., Millar, A.J., Chory, J. & Fankhauser, C. (2003) The *Arabidopsis* SRR1 gene mediates phyB signaling and is required for normal circadian clock function. *Genes Dev.*, **17**, 256–268.

Strayer, C., Oyama, T., Schultz, T.F., Raman, R., Somers, D.E., Mas, P., Panda, S., Kreps, J.A. & Kay, S.A. (2000) Cloning of the *Arabidopsis* clock gene TOC1, an autoregulatory response regulator homolog. *Science*, **289**, 768–771.

Suarez-Lopez, P., Wheatley, K., Robson, F., Onouchi, H., Valverde, F. & Coupland, G. (2001) CON-STANS mediates between the circadian clock and the control of flowering in *Arabidopsis*. *Nature*, **410**, 1116–1120.

Taylor, B.L. & Zhulin, I.B. (1999) PAS domains: internal sensors of oxygen, redox potential, and light. *Microbiol. Mol. Biol. Rev.*, **63**, 479–506.

Tepperman, J.M., Hudson, M.E., Khanna, R., Zhu, T., Chang, S.H., Wang, X. & Quail, P.H. (2004) Expression profiling of phyB mutant demonstrates substantial contribution of other phytochromes to red-light-regulated gene expression during seedling de-etiolation. *Plant J.*, **38**, 725–739.

Toth, R., Kevei, E., Hall, A., Millar, A.J., Nagy, F. & Kozma-Bognar, L. (2001) Circadian clock-regulated expression of phytochrome and cryptochrome genes in *Arabidopsis*. *Plant Physiol.*, **127**, 1607–1616.

Valverde, F., Mouradov, A., Soppe, W., Ravenscroft, D., Samach, A. & Coupland, G. (2004) Photoreceptor regulation of CONSTANS protein in photoperiodic flowering. *Science*, **303**, 1003–1006.

van der Horst, G.T.J., Muijtjens, M., Kobayashi, K., Takano, R., Kanno, S., Takao, M., de Wit, J., Verkerk, A., Eker, A.P.M., van Leenen, D., Buijs, R., Bootsma, D., Hoeijmakers, J.H.J. & Yasui, A. (1999) Mammalian Cry1 and Cry2 are essential for maintenance of circadian rhythms. *Nature*, **398**, 627–630.

Vierstra, R.D. (2003) The ubiquitin/26S proteasome pathway, the complex last chapter in the life of many plant proteins. *Trends Plant Sci.*, **8**, 135–142.

Yanovsky, M.J. & Kay, S.A. (2002) Molecular basis of seasonal time measurement in *Arabidopsis*. *Nature*, **419**, 308–312.

Yanovsky, M.J., Mazzella, M.A. & Casal, J.J. (2000) A quadruple photoreceptor mutant still keeps track of time. *Curr. Biol.*, **10**, 1013–1015.

Zagotta, M.T., Hicks, K.A., Jacobs, C.I., Young, J.C., Hangarter, R.P. & Meeks-Wagner, D.R. (1996) The *Arabidopsis* ELF3 gene regulates vegetative photomorphogenesis and the photoperiodic induction of flowering. *Plant J.*, **10**, 691–702.

5   Photoreceptors and light signalling pathways in plants

Victoria S. Larner, Keara A. Franklin
and Garry C. Whitelam

5.1   Introduction

The action of photoreceptors is required for the regulation of plant growth and development and the entrainment of the circadian clock. Light signals provide a plant with information allowing the synchronization of growth, development and metabolism with the daily light/dark cycle. Light also provides information about the surrounding environment and most critically, energy through photosynthesis. In order for the light signals to be employed by the plant, they must first be detected, then interpreted to produce a wide range of growth and developmental outputs. Higher plants contain a collection of photoreceptors that perceive different wavelengths of light in order that the plant may detect light quality, quantity and duration. It has been apparent for a number of years that plants are able to monitor ultraviolet, blue, red and far-red light (Fankhauser & Staiger, 2002). The phototropins and cryptochromes detect light from the ultraviolet-A (UV-A) and blue (B) regions of the spectrum, and phytochromes monitor primarily red (R) and far-red (FR) light (Wang & Deng, 2004). More recently, the ZEITLUPE family of proteins has been identified as possibly detecting blue/UV-A light (Imaizumi *et al.*, 2003, described in Chapter 4). Additionally, there is evidence for a green light photoreceptor, although its nature is unknown (Folta, 2004).

Following light perception, the photoreceptors initiate signalling cascades leading to entrainment of the clock, regulation of gene expression and a wide range of physiological responses. Developmental processes, including seed germination, the inhibition of elongation growth, the regulation of plant architecture, shade avoidance responses and the transition to flowering are regulated, at least in part, by information from the light environment detected by the photoreceptors. Individual photoreceptors may act alone to control particular processes, but they can also act together, redundantly, synergistically or antagonistically, or in concert with other endogenous signals or environmental cues. In addition, photoreceptors not only mediate light input to the circadian clock but they themselves are also regulated by feedback from the clock.

5.2   Photoreceptor structures

The isolation of phytochrome, cryptochrome and phototropin apoproteins along with the identification of their light-detecting chromophores has given important clues to the structure and function of the photoreceptors. Identification of amino acid motifs within the apoproteins has suggested potential signalling mechanisms employed by the photoreceptors upon detection of light.

5.2.1   Phototropin structure

For many years it has been known that plants are able to respond to the direction of light (Briggs & Chrisite, 2002). This tropic growth is primarily mediated by UV-A and blue light detected by the phototropin family of photoreceptors. In Arabidopsis, there are two phototropins: phototropin 1 (phot1) and phot2. The *PHOTOTROPIN 1* (*PHOT1*) gene (formerly *NPH1*) from Arabidopsis encodes a protein with molecular weight 112 kDa (Briggs *et al.*, 2001). The PHOT1 apoprotein has two distinct domains: a C-terminal serine/threonine kinase domain, and an N-terminal region that contains two LOV domains (Huala *et al.*, 1997). The C-terminal kinase domain has homology with the 11 sequence motifs typical of serine/threonine type kinases. Indeed, phot1 can undergo blue light-dependent autophosphorylation (Christie *et al.*, 1998). The N-terminal part of the protein contains two repeated 107 amino acid regions encoding LOV (light, oxygen, voltage) domains. These LOV domains have been shown to bind flavin mononucleotide (FMN). When PHOT1 LOV domain peptides were expressed in *E. coli* and mixed with FMN, they were seen to bind FMN in stochiometric amounts, hence two FMN molecules are bound per PHOT1 molecule (Christie *et al.*, 1998, 1999). The resulting complex had spectral properties equivalent to those of phototropin action spectra (Christie *et al.*, 1998). When the reconstituted recombinant LOV domains of oat phot1 are irradiated with blue light, they undergo a photocycle characterized by a loss of blue light absorbance in response to light, appearance of a new absorbance peak at 390 nm, followed by a spontaneous recovery of the blue light-absorbing form in the dark (Salomon *et al.*, 2000). Further analyses indicate that the initial photoexcitation event most likely involves the formation of a FMN-cysteinyl adduct within the LOV domain apoprotein, leading to alteration in chromophore conformation. It is speculated that this light-driven reaction leads to a change in the conformation of the LOV domains and that this in turn leads to activation of the kinase domain and so initiates signalling events (Salomon *et al.*, 2000).

The *PHOT2* gene from Arabidopsis was identified due to its homology with *PHOT1* (Jarillo *et al.*, 1998; Sakai *et al.*, 2001). It too has a C-terminal serine/threonine kinase domain and two LOV domains in the N-terminal region which, like PHOT1 LOV domains, bind FMN. The PHOT2 apoprotein shows 67% similarity to PHOT1 along its entire length, but this similarity rises to 82% and 93% in the two conserved LOV domains (Sakai *et al.*, 2001).

5.2.2   Cryptochrome structure

The first blue light photoreceptor from Arabidopsis to be identified followed from the isolation of *hy4*, a mutant that displayed reduced capacity for blue light-mediated inhibition of hypocotyl elongation (Koornneef *et al.*, 1980). The *HY4* locus was later cloned (Ahmad & Cashmore, 1993) through identification of a T-DNA tagged allele, and renamed *CRYPTOCHROME1* (*CRY1*). In Arabidopsis, the *CRY1* gene encodes a protein of 681 amino acids that comprises two distinct regions; an N-terminal domain with around 30% sequence identity with microbial DNA photolyases and a C-terminal region, thought to be involved in nucleocytoplasmic trafficking and protein-protein interactions. Photolyases are enzymes that bind damaged DNA and catalyze repair through a blue or UV-A light-dependent mechanism (Sancar, 1994). Photolyase proteins have a flavin chromophore and a pterin antenna that absorb blue or UV-A light and they repair pyrimidine dimers in damaged DNA through electron transfer. Although the N-terminal domain of Arabidopsis CRY1 associates non-covalently with flavin adenine mononucleotide (FAD) and pterin (5, 10-methenyltetrahydrofolate), MNTF), it lacks photolyase activity (Malhorta *et al.*, 1995). This may in part be due to the CRY1 protein lacking a conserved tryptophan residue found in all bacterial photolyases. The crystal structure of the N-terminal of Arabidopsis cry1 has recently been solved and also suggests reasons why it may be unable to act as a photolyase (Brautigam *et al.*, 2004). Although similar to microbial photolyase proteins in overall structure, cry1 has a mainly negatively charged surface compared with microbial photolyases which are predominantly positive in surface charge. The photolyases also have a positively charged groove which is able to bind to negatively charged DNA, facilitating repair, that is lacking in cry1.

The *CRYPTOCHOME2* (*CRY2*) gene was isolated by screening an Arabidopsis cDNA library with the *CRY1* gene sequence (Hoffman *et al.*, 1996; Lin *et al.*, 1996b). The *CRY2* gene encodes a protein with a high degree of similarity to CRY1 in the chromophore-binding photolyase domain and also binds flavin and pterin. Like CRY1, it lacks the tryptophan residue conserved in bacterial photolyases and also lacks photolyase activity. Arabidopsis CRY1 and CRY2 are 59% identical in the N-terminal domains but only have 13% identity in the C-terminal extension. Indeed, cryptochrome proteins from higher plant species are highly conserved in the chromophore-binding domain but have little similarity in the C-terminal region, with the length of this extension varying greatly in cryptochromes from different species (Lin & Shalitin, 2003). Despite the lack of sequence similarity between C-terminal regions of plant cryptochromes, they do share a set of motifs collectively known as the DAS (DQXVP-acidic-STAES) domain in this part of the protein (Lin & Shalitin, 2003). The role of the C-terminal region in signalling of the blue light perception has been revealed by the construction of fusions of sequences encoding the C-terminal of cry1 or cry2 (assigned CCT) with the *GUS* reporter gene. It was seen that plants expressing a GUS-CCT fusion protein showed constitutive photomorphogenesis, suggesting that the C-terminal region is sufficient to initiate signalling events whilst the N-terminal region is usually required to repress cryptochrome signalling in the dark (Yang *et al.*, 2000).

Cryptochromes are ubiquitous in higher plants and are related to similar photoreceptors from other organisms, including insects and mammals. Higher plant cryptochromes 1 and 2 are more similar to the class I photolyases that repair cyclobutane pyrimidine dimers, whereas the animal crys are more similar to the photolyases that repair 6-4 photoproducts, indicating that they have arisen via separate evolutionary events (Cashmore *et al.*, 1999). Interestingly, cryptochrome does not appear to play a role in photoreception in the mammalian circadian clock. However, it does have a critical role in the molecular mechanism of the mammalian circadian oscillator (Sancar, 2004). In Drosophila, as in plants, cryptochrome acts as a photoreceptor but has also been shown to directly interact with the central oscillator proteins TIM (TIMELESS) and PER (PERIOD) (Stanewsky, 2003). Recently, it has been established that Arabidopsis also possesses a cryptochrome of the type related to the 6-4 photolyases (Brudler *et al.*, 2003). This cryptochrome, called cry3, has been shown to bind DNA (although it does not possess photolyase activity) and carries N-terminal sequences that mediate import into chloroplasts and mitochondria (Kleine *et al.*, 2003). The precise physiological function of cry3 remains to be elucidated.

5.2.3   Phytochrome structure

Phytochrome exists in two interconvertible forms, the inactive Pr form and the biologically active Pfr form (Quail, 1997). Purified phytochrome from dark grown seedlings is blue in colour and absorbs light maximally in the red region of the spectrum (around 660 nm). On absorption of a photon of light, Pr is converted to the Pfr form which absorbs light maximally in the far-red region of the spectrum (around 730 nm). The Pfr form is converted back to Pr upon absorption of far-red light and, in some cases, can be converted to Pr in the absence of light in a process called dark reversion. There are two distinct molecular types of phytochrome, Type I phytochromes where the Pfr form is labile and undergoes rapid proteolysis, and Type II phytochromes where the Pfr form is stable. In Arabidopsis, there are five phytochromes (phyA-E), the apoproteins of which are encoded separate genes (Sharrock & Quail, 1989). Phytochrome A (phyA) is a Type I phytochrome, whereas phytochromes B-E (phyB-phyE) are all Type II phytochromes.

The structure of phytochromes is highly conserved among higher plant species (Montgomery & Lagarias, 2002). Phytochrome is present in cells as a soluble protein covalently attached to a linear tetrapyrole chromophore, and is found as a homodimer. Each monomer of the apoprotein is about 120 kDa in size and each comprises two major domains; an N-terminal chromophore-binding photosensory domain and a C-terminal regulatory domain. These two domains are separated by a protease-sensitive hinge region. The N-terminal region contains a central chromophore-bearing sub-domain along with a short N-terminal extension domain and a so-called PHY sub-domain which are required to maintain the integrity of the Pfr form and to restrict the rate of dark reversion. The C-terminal region contains several conserved sub-domains including the regulatory core, two PAS-related motifs, two dimerisation motifs and a histidine kinase-related domain (HKRD) (Quail, 1997; Montgomery & Lagarias, 2002). The C-terminal region is important for protein-protein interactions, such as the interaction of phytochrome with PIF3 and PIF4 (Kim *et al.*, 2003) and is also required for dimerization and for nuclear import of the phytochromes. The structures of both the N- and C-terminal domains have been further broken down into regions with specific functions by studying the phenotypes of specific phytochrome mutants (Quail, 1997; Matsushita *et al.*, 2003). Domain swapping experiments, where different portions of the *PHYB*, *PHYD* and *PHYE* genes were fused together, showed that the central region, including the PAS-related domains is important in determining the differential functions of each of these phytochromes (Sharrock *et al.*, 2003). Interestingly, the N-terminal region of phyB is apparently sufficient to elicit the full range of responses shown by the intact phytochrome protein, as shown by the finding that a fusion of the N-terminal of PHYB to GUS is a functional photoreceptor and has greater biological activity than intact phyB (Matsushita *et al.*, 2003). This suggests a role for the C-terminal domain in attenuation of phyB activity rather than in signal transduction.

A linear tetrapyrole chromophore, phytochromobilin, synthesised in plastids from haem binds covalently to a conserved cysteine residue in the N-terminal domain of each of the phytochromes (Lagarias & Rapoport, 1980). Absorption of light triggers an isomerisation of the chromophore which induces a structural rearrangement of the protein that initiates signalling.

5.3   Photoreceptor functions

Analysis of mutants lacking or overexpressing one or more functional photoreceptors, mostly using Arabidopsis, has allowed the dissection of the specific roles of each of the photoreceptors. In some processes, a single photoreceptor is responsible for the detection and transduction of a particular light signal, but for many processes a series of photoreceptors, sometimes acting redundantly, are required. Light signals regulate many developmental processes throughout the lifecycle of the plant. The quality, quantity, direction and periodicity of light are all important signals regulating a wide range of physiological responses. The role of individual photoreceptors in the circadian clock is discussed in Chapter 4.

5.3.1   *Phototropin functions*

Developing seedlings and plants must adapt to changes in environmental light conditions. Periods of limiting light reduce photosynthetic activity whereas exposure to excessive light can result in photo-oxidative damage to chloroplasts. Plants possess adaptive strategies to deal with changes in light direction and quantity including phototropism (bending of plant organs towards or away from a light source) and light-induced stomatal opening and chloroplast migration. These responses are all elicited by blue and UV-A light signals and are mediated by the phototropins (Briggs & Christie, 2002). The *non-*phototropic *hypocotyl* 1 (*nph1*) mutant of Arabidopsis that is deficient in phot1 lacks phototropic responsiveness to low fluence rates of unilateral blue light. Genetic studies have shown that phot2 also functions as a photoreceptor mediating hypocotyl phototropism, acting in a conditionally redundant manner with phot1. Thus, whereas *phot2* single mutants exhibit normal phototropic curvature in response to both low and high irradiances of blue light, in the absence of phot1, phot2 is required for phototropism at high irradiances (Sakai *et al.*, 2001). This suggests that phot1 is the primary phototropic receptor. Interestingly, in contrast to *PHOT1*, expression of the *PHOT2* gene is induced by light in dark-grown Arabidopsis seedlings (Jarillo *et al.*, 2001; Kagawa *et al.*, 2001), consistent with its increased role at higher irradiances.

As in the case of phototropism, phot1 and phot2 exhibit partially overlapping functions in mediating light-induced chloroplast movements. Plants are able to control the location of chloroplasts within the cell depending on the light conditions. At low irradiances of light, chloroplasts accumulate at the upper side of the cell to optimize light capture, whereas under high irradiance conditions chloroplasts are aligned on the sides of the cell to minimize light interception and so reduce possible photo-oxidative damage (Sakai *et al.*, 2001). Genetic analyses with the *phot1phot2* double mutant have shown that phot1 and phot2 act together to regulate the chloroplast migration response towards low irradiance blue light (Sakai *et al.*, 2001). In contrast, the avoidance response of chloroplasts to high irradiance light is mediated solely by phot2 (Jarillo *et al.*, 2001; Kagawa *et al.*, 2001).

The phot1 and phot2 photoreceptors act redundantly to regulate blue light-mediated stomatal opening. Although the light-mediated stomatal opening is maintained in either the *phot1* or *phot2* single mutant, the *phot1phot2* double mutant shows an attenuated response (Kinoshita *et al.*, 2001).

5.3.2   Cryptochrome functions

Once an Arabidopsis seed has germinated, light signals act to inhibit hypocotyl growth and promote the opening and expansion of cotyledons, a process termed de-etiolation. Several photoreceptors contribute to the de-etiolation processes depending upon the quality and quantity of ambient light. In response to blue light or UV-A light, the cryptochromes play a major role mediating the inhibition of hypocotyl elongation (Ahmad *et al.*, 1995; Lin *et al.*, 1996a). In addition to having long hypocotyls, *cry1* and *cry2* mutant seedlings also exhibit reduced cotyledon expansion in response to blue light, but have no obvious mutant phenotypes in red or far red light, confirming that the cryptochromes are blue light-specific photoreceptors (Ahmad *et al.*, 1995; Jackson & Jenkins, 1995; Lin *et al.*, 1998). As for the phototropins, the relative importance of each cryptochrome depends upon the irradiance of blue light. Under low irradiances of blue light, cry2 is more important than cry1 in controlling de-etiolation, whereas under higher irradiances cry1 is more important (Lin *et al.*, 1998). The irradiance dependent activity of cry2 may be at least partly due to a rapid blue light-dependent turnover of cry2 that reduces the levels of cry2 under higher irradiances (Shalitin *et al.*, 2002). In Arabidopsis, the cryptochromes are also involved in regulating the timing of reproductive development by controlling photoperiodic induction of flowering. In particular, mutants that are null for cry2 flower late compared to wild type plants, under long-day conditions, indicating a role for cry2 in inducing flowering through this pathway (Koornneef *et al.*, 1991; Guo *et al.*, 1998). An important factor in the photoperiodic control of flowering is the coincidence of a light signal with high levels of expression of *CO* (*CONSTANS*), a transcriptional regulator, whose expression is regulated by the circadian clock. This, in turn, leads to upregulation of expression the flowering regulator gene *FT* (*FLOWERING TIME*), which is required to trigger the transition from vegetative to reproductive development at the apical meristem (Yanovsky & Kay, 2002).

5.3.3   Phytochrome functions

Light signals can influence the timing of seed germination. The role of red and far-red light in regulating germination has been long established and led to the proposal of the Pr/Pfr model of phytochrome photoreversibility (Borthwick *et al.*, 1952). Studies using Arabidopsis null mutants indicate that both phyA and phyB have a role in the R/FR reversible promotion of seed germination (Shinomura *et al.*, 1994, 1996). Analyses of mutants null for multiple photoreceptors subsequently revealed a role for phyE in mediating light-induced germination (Hennig *et al.*, 2002).

  Following germination, phytochromes play a major role in de-etiolation, through detection of red and far-red light. The unique role of phyA in de-etiolation in far-red light was identified by analyzing mutants deficient in phyA in a variety of species including Arabidopsis (Nagatani *et al.*, 1993; Parks & Quail, 1993; Whitelam *et al.*, 1993), tomato (van Tuinen *et al.*, 1995a) and rice (Takano *et al.*, 2001). When grown in continuous FR Arabidopsis, *phyA* mutants have long hypocotyls and are unable to open and expand their cotyledons, thus they resemble wild-type seedlings grown in darkness. In red light, phyB plays the predominant role in de-etiolation. Mutants that are null for phyB have been analyzed in several species including Arabidopsis (Koornneef *et al.*, 1980; Somers *et al.*, 1991), *Brassica rapa* (Devlin *et al.*, 1992), cucumber (López-Juez *et al.*, 1992), tomato (van Tuinen *et al.*, 1995b) and pea (Weller & Reid, 1993). In all these species, *phyB* null mutants display elongated hypocotyls or epicotyls and in the case of Arabidopsis, *B. rapa*, cucumber and tomato, smaller cotyledons compared to wild type controls when grown under either red or white light conditions. The creation of double, triple and quadruple phytochrome-deficient mutants of Arabidopsis, has revealed that although phyA is most important in de-etiolation in far-red light and phyB most important in red light, all the five members of the phytochrome play roles in one or more process of de-etiolation (Reed *et al.*, 1994; Aukerman *et al.*, 1997; Devlin *et al.*, 1998; Franklin *et al.*, 2003a, 2003b; Monte *et al.*, 2003).

The architecture of adult plants is at least in part determined by light signals from the environment detected by the phytochromes. Processes including the size, shape and angle of leaves, plant height and degree of axillary branching are controlled by overlapping actions of individual phytochromes. For instance, in adult Arabidopsis plants, leaves are arranged in a rosette phenotype which serves to maximize the surface area for light capture. Mutant studies have revealed that the functions of phyA, phyB and phyE are all required to maintain this habit (Devlin *et al.*, 1998).

When grown in close proximity to one another, many plants are able to detect the presence of neighbouring vegetation via the reduction in the ratio of red to far-red wavelengths (R:FR ratio) of the light present. Normal daylight has a R:FR ratio of around 1.15, but this value falls significantly when daylight is transmitted through, or reflected from green plants (Smith, 1982; Smith & Whitelam, 1997). This is because chlorophyllous tissues selectively attenuate the red (and blue) regions of the daylight spectrum. Detecting this change in light quality allows plants to adapt to the presence of competing plants in a set of responses known as the shade avoidance syndrome. The shade avoidance syndrome is characterized by multiple changes in plant architecture and pronounced early flowering. These changes include increased stem and petiole elongation, reduced leaf angle from the stem (hyponasty) and increased apical dominance (McLaren & Smith, 1978; Smith & Whitelam, 1997). Such adaptations are thought to increase the chances of light capture and confer increased fitness in a dense stand of plants (Schmitt, 1997). Transient reduction in R:FR ratio results in an increase in elongation growth in a process gated by the circadian clock. A two-hour reduction in R:FR ratio can result in a 30% increase in Arabidopsis hypocotyl length within 24 hours, with maximal elongation occurring following perception of reduced R:FR ratio at subjective dusk (Salter *et al.*, 2003). Unlike the elongation responses, an acceleration of flowering in response to low R:FR is only observed following prolonged exposure to the light signal (Halliday *et al.*, 1994). The detection of R:FR ratio is largely mediated by phyB with additional roles for phyD and phyE (Franklin *et al.*, 2003b). Mutants that are triply null for phyB, phyD and phyE display extreme shade avoidance responses and are no longer able to respond to a reduction in R:FR ratio.

The regulation of flowering time in many species is controlled by measurement of daylength, or photoperiod, in a process that provides an indicator of season thus restricting flowering to a particular time of year (see Chapter 7). Detection of photoperiod in white light is mediated by the phytochromes as well as the cryptochromes (see above). In Arabidopsis, the observation that *phyA* mutant plants flower later than wild-type in long day photoperiods suggested phyA acts as a promoter of flowering in the photoperiodic pathway (Johnson *et al.*, 1994; Reed *et al.*, 1994; Yanovsky & Kay, 2002). Since *phyB* mutants flower earlier that wild-type plants under both long-day and short-day conditions, phytochrome B is thought to be acting as an inhibitor of flowering (e.g. Goto *et al.*, 1991).

5.4   Cellular and sub-cellular localization of photoreceptors

Once the molecular nature of the photoreceptors was known, one of the major questions concerned the cellular and sub-cellular location of the photoreceptors. Knowledge of the regulation of photoreceptor localization has revealed clues about the signalling mechanisms employed to transduce light signals.

5.4.1   Phototropin localization

Before the molecular identity of *PHOT1* was established, it had been noted that blue light causes rapid phosphorylation of a plasma membrane-associated protein that is required for phototropism in etiolated seedlings of several plant species, including Arabidopsis (Reymond *et al.*, 1992). Phosphorylation of this protein was not detectable in the *phot1* mutant and so it was speculated that the protein might be phot1. Despite the fact that it is highly hydrophilic, phot1 was subsequently shown to be associated with the plasma membrane in Arabidopsis (Sakamoto & Briggs, 2002). Precisely how phot1 associates with the plasma membrane is not known, but it seems likely that either it undergoes some post-translational modification or it binds to a protein cofactor to facilitate membrane interaction. After a short exposure to blue light, a proportion of the phot1 protein is lost from the membrane, becomes cytosolic and is eventually degraded (Knieb *et al.*, 2004).

5.4.2   Cryptochrome localization

Both cry1 and cry2 are soluble proteins found throughout all organs of both light- and dark-grown seedlings (Lin *et al.*, 1996a, 1998). In Arabidopsis, the abundance of the cry2 protein is strongly regulated by light, with levels falling to around 10% of those in dark-grown tissue within 15 minutes of light exposure (Shalitin *et al.*, 2002). This regulation occurs at the protein level as *CRY2* mRNA levels are unaffected by light treatment. The blue light-mediated cry2 degradation requires sequences present in the N-terminal portion of the protein (Guo *et al.*, 1999). The sub-cellular location of cry1 is dependent on the light conditions, as it is found in the cytosol in the light, but it is imported into the nucleus in the dark (Guo *et al.*, 1999; Yang *et al.*, 2000). However, cry2 is found in the nucleus regardless of the light conditions, although a proportion of the cry2 pool is localized in the cytoplasm (Cashmore *et al.*, 1999; Guo *et al.*, 1999; Kleiner *et al.*, 1999). The C-terminal portion of the cry2 protein, which contains a classical basic bipartite nuclear localization signal, is both necessary and sufficient for the nuclear import of cry2 (Guo *et al.*, 1999).

5.4.3   Phytochrome localization

Phytochromes can change their nucleocytoplasmic partitioning in response to light in a waveband and irradiance-dependent process (Gil *et al.*, 2000; Kim *et al.*, 2000;

Nagy *et al.*, 2000). Light triggers the translocation of a pool of phyA and phyB from the cytoplasm, where the proteins are synthesized and chromophorylated, to the nucleus. Once inside the nucleus, both phyA and phyB can form speckles (also called foci or nuclear bodies) depending on the light conditions. Using constitutively expressed fusions between phyA or phyB and GFP, the sub-cellular and sub-nuclear localization of the phytochromes has been extensively studied (reviewed by Nagatani, 2004). The kinetics of translocation and the quality of light that brings about the movement differs for each phytochrome. The nuclear translocation of phyA is rapid and occurs within 15 minutes of exposure to red, far-red or blue light. Continuous far-red or blue light, but not red light, causes phyA to rapidly accumulate in nuclear speckles (Kim *et al.*, 2000; Hisada *et al.*, 2000). The movement of phyB from the cytoplasm to the nucleus upon exposure to red (but not far-red or blue light) takes much longer than that of phyA, requiring around two hours of continuous irradiation light. However, even under optimal light conditions, a significant pool of phytochrome remains cytosolic and only a small fraction of the total phyA and phyB within the cell moves into the nucleus (Kircher *et al.*, 2002). All five Arabidopsis phytochromes have a cytosolic localization in dark-grown seedlings and are translocated into the nucleus in a light-dependent manner. Once in the nucleus, all five phytochromes also form nuclear speckles (Yamaguchi *et al.*, 1999; Gil *et al.*, 2000; Kircher *et al.*, 2002). The formation of phytochrome nuclear speckles is under circadian as well as light regulation. In seedlings grown under light/dark cycles, the appearance of nuclear speckles of all the phytochromes begins about 10 minutes before dawn suggesting circadian regulation, with maximal speckle accumulation occurring during the light phase and then falling during the dark phase (Kircher *et al.*, 2002). The regions of the phytochrome molecule required for nuclear localization have been dissected by studying phyB N-terminal and C-terminal fusion proteins (Sakamoto & Nagatani, 1996; Matsushita *et al.*, 2003). The phyB N-terminal fusion was seen to be constitutively localized to the cytoplasm regardless of light conditions, whereas the C-terminal fusion retained nuclear localization activity similar to that of the full-length protein. Despite the nuclear localization of the C-terminal alone, it was unable to form nuclear speckles. This led to the hypothesis that in darkness, the N-terminal of the Pr form of phytochrome functions to retain the whole molecule in the cytoplasm by suppressing the nuclear localization activity of the C-terminal. In the light, the N-terminal of the Pfr phytochrome form is unable to suppress nuclear localization and the whole molecule is translocated into the nucleus where it forms speckles (Matsushita *et al.*, 2003). Further dissection of the portions of the phytochrome molecule and light conditions required for the formation of nuclear speckles was carried out by the analysis of different *phyB* mutant alleles (Kircher *et al.*, 2002; Chen *et al.*, 2003). The formation of nuclear speckles depends on the proportion of Pfr phytochrome in the cell, with more and larger speckles being formed when more Pfr is present. Nuclear localization of phyB alone is not sufficient for the formation of nuclear speckles, as fusions containing just the phyB C-terminal are able to move to the nucleus but not form nuclear speckles, regardless of light conditions. Several mutations that reduce the activity of phyB have been shown to display reduced nuclear speckle formation, leading to the notion that speckle formation is required for phyB function. However, the observation that the phyB N-terminal alone, when fused to GUS and a nuclear localization signal, does localize to the nucleus but without forming speckles and yet has greater activity than the full-length phyB, suggests the function of nuclear speckles is not so simple (Oka *et al.*, 2004). It is possible that nuclear speckles are a site for sequestration of photoreceptors and may act as a means of desensitizing active phytochrome under higher irradiances. It is also possible that nuclear speckles are a site of photoreceptor association with transcription factors and other proteins. Indeed, cry2, COP1 and PIF3 have all been identified co-localized with phytochromes in nuclear speckles (Mas *et al.*, 2000; Seo *et al.*, 2004; Bauer *et al.*, 2004). In addition, TOC1 produces similar nuclear speckles, although co-localization in these speckles with phytochromes has not been confirmed (Strayer *et al.*, 2000). Similar speckling is also seen for the Drosophila dCRY (Ceriani *et al.*,1999). The biological significance of these nuclear speckles remains uncertain.

5.5   Photoreceptor signalling

Although the nature of plant photoreceptors has been long established, the mechanisms by which light signals are transduced leading to physiological changes have been more difficult to establish. Over recent years, studies have begun to allow the elucidation of signalling processes. Photoreceptors perceive light signals from the environment but must transmit these signals throughout the cell to bring about changes in gene expression and physiology required by the plant to make use of the light information. To do this, the photoreceptors themselves may undergo conformational changes and changes in sub-cellular location. They may also amplify the light signal via cytosolic second messengers or directly interact with the cell's transcriptional or post-transcriptional machinery to alter gene expression. Over recent years, it has emerged that photoreceptors use a variety of mechanisms to transmit light signals.

5.5.1   Post-translational Phosphorylation

Phosphorylation and dephosphorylation are mechanisms widely used by organisms in signalling cascades. The presence of putative kinase domains within photoreceptor proteins has suggested a role for phosphorylation in light signalling. Blue light causes the rapid phosphorylation of a 120 kDa membrane-associated protein in etiolated seedlings. It has been shown that a mutant deficient in phototropism (later shown to be *phot1*) is also deficient in phosphorylation of the 120kDa protein. Phosphorylation of phot1 is therefore involved in the early stage of phot1 signal transduction, at least for phototropism (Reymond *et al.*, 1992).

It has been shown that both cry1 and cry2 are phosphorylated upon expose to blue light, and that this phosphorylation decreases upon return to darkness. An *in vitro* assay, where cry1 was expressed in an insect cell system, showed that cry1 was able to be phosphorylated in a blue light-dependent manner in the absence of an additional protein kinase (Shalitin *et al.*, 2003). This suggests that cry1 is able to undergo blue light-dependent autophosphorylation. Mutant alleles of *cry1* and *cry2* that lacked function were also found to lack phosphorylation. In addition, constitutively active C-terminal cry2 fusions were shown to always be constitutively phosphorylated (Shalitin *et al.*, 2002). These results suggest that blue light-mediated phosphorylation of cry1 and cry2 is associated with their function.

It has been suggested for a number of years that phytochrome may act as a light-regulated kinase. Indeed, the C-terminal domain of the phytochrome molecule contains a region of sequence with homology to a histidine kinase. Several potential phosphorylation sites have been identified in oat phyA, most of which show autophosphorylation in a light-dependent manner, suggesting phosphorylation may act as a biological switch in phytochrome signalling (Park *et al.*, 2000). Recent research has shown that phosphorylation of a serine residue in the phyA hinge region controls the interaction of phytochrome with putative signal transducers (Kim *et al.*, 2004). Phosphorylation of Ser[598] inhibits the interaction between phyA and PIF3, thus acting as a regulatory mechanism in phytochrome signalling. In addition to autophosphorylation, phyA and phyB also phosphorylate the protein PKS1 (PHYTOCHROME KINASE SUBSTRATE 1) in a light-dependent manner (Fankhauser *et al.*, 1999). The phosphorylation of PKS1 acts to negatively regulate phytochrome signalling and suggests that the serine-threonine kinase activity of phytochrome has an important function in light signalling.

5.5.2   *Cytosolic signalling mechanisms*

The levels and location of Ca^{2+} within a cell is used as a signalling mechanism that can couple a wide range of extracellular signals to specific intracellular responses (for the role of Ca^{2+} in the circadian clock, see Chapter 8). The speed and amplitude, as well as the spatial location, of the Ca^{2+} messenger induce differing responses (Sanders *et al.*, 1999). The use of plants containing the luminescent Ca^{2+} sensitive marker protein aequorin has shown that pulses of blue light induce an increase in cytosolic free Ca^{2+} (Baum *et al.*, 1999). The patterns of Ca^{2+} sub-cellular location in response to a pulse of blue light differs in *phot1* mutant seedlings compared with the wild type, suggesting a role for this photoreceptor in mediating the response. In addition, it was shown that the wavelengths of light responsible for stomatal opening and phototropic curvature overlap with those at which cytosolic free Ca^{2+} levels are induced. These data suggested that physiological responses regulated by phot1, and perhaps phot2, might be transduced through changes in cytosolic free Ca^{2+} levels (Baum *et al.*, 1999). The specific roles of phot1 and phot2 in controlling blue light-dependent increases in cytosolic Ca^{2+} levels have since been elucidated. Both phot1 and phot2 can induce Ca^{2+} influx into the cytosol from the apoplast (extracellular) through Ca^{2+} channels in the plasma membrane (Harada *et al.*, 2003). However, phot2 alone also leads to release of Ca^{2+} from internal stores in response to blue light. This difference in mechanism may account for at least some of the differences in function of phot1 and phot2 (Liscum *et al.*, 2003).

Calcium may also be employed as a messenger following light perception by cry1 and cry2. One line of evidence for this is the isolation of the SUB1 (SHORT UNDER BLUE LIGHT) calcium-binding protein of Arabidopsis which acts downstream of cryptochromes in blue light signalling (Guo *et al.*, 2001). Analysis of double mutants between *cry1/cry2* and *sub1* suggest that *SUB1* acts as part of the cryptochrome signalling mechanism that suppresses the light-dependent accumulation of the transcriptional regulator HY5.

The blue light-mediated inhibition of hypocotyl elongation mediated by both the cryptochromes and phototropins involves both a rapid- and a slow-phase membrane depolarization. This occurs via blue light-dependent activation of anion channels in the hypocotyl cells resulting in membrane depolarization and inhibition of cell elongation (Sanders *et al.*, 1999). The rapid phase of growth inhibition is caused by light activation of anion channels by phototropin-mediated blue light perception (Folta & Spalding, 2001). After around 30 minutes of blue light, depolarization by cry1 and cry2 signalling cascades replaces phototropin mediated inhibition.

It has also been reported that phytochrome signal transduction is partly mediated via cytosolic free Ca^{2+} and also by membrane-associated cyclic GMP pathways (Bowler *et al.*, 1994; Bowler & Chua, 1994). Microinjection experiments have shown that the development of chloroplasts and biosynthesis of anthocyanin are regulated by light signals transduced via cytosolic free Ca^{2+} levels and cyclic GMP signalling. These proposed cytosolic signalling processes may suggest a role for the active Pfr phytochrome retained in the cytoplasm even in light conditions optimal for nuclear translocation of phytochrome (Nagy & Schäfer, 2002).

5.5.3   Regulation of protein degradation

One general mechanism by which light regulates photomorphogenesis is via the specific targeting of proteins for ubiquitination and proteasome-mediated degradation. One of the key regulators of this process is the COP1 (CONSTITUTIVE PHOTOMORPHOGENESIS 1) E3 ubiquitin protein ligase which acts downstream of both phytochromes and cryptochromes (Ang and Deng, 1994). The COP1 protein is a member of the COP/DET/FUS group of proteins which act as regulators of photomorphogenesis. The COP1 protein, along with DET1 (DE-ETIOLATED 1), is localized to the nucleus in the dark. In the nucleus, COP1 associates with the COP9 signalosome, a 12 subunit complex which is constitutively localized in the nucleus. In the light, COP1 moves out of the nucleus. When COP1 leaves the nucleus, proteins involved in the positive regulation of photomorphogenesis such as HY5 are no longer degraded. Consequently, their levels rise and photomorphogenesis occurs. Several photoreceptors have been identified as part of the mechanism involved in moving COP1 from the nucleus to the cytoplasm in the light. For example, the *cop1* mutation is epistatic to *cry1* (Ang & Deng, 1994) and cry1 is involved in triggering the migration of cop1 from the nucleus to the cytoplasm on exposure to light

(Osterlund & Deng, 1998). In addition to the re-localisation of COP1, blue light also causes a change in conformation of COP1 and thereby, a change in COP1 activity (Yang *et al.*, 2001). Both cry1 and cry2 have been shown to physically interact with COP1 in a light-dependent manner and so act to repress COP1 activity through direct protein-protein interactions (Yang *et al.*, 2000; Wang *et al.*, 2001). It has also been shown that COP1 interacts with phyA and phyB leading to their ubiquitination, suggesting COP1 can act to regulate phytochrome degradation (Wang *et al.*, 2001; Seo *et al.*, 2004). This feedback acts to negatively regulate phytochrome signalling activity.

5.5.4 *Photoreceptor interacting proteins*

In order to identify proteins that interact directly with phyB, and are therefore likely to be early components in the phyB-mediated light signalling pathway, a yeast two-hybrid screen was carried out using the C-terminal region of phyB as bait (Ni *et al.*, 1998). One of the proteins identified, named PIF3 (PHYTOCHROME INTER-ACTING FACTOR 3) belongs to the basic helix-loop-helix (bHLH) superfamily of transcription factors. Basic helix-loop-helix proteins are found ubiquitously in eukaryotes and function as regulatory components in transcriptional networks controlling many diverse processes (Heim *et al.*, 2003; Toledo-Ortiz *et al.*, 2003). In Arabidopsis, bHLH proteins form one of the largest families of transcription factors, currently estimated to contain 162 members (Bailey *et al.*, 2003). The bHLH signature domain consists of around 60 amino acids with two functionally distinct regions. The N-terminal end of the domain (the basic region) has around 15 basic amino acid residues and is usually involved in DNA binding. The C-terminal end of the domain has mainly hydrophobic residues that are predicted to form two amphipathic α-helices separated by an intervening loop region of variable sequence and length. The C-terminal region of the bHLH motif is involved in dimerization allowing the formation of either homo- or heterodimers with other bHLH proteins. A single bHLH protein may be able to form dimers with several different partners with the basic region of each protein binding half of the DNA recognition site. This allows precise control in signalling pathways. The DNA sequence motif recognized by most bHLH transcription factors is called the E-box, a hexameric sequence CAN-NTG. In Arabidopsis, the most commonly recognized type of E-box is a called the G-box (CACGTG) (Toledo-Ortiz *et al.*, 2003). PIF3 is a member of the Arabidopsis bHLH subfamily 15 which consists of 15 members in total (Toledo-Ortiz *et al.*, 2003).

PIF3 binds to full-length phyB in a specific and photo-reversible manner (Ni *et al.*, 1999). When phyB is in its photoactivated Pfr form it is able to bind PIF3, but upon reversion to Pr, phyB dissociates from the complex. In addition, PIF3 is able to bind the Pfr form of phyB whilst bound to its target DNA G-box motif (Martínez-García *et al.*, 2000). PIF3 also binds selectively and reversibly to the Pfr form of phyA although with far less affinity than it binds to phyB (Zhu *et al.*, 2000). Binding occurs in stochiometric amounts with the N-terminal region of the phytochrome molecule being required for the interaction. This is interesting, as the C-terminal of phytochrome alone was used as the bait to identify PIF3 in the initial yeast two-hybrid screens (Ni *et al.*, 1998). Recently, it has been shown that the interaction of PIF3 and phyB occurs within the speckles in the nucleus (Bauer, *et al.*, 2004). This evidence suggests that phyB may signal directly to target genes in the nucleus by physically interacting with PIF3 in a conformation-dependent way, acting as a light-switchable component of transcriptional regulation. The nature of this regulation has been difficult to resolve due to several contradictory reports. Initially, it was thought that PIF3 acted as a positive factor in phytochrome signalling. This is because a mutant thought to be overexpressing the *PIF3* gene was in fact lacking functional transcript (Ni *et al.*, 1998; Halliday *et al.*, 1999; Martínez-García *et al.*, 2000). It has since been shown that PIF3 can act as either a positive or negative regulator of phytochrome signalling, depending on the process under control. Seedlings overexpressing *PIF3* are hyposensitive to red and far-red light in terms of the inhibition of hypocotyl elongation and cotyledon opening and expansion, whilst *pif3* mutants are hypersensitive to red and far-red light for these responses. This indicates that PIF3 acts to negatively regulate seedling de-etiolation (Kim *et al.*, 2003). However, PIF3 is required for normal greening and chloroplast development during early de-etiolation and plants overexpressing *PIF3* have increased levels of anthocyanin, indicating a positive role in phytochrome signal transduction in these processes (Kim *et al.*, 2003; Monte *et al.*, 2004). Binding of phyB by PIF3 is also important in regulating PIF3 levels within the seedling. PIF3 accumulates in the nucleus of all cells in etiolated seedlings (Bauer *et al.*, 2004; Monte *et al.*, 2004). Upon transfer to red (phyB- and phyD-mediated) or far-red (phyA-mediated) light, PIF3 levels fall rapidly to a low steady state of about 20% of dark levels within one hour of illumination. This fall in PIF3 levels is mediated by the ubiquitination of PIF3 leading to its subsequent degradation by the 26S proteasome (Park *et al.*, 2004). Levels of PIF3 rise again during a subsequent dark period. Thus plants grown in light/dark cycles will display diurnal fluctuations in PIF3 levels, although this appears to be regulated by light and dark rather than the circadian clock (Monte *et al.*, 2004).

In addition to PIF3, several other members of the bHLH family of transcription factors have been implicated in phytochrome signalling. Each bHLH involved in phytochrome signalling has specific roles in regulation of a subset of phytochrome-mediated responses. PIF1 acts to negatively regulate chlorophyll biosynthesis during early seedling development (Huq *et al.*, 2004). Seedlings that are null for PIF1 accumulate excess free protochlorophyllide, a precursor of chlorophyll, causing lethal photo-bleaching upon transfer to the light. PIF1 localizes to the nucleus where it is able to form heterodimers with PIF3, but not whilst PIF3 is bound to phytochrome. The *PIF4* locus was identified by the mutant known as *srl2* (*short under red light 2*) as mutants are hypersensitive to red light (Huq & Quail, 2002). PIF4 can bind phyB and acts negatively in phyB-mediated cell expansion regulation, but is not required for chloroplast development. PIF4 can bind to the G-box promoter element in light-regulated genes but not when it is bound to phyB. PIL5 has a role in the negative regulation of phytochrome-mediated seed germination and is also involved in the inhibition of phyA-mediated hypocotyl agravitropism and the inhibition of hypocotyl elongation (Oh *et al*., 2004). PIL6 (also known as PIF5) plays a similar role to PIF3 in the regulation of phytochrome-mediated de-etiolation, acting as a negative regulator of red light-mediated responses (Fujimori *et al*., 2004). Two members of the bHLH subfamily 15, HFR1 (LONG HYPOCOTYL IN FAR-RED) and PIL1 do not directly bind phytochrome but do have a role in phytochrome signal transduction (Fairchild *et al*., 2000; Salter *et al*., 2003). HFR1 is required for a subset of phyA-mediated light responses (Duek & Fankhauser, 2003), whilst PIL1 is required for the proper circadian gating of the shade avoidance response mediated by phyB, phyD and phyE (Salter *et al*., 2003). Although HFR1 does not bind directly to phytochrome, it is able to form heterodimers with PIF3, which can in turn bind phyB (Fairchild *et al*., 2000). This raises the possibility of a whole range of interactions between individual members of the bHLH family in different combinations under different conditions, and indeed PIF3 and PIF4 have been shown to bind strongly to one another (Toledo-Ortiz *et al*., 2003). In addition to binding to other members of the bHLH family, several members have been shown to interact with TOC1, a component of the central oscillator of the circadian clock (Makino *et al*., 2002; Yamashino *et al*., 2003). However, none of the PIF proteins appear to have a direct role in the function of the circadian clock itself (Makino *et al*., 2002; Yamashino *et al*., 2003; Oda *et al*., 2004).

Members of the bHLH subfamily 15 which can bind phyB (PIF1, PIF3, PIF4, PIF5/PIL6 and PIF6/PIL2) all share a region of conserved sequence identified as the active phytochrome binding motif (APB) (Huq & Quail, 2002; Yamashino *et al*., 2003; Huq *et al*., 2004; Khanna *et al*., 2004). The APB is found in the N-terminal region of the PIF proteins and is conserved in all phytochrome-interacting bHLH proteins but not other members of the family. Two further bHLH proteins from subfamily 15 contain the conserved APB but do not bind to phyB in *in vitro* binding assays – bHLH023 and PIL1. The bHLH023 protein lacks some conserved residues found in the APB of all phyB binding bHLH proteins whilst the PIL1 N-terminal region, but not full-length protein, binds weakly to phyB in a photoreversible manner.

The NDPK2 (NUCLEOSIDE DIPHOSPHATE KINASE 2) protein of Arabidopsis binds preferentially to light-activated phytochrome and acts in the downstream signalling of red and far-red light (Choi *et al*., 1999). The interaction between phytochrome and NDPK2, which is mediated by the C-terminal regions of each protein, results in an increase in NDPK2 activity (Im *et al*., 2004; Shen *et al*., 2004). The observation that loss of *NDPK2* function results in reduced responses to red and far-red light, indicates that NDPK2 is a positive-acting component of the phytochrome signalling pathway (Choi *et al*., 1999).

Several proteins that act downstream of phot1 and phot2 in signalling blue light have been identified. These include the related proteins RPT2 (root phototropism 2) and NPH3 (non-phototropic hypocotyl 3) (Motchoulski & Liscum, 1999; Sakai *et al*., 2000). Both *RPT2* and *NPH3* encode proteins with BTB/POZ (broad complex, tram track, bric à brac/pox virus, zinc finger) domains at the N-terminal and a coiled-coil domain at the C-terminal, both thought to be involved in protein-protein interactions (Inada *et al.*, 2004). RPT2 and NPH3 have been shown to interact with phot1 possibly at the plasma membrane in a blue light-dependent manner and are required to transduce signals from the phototropins (Inada *et al.*, 2004). NPH3 and RPT2 belong to a family of plant-specific proteins named NRL (NPH3/RPT2-Like) leading to the possibility that other proteins interact in the same complex with phot1 and phot2 (Celaya & Liscum, 2004).

5.6   Crosstalk in photoreceptor signalling

Seldom in nature do plants receive environmental cues from just one source and plants have receptors to detect signals from a variety of different sources. In order to produce a range of physiological responses, plants are able to integrate signals from several environmental sources including gravity and temperature along with those from endogenous sources such as the circadian clock with light signals to regulate development.

5.6.1   Integration of environmental signals

Crosstalk between red and blue light sensing photoreceptors occurs at all stages of plant development. Although the exact nature of co-action is not fully understood, it is known that blue light-mediated de-etiolation involves interaction of both phytochrome and cryptochrome signalling (Yanovsky *et al.*, 1995; Ahmad & Cashmore, 1997; Casal & Mazzella, 1998). Studies of mutants lacking multiple combinations of phyA, phyB, phyC, phyD, cry1 and cry2 revealed numerous genetic and physical interactions between these photoreceptors during seedling and plant development (Ahmad *et al.*, 1998; Casal & Mazzella, 1998; Neff & Chory, 1998; Hennig *et al.*, 1999; Mas *et al.*, 2000; Franklin *et al.*, 2003a). The integration of light and gravity signals allows plants to orient themselves and adjust their architecture to enable maximal photosynthetic activity. Gravity provides a unidirectional signal which is integrated with light signals from phytochromes and phototropins. In white light and blue light, Arabidopsis roots display negative phototropism mediated by phototropins (Okada & Shimura, 1992; Briggs & Christie, 2002). In contrast, red light has been shown to induce a weak positive phototropism response in roots mediated largely by phyA and phyB (Ruppel *et al.*, 2001; Kiss *et al.*, 2003). The integration of light and temperature signals provides the plant with important information concerning the season. Periods of cold temperature and shortened daylength provide plants with a measure of seasonal progression which are important in regulating the transition to flowering (Simpson & Dean, 2002).

5.6.2   Integration of light and endogenous signals

In addition to multiple environmental signals combining to elicit physiological responses, photoreceptor action is also regulated by the endogenous circadian clock within the plant itself. The circadian clock can be broadly divided into three parts, defined as input pathways, central oscillator and output pathways. The input pathways perceive environmental cues, especially light and temperature and transduce these signals to the central oscillator. The oscillator consists of genes and proteins the expression of which is interlinked in a negative feedback loop which serves to produce the rhythm. This molecular rhythm acts to control the timing of many processes including gene expression, physiological and developmental responses. Studies have shown that all the photoreceptors tested can serve to input light signals of distinct wavelengths to entrain the central oscillator (Somers *et al.*, 1998). However, rhythmic output signals from the clock also feed back on the light input pathway. For example, the expression of cryptochrome and phytochrome genes (except *PHYC*) is controlled by the circadian clock at the transcriptional level (Bognár *et al.*, 1999; Hall *et al.*, 2001; Tóth *et al.*, 2001). Under light/dark cycles the mRNA levels of the cryptochrome and phytochrome genes begin to rise before dawn, thus anticipating the onset of the light period. Levels of mRNA reach a peak during the light phase, making the plant maximally sensitive to the predicable light signals required to reset the circadian clock each day. However, protein levels of the phytochromes do not fluctuate, although the appearance of speckles prior to down suggests the clock my regulate speckle formation (Bognár *et al.*, 1999; Kircher *et al.*, 2002). In addition, microarray experiments have shown that *PHOT1* mRNA levels are also under the control of the circadian clock together with a number of genes involved in light signal transduction, including *SPA1* and *RPT2* (Harmer *et al.*, 2000).

The circadian clock also regulates its own sensitivity to light input signals via the so-called gating pathway mediated by *EARLY FLOWERING3* (*ELF3*) (McWatters *et al.*, 2000). The *ELF3* protein is expressed rhythmically with peak expression around dusk and is essential for the normal entrainment of the circadian clock under long photoperiods and in continuous light conditions. *ELF3* has been shown to interact directly with phyB in a yeast two-hybrid assay leading to the hypothesis that this interaction inhibits phyB signalling (Liu *et al.*, 2001, discussed further in Chapter 4).

5.7   Conclusions

Identification of the phytochrome, cryptochrome and phototropin families of plant photoreceptors has allowed the nature of regulation of some processes mediated by light to be elucidated. Studies of mutant plants lacking the function of one or more photoreceptors have given useful information concerning the roles of individual photoreceptors in determining many aspects of plant development coupled to the light environment to allow the most efficient use of available resources. Recently, steps have been taken towards understanding the signalling mechanisms employed by the photoreceptors. Although much progress is still to be made in this direction several general themes have begun to emerge, not least the nucleocytoplasmic partitioning of active phytochromes and direct interaction of phytochrome molecules with transcriptional regulators. Use of microarray technology and further investigation of the targets of active photoreceptors should allow a more detailed picture of the integrated signalling networks employed by the plant photoreceptors to be determined.

References

Ahmad, M. & Cashmore, A.R. (1993) HY4 gene of *A. thaliana* encodes a protein with characteristics of a blue-light photoreceptor. *Nature*, **366**, 162–166.

Ahmad, M., Lin, C. & Cashmore, A.R. (1995) Mutations throughout an *Arabidopsis* blue-light photoreceptor impair blue-light-responsive anthocyanin accumulation and inhibition of hypocotyl elongation. *Plant J.*, **8**, 653–658.

Ahmad, M. & Cashmore, A.R. (1997) The blue light receptor cryptochrome 1 shows functional dependence on phytochrome A or phytochrome B in *Arabidopsis thaliana*. *Plant J.*, **11**, 421–427.

Ahmad, M., Jarillo, J., Smirnova, O. & Cashmore, A.R. (1998) The CRY1 blue light photoreceptor of Arabidopsis interacts with phytochrome A *in vitro*. *Mol. Cell*, **1**, 939–948.

Ang, L.H. & Deng, X.W. (1994) Regulatory hierarchy of photomorphogenic loci: allele-specific and light-dependent interaction between HY5 and COP1 loci. *Plant Cell*, **6**, 613–628.

Aukerman, M.J., Hirschfeld, M., Webster, L. *et al.* (1997) A deletion in the *PHYD* gene of the Arabidopsis Wassilewskija ecotype defines a role for phytochrome D in red/far-red light sensing. *Plant Cell*, **9**, 1317–1326.

Bailey, P.C., Martin, C., Toledo-Ortiz, G. *et al.* (2003) Update on the basic helix-loop-helix transcription factor gene family in Arabidopsis thaliana. *Plant Cell*, **15**, 2497–2501.

Bauer, D., Viczián, A., Kircher, S. *et al.* (2004) Constitutive photomorphogenesis 1 and multiple photoreceptors control degradation of phytochrome interacting factor 3, a transcription factor required for light signaling in Arabidopsis. *Plant Cell*, **16**, 1433–1445.

Baum, G., Long, J.C., Jenkins, G.I. & Trewavas, A.J. (1999) Stimulation of the blue light phototropic receptor NPH1 causes a transient increase in cytosolic Ca^{2+}. *Proc. Natl. Acad. Sci. USA*, **96**, 13554–13559.

Bognár, L.K., Hall, A., Ádám, E., Thain, S.C., Nagy, F. & Millar, A.J. (1999) The circadian clock controls the expression pattern of the circadian input photoreceptor, phytochrome B. *Proc. Natl. Acad. Sci. USA*, **96**, 14652–14657.

Borthwick, H.A., Hendricks, S.B., Parker, M.W., Toole, E.H. & Toole, V.K. (1952) A reversible photoreaction controlling seed germination. *Proc. Natl. Acad. Sci. USA*, **38**, 662–666.

Bowler, C. & Chua, N.H. (1994) Emerging themes of plant signal transduction. *Plant Cell*, **6**, 1529–1541.

Bowler, C., Neuhaus, G., Yamagata, H. & Chua, N.H. (1994) Cyclic GMP and calcium mediate phytochrome phototransduction. *Cell*, **77**, 73–81.

Brautigam, C.A., Smith, B.S., Ma, Z. *et al.* (2004) Structure of the photolyase-like domain of cryptochrome 1 from *Arabidopsis thaliana*. *Proc. Natl. Acad. Sci. USA*, **101**, 12142–12147.

Briggs, W.R., Beck, C.F., Cashmore, A.R. *et al.* (2001) The phototropin family of photoreceptors. *Plant Cell*, **13**, 993–997.

Briggs, W.R. & Christie, J.M. (2002) Phototropins 1 and 2: versatile plant blue-light receptors. *Trends Plant Sci.*, **7**, 204–210.

Brudler, R., Hitomi, K., Daiyasu, H. *et al.* (2003) Identification of a new cryptochrome class: structure, function, and evolution. *Mol. Cell*, **11**, 59–67.

Casal, J.J. & Mazzella, M.A. (1998) Conditional synergism between cryptochrome 1 and phytochrome B is shown by analysis of *phyA*, *phyB*, *hy4* simple, double and triple mutants in *Arabidopsis*. *Plant Physiol.*, **118**, 19–25.

Cashmore, A.R., Jarillo, J.A., Wu, Y.J. & Liu, D. (1999) Cryptochromes: blue light receptors for plants and animals. *Science*, **284**, 760–765.

Celaya, R.B. & Liscum, E. (2004) Phototropins and associated signaling: providing the power of movement in higher plants. *Photoch. Photobiol.*, e-published ahead of print.

Ceriani, M.F., Darlington, T.K., Staknis, D., Mas, P., Petti, A.A., Weitz, C.J. & Kay, S.A. (1999) Light-dependent sequestration of TIMELESS by CRYPTOCHROME. *Science*, **285**, 506–507.

Chen, M., Schwab, R. & Chory, J. (2003) Characterization of the requirements for localization of phytochrome B to nuclear bodies. *Proc. Natl. Acad. Sci. USA*, **100**, 14493–14498.

Choi, G., Yi, H., Lee, J. *et al.* (1999) Phytochrome signalling is mediated through nucleoside diphosphate kinase 2. *Nature*, **401**, 610–313.

Christie, J.M., Reymond, P., Powell, G.K. *et al.* (1998) *Arabidopsis* NPH1: a flavoprotein with the properties of a photoreceptor for phototropism. *Science*, **282**, 1698–1701.

Christie, J.M., Salomon, M., Nozue, K., Wada, M. & Briggs, W.R. (1999) LOV (light, oxygen or voltage) domains of the blue-light photoreceptor phototropin (nph1): binding sites for the chromophore flavin mononucleotide. *Proc. Natl. Acad. Sci. USA*, **96**, 8779–8783.

Devlin, P.F., Rood, S.B., Somers, D.E., Quail, P.H. & Whitelam, G.C. (1992) Photophysiology of the *elongated internode* (*ein*) mutant of *Brassica rapa*: *ein* mutant lacks a detectable phytochrome B-like protein. *Plant Physiol.*, **100**, 1442–1447.

Devlin, P.F., Patel, S.R. & Whitelam, G.C. (1998) Phytochrome E influences internode elongation and flowering time in *Arabidopsis*. *Plant Cell*, **10**, 1479–1487.

Duek, P.D. & Fankhauser, C. (2003) HFR1, a putative bHLH transcription factor, mediates both phytochrome A and cryptochrome signaling. *Plant J.*, **34**, 827–836.

Fairchild, C.D., Schumaker, M.A. & Quail, P.H. (2000) HFR1 encodes an atypical bHLH protein that acts in phytochrome A signal transduction. *Genes Dev.*, **14**, 2377–2391.

Fankhauser, C., Yeh, K.C., Lagarias, J.C., Zhang, H., Elich, T.D. & Chory, J. (1999) PKS1, a substrate phosphorylated by phytochrome that modulates light signalling in Arabidopsis. *Science*, **284**, 1539–1541.

Fankhauser, C. & Staiger, D. (2002) Photoreceptors in Arabidopsis thaliana: light perception, signal transduction and entrainment of the endogenous clock. *Planta*, **216**, 1–16.

Folta, K.M. & Spalding, E.P. (2001) Unexpected roles for cryptochrome 2 and phototropin revealed by high-resolution analysis of blue light-mediated hypocotyl growth inhibition. *Plant J.*, **26**, 471–478.

Folta, K.M. (2004) Green light stimulates early stem elongation, antagonizing light-mediated growth inhibition. *Plant Physiol.*, **135**, 1407–1416.

Franklin, K.A., Davis, S.J., Stoddard, W.M., Vierstra, R.D. & Whitelam, G.C. (2003a) Mutant analyses define multiple roles for phytochrome C in Arabidopsis thaliana photomorphogenesis. *Plant Cell*, **15**, 1981–1989.

Franklin, K.A., Praekelt, U., Stoddart, W.M., Billingham, O.E., Halliday, K.J. & Whitelam, G.C. (2003b) Phytochromes B, D and E act redundantly to control multiple physiological responses in Arabidopsis. *Plant Physiol.*, **131**, 1340–1346.

Fujimori, T., Yamashino, T., Kato, T. & Mizuno, T. (2004) Circadian-controlled basic/helix-loop-helix factor, PIL6, implicated in light-signal transduction in Arabidopsis thaliana. *Plant Cell Physiol.*, **45**, 1078–1086.

Gil, P., Kircher, S., Adam, E. *et al.* (2000) Photocontrol of subcellular partitioning of phytochrome-B:GFP fusion protein in tobacco seedlings. *Plant J.*, **22**, 135–145.

Goto, N., Kumagai, T. & Koornneef, M. (1991) Flowering responses to light-breaks in photomorphogenic mutants of Arabidopsis thaliana, a long-day plant. *Physiol. Plant.*, **83**, 209–215.

Guo, H., Yang, H., Mockler, T.C. & Lin, C. (1998) Regulation of flowering time by Arabidopsis photoreceptors. *Science*, **279**, 1360–1363.

Guo, H., Duong, H., Ma, N. & Lin, C. (1999) The Arabidopsis blue light receptor cryptochrome 2 is a nuclear protein regulated by a blue light-dependent post-transcriptional mechanism. *Plant J.*, **13**, 279–287.

Guo, H., Mockler, T., Duong, H. & Lin, C. (2001) SUB1, an *Arabidopsis* Ca^{2+}-binding protein involved in cryptochrome and phytochrome coaction. *Science*, **291**, 487–490.

Hall, A., Bognár, L.K., Tóth, R., Nagy, F. & Millar, A.J. (2001) Conditional circadian regulation of *PHYTOCHROME A* gene expression. *Plant Physiol.*, **127**, 1808–1818.

Halliday, K.J., Koornneef, M. & Whitelam, G.C. (1994) Phytochrome B and at least one other phytochrome mediate the accelerated flowering response of Arabidopsis thaliana L. to low red/far-red ratio. *Plant Physiol.*, **104**, 1311–1315.

Halliday, K.J., Hudson, M., Ni, M., Qin, M. & Quail, P.H. (1999) poc1: an Arabidopsis mutant perturbed in phytochrome signalling because of a T-DNA insertion in the promoter of PIF3, a gene encoding a phytochrome-interacting bHLH protein. *Proc. Natl. Acad. Sci. USA*, **96**, 5832–5837.

Harada, A., Sakai, T. & Okada, K. (2003) phot1 and phot2 mediate blue light-induced transient increases in cytosolic Ca^{2+} differently in Arabidopsis leaves. *Proc. Natl. Acad. Sci. USA*, **100**, 8583–8588.

Harmer, S.L., Hogenesch, L.B., Straume, M. *et al.* (2000) Orchestrated transcription of key pathways in Arabidopsis by the circadian clock. *Science*, **290**, 2110–2113.

Heim, M.A., Jakoby, M., Werber, M., Martin, C., Weisshaar, B. & Bailey, P.C. (2003) The basic helix-loop-helix transcription factor family in plants: a genome-wide study of protein structure and function diversity. *Mol. Biol. Evol.*, **20**, 735–747.

Hennig, L., Funk, M., Whitelam, G.C. & Schäfer, E. (1999) Functional interaction of cryptochrome 1 and phytochrome D. *Plant J.*, **20**, 289–294.

Hennig, L., Stoddart, W.M., Dieterle, M., Whitelam, G.C. & Schäfer, E. (2002) Phytochrome E controls light-induced germination of Arabidopsis. *Plant Physiol.*, **128**, 194–200.

Hisada, A, Hanrawa, H., Waller, J.L., Nagatani, A., Reid, J.B. & Furuya, M. (2000) Light induced nuclear translocation of endogenous pea phytochrome A visualized by immunocytochemical procedures. *Plant Cell*, **12**, 1063–1078.

Hoffman, P.D., Batschauer, A. & Hays, J.B. (1996) PHH1, a novel gene from Arabidopsis thaliana that encodes a protein similar to plant blue-light photoreceptors and microbial photolyases. *Mol. Gen. Genet.*, **253**, 259–265.

Huala, E., Oeller, P.W., Liscum, E., Han, I.S., Larsen, E. & Briggs, W.R. (1997) Arabidopsis NPH1: a protein kinase with a putative redox-sensing domain. *Science*, **278**, 2120–2123.

Huq, E. & Quail, P.H. (2002) PIF4, a phytochrome-interacting bHLH factor, functions as a negative regulator of phytochrome B signaling in *Arabidopsis*. *EMBO J.*, **21**, 2441–2450.

Huq, E., Al-Sady, B., Hudson, M., Kim, C., Apel, K. & Quail, P.H. (2004) Phytocrome-interacting factor 1 is a critical bHLH regulator of chlorophyll biosynthesis. *Science*, **305**, 1937–1941.

Im, Y.J., Kim, J.I., Shen, Y. *et al.* (2004) Structural analysis of Arabidopsis thaliana nucleoside diphosphate kinase-2 for phytochrome-mediated light signaling. *J. Mol. Biol.*, **343**, 659–670.

Imaizumi, T., Tran, H.G., Swartz, T.E., Briggs, W.R. & Kay, S.A. (2003) FKF1 us essential for photoperiodic-specific light signalling in Arabidopsis. *Nature*, **426**, 302–306.

Inada, S., Ohgishi, M., Mayama, T., Okada, K. & Sakai, T. (2004) RPT2 is a signal transducer involved in phototropic response and stomatal opening in association with phototropin 1 in Arabidopsis thaliana. *Plant Cell*, **16**, 887–896.

Jackson, J.A. & Jenkins, G.I. (1995) Extension-growth responses and expression of flavonoid biosynthesis genes in the *Arabidopsis hy4* mutant. *Planta*, **197**, 233–239.

Jarillo, J.A., Ahmad, M. & Cashmore, A.R. (1998) NPL1 (Accession no. AF053941): a second member of the NPH serine/threonine kinase family of Arabidopsis. *Plant Physiol.*, **117**, 719.

Jarillo, J.A., Gabrys, H., Capel, J., Alonso, J.M., Ecker, J.R. & Cashmore, A.R. (2001) Phototropin-related NPL1 controls chloroplast relocation mediated by blue light. *Nature*, **410**, 952–954.

Johnson, E., Bradley, J.M., Harberd, N.P. & Whitelam, G.C. (1994) Photoresponses of light-grown *phyA* mutants of Arabidopsis: phytochrome A is required for the perception of daylength extensions. *Plant Physiol.*, **105**, 141–149.

Kagawa, T., Sakai, T., Suetsugu, N. *et al.* (2001) Arabidopsis NPL1: a phototropin homolog controlling the chloroplast high light avoidance response. *Science*, **291**, 2138–2141.

Khanna, R., Huq, E., Kikis, E.A., Al-Sady, B., Lanzatella, C. & Quail, P.H. (2004) A novel molecular recognition motif necessary for targeting photoactivated phytochrome signaling to specific basic helix-loop-helix transcription factors. *Plant Cell*, **16**, 3033–3044.

Kim, L., Kircher, S., Tóth, R., Adam, E., Schäfer, E. & Nagy, F. (2000) Light-induced nuclear import of phytochrome-A:GFP fusion proteins is differentially regulated in transgenic tobacco and Arabidopsis. *Plant J.*, **22**, 125–133.

Kim, J., Yi, H., Choi, G., Shin, B., Song, P.S. & Choi, G. (2003) Functional characterization of phytochrome interacting factor 3 in phytochrome-mediated light signal transduction. *Plant Cell*, **15**, 2399–2407.

Kim, J., Shen, Y., Han, Y.J. *et al.* (2004) Phytochrome phosphorylation modulates light signaling by influencing the protein-protein interaction. *Plant Cell*, **16**, 2629–2640.

Kinoshita, T., Doi, M., Suetsugu, N., Kagawa, T., Wada, M. & Shimazaki, K. (2001) phot1 and phot2 mediate blue light regulation of stomatal opening. *Nature*, **414**, 656–660.

Kircher, S., Gil, P., Bognár, L.K. *et al.* (2002) Nucleocytoplasmic partitioning of the plant photoreceptors phytochrome A, B, C, D, and E is regulated differentially by light and exhibits a diurnal rhythm. *Plant Cell*, **14**, 1541–1555.

Kiss, J.Z., Mullen, J.L., Correll, M.J. & Hangarter, R.P. (2003) Phytochromes A and B mediate red-light-induced positive phototropism in roots. *Plant Physiol.*, **131**, 1411–1417.

Kleine, T., Lockhart, P. & Batschauer, A. (2003) An Arabidopsis protein closely related to *Synechocystis* cryptochrome is targeted to organelles. *Plant J.*, **93**, 93–103.

Kleiner, O., Kircher, S., Harter, K. & Batschauer, A. (1999) Nuclear localization of the Arabidopsis blue light receptor cryptochrome 2. *Plant J.*, **19**, 289–296.

Knieb, E., Salomon, M. & Rüdiger, W. (2004) Tissue-specific and subcellular localization of phototropin determined by immuno-blotting. *Planta*, **218**, 843–851.

Koornneef, M., Rolff, E. and Spruitt, C.J.P. (1980) Genetic control of light-inhibited hypocotyl elongation in Arabidopsis thaliana L. Heyhh. *Z. Pflanzenphysiol.*, **100**, 147–160.

Koornneef, M., Hanhart, C.J. & van der Veen, J.H. (1991) A genetic and physiological analysis of late flowering mutants in Arabidopsis thaliana. *Mol. Gen. Genet.*, **229**, 57–66.

Lagarias, J.C. & Rapoport, H. (1980) Chromopeptides from phytochrome. The structure and linkage of the Pfr form of the phytochrome chromophore. *J. Am. Chem. Soc.*, **102**, 4821–4828.

Lin, C., Ahmad, M. & Cashmore, A. (1996a) Arabidopsis cryptochrome 1 is a soluble protein mediating blue light-dependent regulation of plant growth and development. *Plant J.*, **10**, 893–902.

Lin, C., Ahmad, M., Chan, J. & Cashmore, A.R. (1996b) CRY2: a second member of the Arabidopsis cryptochrome gene family (Accession no. U43397). *Plant Physiol.*, **110**, 1047–1048.

Lin, C., Yang, H., Guo, H., Mockler, T., Chen, J. & Cashmore, A.R. (1998) Enhancement of blue-light sensitivity of Arabidopsis seedlings by a blue light receptor cryptochrome 2. *Proc. Natl. Acad. Sci. USA*, **95**, 2686–2690.

Lin, C. & Shalitin, D. (2003) Cryptochrome structure and signal transduction. *Annu. Rev. Plant Biol.*, **54**, 469–496.

Liscum, E., Hodgson, D.W. & Campbell, T.J. (2003) Blue light signaling through the cryptochromes and phototropins. So that's what the blues is all about. *Plant Physiol.*, **133**, 1429–1436.

Liu, X.L., Covington, M.F., Fankhauser, C., Chory, J. & Wagner, D.R. (2001) *ELF3* encodes a circadian clock-regulated nuclear protein that functions in an Arabidopsis *PHYB* signal transduction pathway. *Plant Cell*, **13**, 1293–1304.

López-Juez, E., Nagatani, A., Tomizawa, K.I. *et al.* (1992) The cucumber long hypocotyl mutant lacks a light-stable PHYB-like phytochrome. *Plant Cell*, **4**, 241–251.

Makino, S., Matsushika, A., Kojima, M., Yamashino, T. & Mizuno, T. (2002) The APRR1/TOC1 quintet implicated in circadian rhythms of Arabidopsis thaliana: I. Characterization with APRR1-overexpressing plants. *Plant Cell Physiol.*, **43**, 58–69.

Malhorta, K., Kim, S.T., Batschauer, A., Dawut, L. & Sancar, A. (1995) Putative blue-light photoreceptors from Arabidopsis thaliana and Sinapis alba with a high degree of sequence homology to DNA photolyase contain the two photolyase cofactors but lack DNA photolyase activity. *Biochemistry*, **34**, 6892–6899.

Martínez-García, J.F., Huq, E. & Quail, P.H. (2000) Direct targeting of light signals to a promoter element-bound transcription factor. *Nature*, **288**, 859–863.

Mas, P., Devlin, P.F., Panda, S. & Kay, S.A. (2000) Functional interaction of phytochrome B and cryptochrome 2. *Nature*, **408**, 207–211.

Matsushita, T., Mochizuki, N. & Nagatani, A. (2003) Dimers of the N-terminal domain of phytochrome B are functional in the nucleus. *Nature*, **424**, 571–574.

McLaren, J.S. & Smith, H. (1978) The function of phytochrome in the natural environment. VI. Phytochrome control of the growth and development of *Rumex obtusifolius* under simulated canopy light environments. *Plant Cell Env.*, **1**, 61–67.

McWatters, H.G., Bastow, R.M., Hall, A. & Millar, A.J. (2000) The ELF3 zeitnehmer regulates light signaling to the circadian clock. *Nature*, **7**, 716–720.

Monte, E., Alonso, J.M., Ecker, J.R. *et al.* (2003) Isolation and characterization of *phyC* mutants in Arabidopsis reveals complex crosstalk between phytochrome signaling pathways. *Plant Cell*, **15**, 1962–1980.

Monte, E., Tepperman, J.M., Al-Sady, B. *et al.* (2004) The phytochrome-interacting transcription factor, PIF3, acts early, selectively, and positively in light-induced chloroplast development. *Proc. Natl. Acad. Sci. USA*, **16**, 16091–16098.

Montgomery, B.L. & Lagarias, J.C. (2002) Phytochrome ancestry: sensors of bilins and light. *Trends Plant Sci.*, **7**, 357–366.

Motchoulski, A. & Liscum, E. (1999) Arabidopsis NPH3: a NPH1 photoreceptor-interacting protein essential for phototropism. *Science*, **286**, 961–964.

Nagatani, A., Reed, J.W. & Chory, J. (1993) Isolation and initial characterization of Arabidopsis mutants that are deficient in functional phytochrome A. *Plant Phys.*, **102**, 269–277.

Nagatani, A. (2004) Light-regulated nuclear localization of phytochromes. *Curr. Opin. Plant Biol.*, **7**, 708–711.

Nagy, F. Kircher, S. & Schäfer, E. (2000) Nucleo-cytoplasmic partitioning of the plant photoreceptors phytochromes. *Semin. Cell Dev. Biol.*, **11**, 505–510.

Nagy, F. & Schäfer, E. (2002) Phytochromes control photomorphogenesis by differentially regulated, interacting signaling pathways in higher plants. *Annu. Rev. Plant Biol.*, **53**, 329–355.

Neff, M.M. & Chory, J. (1998) Genetic interaction between phytochrome A, phytochrome B and cryptochrome 1 during Arabidopsis development. *Plant Physiol.*, **104**, 1027–1032.

Ni, M., Tepperman, J.M. & Quail, P.H. (1998) PIF3, a phytochrome-interacting factor necessary for normal photoinduced signal transduction, is a novel basic helix-loop-helix protein. *Cell*, **95**, 657–667.

Ni, M., Tepperman, J.M. & Quail, P.H. (1999) Binding of phytochrome B to its nuclear signaling partner PIF3 is reversibly induced by light. *Nature*, **400**, 781–784.

Oda, A., Fujiwara, S., Kamada, H., Coupland, G. & Mizoguchi, T. (2004) Antisense suppression of the *Arabidopsis PIF3* gene does not affect circadian rhythms but causes early flowering and increases *FT* expression. *FEBS Lett.*, **557**, 259–264.

Oh, E., Kim, J., Park, E., Kim, J., Kang, C. & Choi, G. (2004) PIL5, a phytochrome-interacting basic helix-loop-helix protein, is a key negative regulator of seed germination in Arabidopsis thaliana. *Plant Cell*, **16**, 3045–3058.

Oka, Y., Matsushita, T., Mochizuki, N., Suzuki, T., Tokutomi, S. & Nagatani, A. (2004) Functional analysis of a 450-amino acid N-terminal fragment of phytochrome B in Arabidopsis. *Plant Cell*, **16**, 2104–2116.

Okada, K. & Shimura, Y. (1992) Aspects of recent developments in mutational studies of plant signaling in Arabidopsis. *Cell*, **7**, 369–372.

Osterlund, M.T. & Deng, X.W. (1998) Multiple photoreceptors mediate the light-induced reduction of GUS-COP1 from Arabidopsis hypocotyl nuclei. *Plant J.*, **16**, 201–208.

Park, E.M., Bhoo, S.H. & Song, P.S. (2000) Inter-domain cross-talk in the phytochrome molecules. *Semin. Cell. Dev. Biol.*, **11**, 449.456.

Park, E., Kim, J., Lee, Y. *et al.* (2004) Degradation of phytochrome interacting factor 3 in phytochrome-mediated light signaling. *Plant Cell Physiol.*, **136**, 968–975.

Parks, B.M. & Quail, P.H. (1993) *hy8*, a new class of Arabidopsis long hypocotyl mutants deficient in functional phytochrome A. *Plant Cell*, **3**, 39–48.

Quail, P.H. (1997) An emerging molecular map of the phytochromes. *Plant Cell Env.*, **20**, 657–665.

Reed, J.W., Nagatani, A., Elich, T., Fagan, M. & Chory, J. (1994) Phytochrome A and phytochrome B have overlapping but distinct functions in Arabidopsis development. *Plant Physiol.*, **104**, 1139–1149.

Reymond, P., Short, T.W., Briggs, W.R. & Poff, K.L. (1992) Light-induced phosphorylation of a membrane protein plays an early role in signal transduction for phototropism in Arabidopsis thaliana. *Proc. Natl. Acad. Sci. USA*, **89**, 4718–4721.

Ruppel, N.J., Hangarter, R.P. & Kiss, J.Z. (2001) Red-light-induced positive phototropism in Arabidopsis roots. *Planta*, **212**, 424–430.

Sakai, T., Wada, T., Ishiguro, S. & Okada, K. (2000) RPT2: A signal transducer of the phototropic response in Arabidopsis. *Plant Cell*, **12**, 225–236.

Sakai, T., Kagawa, T., Kasahara, M. *et al.* (2001) *Arabidopsis* nph1 and npl1: blue light receptors that mediate both phototropism and chloroplast relocation. *Proc. Natl. Acad. Sci. USA*, **98**, 6969–6974.

Sakamoto, K. & Briggs, W.R. (2002) Cellular and subcellular localization of phototropin 1. *Plant Cell*, **14**, 1723–1735.

Sakamoto, K. & Nagatani, A. (1996) Nuclear localization activity of phytochrome B. *Plant J.*, **10**, 859–868.

Salomon, M., Christie, J.M., Knieb, E., Lempert, U. & Briggs, W.R. (2000) Photochemical and mutational analysis of the FMN-binding domains of the plant blue light receptor, phototropin. *Biochemistry*, **39**, 9401–9410.

Salter, M.G., Franklin, K.A. & Whitelam, G.C. (2003) Gating of the rapid shade avoidance responses by the circadian clock in plants. *Nature*, **11**, 680–683.

Sancar, A. (1994) Structure and function of DNA photolyase. *Biochemistry*, **33**, 2–9.

Sancar, A. (2004) Regulation of the mammalian circadian clock by cryptochrome. *J. Biol. Chem.*, **279**, 34079–34082.

Sanders, D., Brownlee, C. & Harper, J.F. (1999) Communicating with calcium. *Plant Cell*, **11**, 691–706.

Schmitt, J. (1997) Is photomorphogenic shade avoidance adaptive? Perspectives from population biology. *Plant Cell Env.*, **20**, 826–830.

Seo, H.S., Watanabe, E., Tokutomi, S., Nagatani, A. and Chua, N.H. (2004) Photoreceptor ubiquitination by COP1 E3 ligase desensitizes phytochrome A signaling. *Genes Dev.*, **18**, 617–622.

Shalitin, D., Yang, H., Mockler, T. *et al.* (2002) Regulation of Arabidopsis cryptochrome 2 by blue-light-dependent phosphorylation. *Nature*, **417**, 763–767.

Shalitin, D., Yu, X., Maymon, M., Mockler, T. & Lin, C. (2003) Blue light-dependent in vivo and in vitro phosphorylation of Arabidopsis cryptochrome 1. *Plant Cell*, **15**, 2421–2429.

Sharrock, R.A. & Quail, P.H. (1989) Novel phytochrome sequences in Arabidopsis thaliana: structure, evolution, and differential expression of a plant regulatory photoreceptor family. *Genes Dev.*, **3**, 8129–8133.

Sharrock, R.A., Clack, T. & Goosey, L. (2003) Signaling activities among the Arabidopsis phyB/D/E-type phytochromes: a major role for the central region of the apoprotein. *Plant J.*, **34**, 317–326.

Shen, Y., Kim, J.I. & Song, P.S. (2004) NDPK2 as a signal transducer in the phytochrome-mediated light signaling. *J. Biol. Chem.*, e-publication.

Shinomura, T., Nagatani, A., Chory, J. & Furuya, M. (1994) The induction of seed germination in Arabidopsis thaliana is regulated principally by phytochrome B and secondarily by phytochrome A. *Plant Physiol.*, **104**, 363–371.

Shinomura, T., Nagatani, A., Manzawa, H., Kubota, M., Watanabe, M. & Furuya, M. (1996) Action spectra for phytochrome A and B-specific photoinduction of seed germination in Arabidopsis thaliana. *Proc. Natl. Acad. of Sci. USA*, **93**, 8129–8133.

Simpson, G.G. & Dean, C. (2002) Arabidopsis, the Rosetta stone of flowering time? *Science*, **296**, 285–289.

Smith, H. (1982) Light quality, photoperception and plant strategy. *Annu. Rev. Plant Physiol.*, **33**, 285–289.

Smith, H. & Whitelam, G.C. (1997) The shade avoidance syndrome: multiple responses mediated by multiple phytochromes. *Plant Cell Env.*, **20**, 840–844.

Somers, D.E., Devlin, P.F. & Kay, S.A. (1998) Phytochromes and cryptochromes in the entrainment of the Arabidopsis circadian clock. *Science*, **282**, 1488–1490.

Somers, D.E., Sharrock, R.A., Tepperman, J.M. & Quail, P.H. (1991) The *hy3* long hypocotyl mutant of Arabidopsis is deficient in phytochrome B. *Plant Cell*, **3**, 1263–1274.

Stanewsky, R. (2003) Genetic analysis of the circadian system in Drosophila melonogaster and mammals. *J. Neurobiol.*, **54**, 111–147.

Strayer, C., Oyama, T., Schultz, T.F., Raman, R., Somers, D.E., Mas, P., Panda, S., Krepps, J.A. & Kay, S.A. (2000) Cloning of the Arabidopsis clock gene TOC1, an autoregulatory response regulator homolog. *Science*, **289**, 768–771.

Takano, M., Kanegae, H., Shinomura, T., Miyao, A., Hirochika, H. & Furuya, M. (2001) Isolation and characterization of rice phytochrome A mutants. *Plant Cell*, **13**, 521–534.

Toledo-Ortiz, G., Huq, E. & Quail, P.H. (2003) The Arabidopsis basic/helix-loop-helix transcription factor family. *Plant Cell*, **15**, 1749–1770.

Tóth, R., Kevei, E., Hall, A., Millar, A.J., Nagy, F. & Bognár, L.K. (2001) Circadian clock-regulated expression of phytochrome and cryptochrome genes in Arabidopsis. *Plant Physiol.*, **127**, 1607–1616.

van Tuinen, A., Kerckhoffs, L.H.J., Nagatani, A., Kendrick, R.E. & Koornneef, M. (1995a) A temporarily red light-insensitive mutant of tomato lacks a light-stable B-like phytochrome. *Plant Physiol.*, **108**, 939–947.

van Tuinen, A., Kerckhoffs, L.H.J., Nagatani, A., Kendrick, R.E. & Koornneef, M. (1995b) Far–red light insensitive, phytochrome A-deficient mutants of tomato. *Mol. Gen. Genet.*, **246**, 133–141.

Wang, H., Ma, L.G., Li, J.M., Zhao, H.Y. & Deng, X.W. (2001) Direct interaction of Arabidopsis cryptochromes with COP1 in light control development. *Science*, **294**, 154–158.

Wang, H. & Deng, X.W. (2004) Phytochrome signaling mechanism. In The Arabidopsis Book, Somerville, C. and Meyerowitz E.M., Eds. (American Society of Plant Biologists) http://aspb.org/publications/arabidopsis.

Weller, J.L. & Reid, J. (1993) Photoperiodism and photocontrol of stem elongation in two photomorphogenic mutants of *Pisum sativum L. Planta*, **189**, 15–23.

Whitelam, G.C., Johnson, E., Peng, J. *et al.* (1993) Phytochrome A null mutants of Arabidopsis display a wild-type phenotype in white light. *Plant Cell*, **5**, 757–768.

Yamaguchi, R., Nakamura, M., Mochizuki, N., Kay, S.A. & Nagatani, A. (1999) Light-dependent translocation of a phytochrome B-GFP fusion protein to the nucleus in transgenic Arabidopsis. *J. Cell Biol.*, **3**, 437–445.

Yamashino, T., Matsushika, A., Fujimori, T. *et al.* (2003) A link between circadian-controlled bHLH factors and the APRR1/TOC1 quintet in Arabidopsis thaliana. *Plant Cell Physiol.*, **44**, 619–629.

Yang, H.Q., Wu, Y.J., Tang, R.H., Liu, D., Liu, Y. & Cashmore, A. (2000) The C termini of Arabidopsis cryptochromes mediate a constitutive light response. *Cell*, **103**, 815–827.

Yang, H.Q., Tang, R.H. & Cashmore, A.R. (2001) The signaling mechanism of Arabidopsis CRY1 involves direct interaction with COP1. *Plant Cell*, **13**, 2573–2587.

Yanovsky, M.J., Casal, J.J. & Whitelam, G.C. (1995) Phytochrome A, phytochrome B and HY4 are involved in hypocotyl growth responses to natural radiation in Arabidopsis: weak de-etiolation of the *phyA* mutant under dense canopies. *Plant Cell Env.*, **18**, 788–794.

Yanovsky, M.J. & Kay, S.A. (2002) Molecular basis of seasonal time measurements in Arabidopsis. *Nature*, **419**, 308–312.

Zhu, Y., Tepperman, J.M., Fairchild, C.D. & Quail, P.H. (2000) Phytochrome B binds with greater apparent affinity than phytochrome A to the basic helix-loop-helix factor PIF3 in a reaction requiring the PAS domain of PIF3. *Proc. Natl. Acad. Sci. USA*, **97**, 13419–13424.

6 Circadian regulation of global gene expression and metabolism

Stacey L. Harmer, Michael F. Covington, Oliver Bläsing and Mark Stitt

It's not true that life is one damn thing after another; it is one damn thing over and over
—Edna St. Vincent Millay

Although this remark on the cyclical nature of life was made in modern times, observers of the natural world have long known that organisms show cyclical patterns in behavior and physiology. Many of these are daily rhythms regulated by the circadian clock and can be considered to be clock outputs. The study of these rhythmic outputs has been very fruitful and has led to the discovery of the existence of circadian clocks, the creation of tools for studying clock function and physiology, and the identification of clock-associated genes. In recent years, much attention has focused on circadian regulation of gene expression and metabolism. In this chapter, we discuss genes regulated by the plant circadian clock and their likely roles in plant growth and metabolism.

6.1 Circadian rhythms in transcription

In recent years, much attention has been paid to outputs controlled at the cellular level. As early as the middle of the 20th century, it was shown that regulation of DNA and RNA metabolism has a circadian component (Halberg *et al.*, 1959). The first description of rhythmic expression of genes encoding specific plant proteins came in 1985 when Kloppstech demonstrated that the abundance of three message encoding proteins involved in photosynthesis was under circadian regulation (Kloppstech, 1985). Many additional genes were found to be clock-regulated at the steady-state mRNA level, both through serendipity and directed approaches such as differential display (Carpenter *et al.*, 1994; Heintzen *et al.*, 1994a, 1994b; Sage-Ono *et al.*, 1998; Kreps *et al.*, 2000; Taybi *et al.*, 2000). Circadian regulation of message levels is often due to rhythmic changes in gene expression, but can clearly also be affected by changes in the rate of transcript degradation. A number of clock-regulated plant genes have markedly unstable transcripts (Gutierrez *et al.*, 2002).

6.1.1 Monitoring rhythms of transcription

The realization that transcription is clock-regulated has allowed investigators to use Northern blots to monitor circadian function in plants. For example, plants with defects in photomorphogenesis and photoperiodism were found also to display arrhythmic expression of multiple clock-regulated genes, leading to the identification of the Myb-like transcription factors *CCA1* (*CIRCADIAN CLOCK ASSOCIATED 1*) and *LHY* (*LATE ELONGATED HYPOCOTYL*) as genes closely associated with the circadian clock (Wang & Tobin, 1998; Schaffer *et al.*, 1998). But despite the power of this approach, there are considerable drawbacks. Collecting tissue samples and performing the blotting protocols are laborious and time consuming. Perhaps more importantly, throughput is low and the samples are destroyed in the process. These considerations have made Northern blot analysis unsuitable for genetic screens – a serious drawback, since most clock-associated genes have been originally identified using forward genetic methods (Young & Kay, 2001).

A method well suited to high throughput analysis was developed by Kay and colleagues in the early 1990s (Millar *et al.*, 1992). These workers generated transgenic plants expressing firefly luciferase under the control of a clock-regulated promoter. When supplied with the substrate luciferin, these plants show a circadian pattern of bioluminescence, reflecting circadian changes in promoter activity and allowing the activity of the central clock to be monitored (Fig. 6.1). New bioinformatics tools were developed to allow determination of the period, amplitude, and robustness

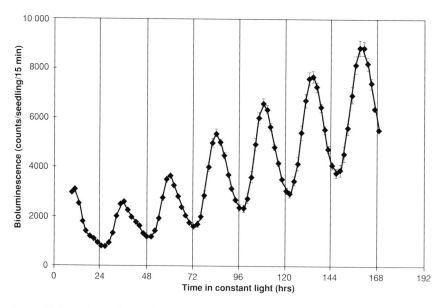

Figure 6.1 Circadian regulation of luciferase activity. Transgenic Arabidopsis seedlings, expressing luciferase under the control of the promoter of the *CCR2/AtGRP7* gene (Strayer *et al.*, 2000), were entrained in light/dark cycles and then transferred to constant light. Luminescence of individual seedlings was recorded at two-hour intervals using a CCD camera. The average luminescence of 84 plants, ± SEM, is shown.

of these oscillations (Plautz *et al.*, 1997). In contrast to Northern blot analysis, luciferase imaging is not very labor intensive and its non-destructive nature allows promoter activity in individual plants to be monitored repeatedly. Given these qualities, it is not surprising that it has been used as a circadian marker in a number of successful forward genetic screens (Millar *et al.*, 1995; Onai *et al.*, 2004).

The high time resolution provided by luciferase imaging has also allowed the reliable characterization of mutants with subtle clock phenotypes and the evaluation of clock function in different environmental conditions (Somers *et al.*, 1998; Eriksson *et al.*, 2003; Michael *et al.*, 2003). In addition, tissue specific, clock-regulated luciferase reporters have revealed that different tissues have circadian rhythms with distinct properties (Thain *et al.*, 2002). The recent development of luciferase reporters that emit at different wavelengths, together with single cell imaging technology, should in future allow the promoter activity of two genes to be measured simultaneously in the same cell. This would allow the investigation of multiple independent oscillators that may function in a single cell (Michael *et al.*, 2003). Enhancer trap experiments using a luciferase reporter have also led to the identification of new clock-regulated genes (Michael & McClung, 2003).

Despite the power of this technology, there are important drawbacks to the use of luciferase as a circadian marker: transgenic plants must be generated; expensive, state-of-the-art low-light imaging technology is required to monitor luminescence; the luciferin substrate is expensive; and finally, the activity of only one promoter is monitored at a time. An alternative tool, DNA microarrays, has instead been used to monitor circadian regulation of gene expression in a highly parallel manner.

6.1.2 Circadian regulation of the transcriptome

DNA microarrays allow the steady-state abundance of thousands of gene transcripts to be determined simultaneously. This relatively unbiased examination of the genome has allowed the identification of genes or pathways previously not suspected to be under circadian regulation. However, the manufacture of these arrays requires extensive genomic resources, they are often expensive, and the design and interpretation of microarray experiments are not trivial. A further drawback to all microarray platforms is their limited sensitivity and dynamic range. An important recent advance has been the development of highly multiparallel real time RT-PCR (Czechowski *et al.*, 2004). This platform was developed to allow the parallel determination of transcript levels for over 1400 transcription factors in *Arabidopsis*. Comparison with measurements of transcripts using ATH1 Affymetrix microarrays revealed that the latter did not provide quantitatively reliable results for transcripts with either very low or very high expression levels, due to background hybridization and signal attenuation, respectively (Czechowski *et al.*, 2004; Scheible *et al.*, 2004). Multiparallel RT-PCR thus provides an important adjunct to array techniques.

Despite these limitations, the use of microarrays has provided new insights into circadian rhythms in many different species. To date, over a dozen circadian microarray experiments have been carried out in organisms including *Arabidopsis*, *Drosophila*, mouse, *Neurospora*, and even a dinoflagellate (see Sato *et al.* (2003) and references therein and Okamoto and Hastings (2003)). Remarkably, almost each of these studies has concluded that between 2% and 10% of the genes in any given tissue sample show circadian variation in steady-state mRNA levels; this despite the fact that these experiments were done in organisms ranging from fungus to plants to animals, and used a variety of experimental and analytical protocols. After identifying genes of interest, it is of course important to integrate all available information about their roles in biological processes. This is undoubtedly the most challenging part of transcriptome analysis, but is being aided by improvements in gene annotation and the development of bioinformatics tools that allow the user to display results in their biological context (Thimm *et al.*, 2004). We will return to the nature of the genes under circadian regulation in plants later in this chapter.

6.2   Post-transcriptional circadian regulation

Although circadian regulation of transcription is ubiquitous and is thought to play a role in the functioning of the central clock in many organisms (Young & Kay, 2001), post-transcriptional regulation is clearly important as well. Detailed studies of a few genes implicated in plant circadian function have revealed several modes of post-transcriptional regulation. For example, levels of transcript encoding the F-box protein ZEITLUPE (ZTL) do not show circadian variation, but ZTL protein abundance is clock-regulated via circadian modulation of protein stability (Kim *et al.*, 2003b). Differential splicing of transcripts can also regulate protein abundance: the RNA-binding protein COLD CIRCADIAN RHYTHM 2 (CCR2)/ARABIDOPSIS THALIANA GLYCINE-RICH PROTEIN 7 (AtGRP7) specifically binds to its own transcript, regulating splice site selection and mRNA stability (Staiger *et al.*, 2003). Finally, modification of proteins can affect their function in the clock, as has been shown for phosphorylation of CCA1 (Daniel *et al.*, 2004). (See Chapters 4, 3 and 1 for more information on regulation of ZTL, CCR2/AtGRP7, and CCA1, respectively.)

The above examples make it clear that, although changes in transcript levels provide an excellent read-out of changes in signaling, it is important to systematically examine protein abundance and function as well. Such studies will not only reveal regulatory mechanisms that are superimposed upon transcriptional regulation as described above, but will also reveal instances where there is limited correlation between transcriptional regulation and protein activity. In the past, the study of changes in protein abundance and function has been a slow and case-by-case business, involving the use of dedicated assays or the production of specific antibodies. Similar limitations have hindered the study of metabolism and physiological processes. However, new and emerging technologies are making it possible to monitor proteins and metabolites in a more systematic and rapid manner. These new technologies do, nonetheless, have limitations and handicaps that will be briefly outlined below.

6.2.1 Proteomics

The goal of proteomics is to measure the abundance of every protein within a cell, although this has not yet been achieved. Indeed, most proteomic approaches provide only qualitative information, although some methods do allow for the highly parallel analysis of protein levels. Promising developments include gel-free proteomics using chromatographic separation of peptides, the use of heavy isotope labeled standards, and improvements in instrumentation and software (Chelius et al., 2003; Weckwerth et al., 2004; Zhang et al., 2004). It is only a matter of time before the broad-based analysis of circadian changes in protein levels becomes feasible.

A related, although more restricted, approach is to measure the activities of large numbers of enzymes in optimized assays as a proxy for determination of protein levels (Gibon et al., 2004a). A robotized platform using highly sensitive and optimized assays allows the activities of 23 enzymes involved in central carbon and nitrogen metabolism to be measured very precisely, with low operating costs and high throughput.

6.2.2 Metabolomics

The systematic determination of metabolite levels within a cell or organism (metabolomics) is technically even more demanding than measurements of proteins (Fiehn, 2002). There are probably more than 20 000 metabolites in any one tissue, with an enormous range of chemical structures that require different methods for extraction, stabilization and analysis. Unfortunately, there is no practical way to generate a list of all predicted metabolites in an organism, unlike the complete inventory of transcripts and proteins afforded by annotations of a full genome sequence. Furthermore, the methods used to extract and assay metabolites are quite sensitive to interference from other things in the extract. Traditionally, metabolites are measured by a wide range of specialized methods that are each linked to a wide range of detection strategies. The result is an enormous range of isolated analytic platforms, each of which can be highly sensitive and specific, but only able to detect a small number of metabolites (Fernie et al., 2005). This can be inefficient in providing a broad overview of responses and interactions in a system. A recent major advance has been the coupling of the separation power of gas chromatography and liquid chromatography with the separation, detection and identification power of mass spectrometry (Fiehn, 2002; Kopka et al., 2004; Fernie et al., 2005). In principle, this should allow the parallel determination of hundreds of metabolites. However, these powerful analytic platforms still have some important technical limitations that must be resolved. The deconvolution and evaluation of the mass chromatographic spectra is a slow and technically demanding process, and the algorithms and chemical databases that allow this process to be automated or semi-automated are still being developed. In addition, the stability and running time of a particular metabolite can vary depending on the apparatus and site. Finally, the vast majority of the analytes detected by these technologies remain to be identified, a time consuming process that requires specialized chemical expertise and infrastructure.

Despite these caveats, we are optimistic that broad proteomic and metabolomic approaches will soon be applied to plant circadian biology, giving us a more thorough understanding of the circadian system and the physiological processes that it governs.

6.3 Circadian regulation of transcription – what kinds of genes, and why?

Over 10% of the genes detectably expressed in *Arabidopsis* seedlings are circadian-regulated at the steady-state mRNA level (Harmer *et al.*, 2000; Covington *et al.*, 2005). An even higher fraction of genes is estimated to be clock-regulated at the transcriptional level, based on enhancer trap experiments (Michael & McClung, 2003). These clock-regulated genes encode proteins with a variety of predicted functions, ranging from cytoskeletal components to metabolic enzymes to transcriptional regulators. This observation prompts the obvious questions: What is the physiological significance of rhythmic gene expression? How frequently does a circadian change in gene expression cause a circadian change in protein level or function? And finally, how is clock-regulation of gene expression and protein function modulated in real-world conditions (i.e. light/dark cycles)? We will attempt to address these questions in the remainder of this chapter.

6.3.1 Clock-associated genes

One important category of clock-regulated genes in *Arabidopsis* includes those involved in the functioning of the clock itself. These include genes that are part of the putative core oscillator (*CCA1*, *LHY*, and *TOC1* (*TIMING OF CAB EXPRESSION 1*)); genes thought to modulate clock input pathways (*GIGANETA (GI)*, *EARLY FLOWERING 3 (ELF3)*, and the phytochrome and cryptochrome photoreceptors); and genes that play ill-defined roles in clock function (the *PSEUDORESPONSE REGULATOR*s, and *EARLY FLOWERING 4 (ELF4)*). (See Chapters 1, 2 and 3, for more information on these classes of genes.) Many of the protein products of these genes are likely to be clock-regulated, as has indeed been shown for *CCA1*, *LHY*, *TOC1*, and *ELF3* (Wang & Tobin, 1998; Liu *et al.*, 2001b; Kim *et al.*, 2003a; Mas *et al.*, 2003). However, some genes such as the phytochromes show robust rhythms in levels of steady state mRNA and synthesis of new proteins, but no evidence of cycling at the level of total protein (Bognar *et al.*, 1999; Toth *et al.*, 2001). In these last cases, it is possible that there is physiological significance in the clock-regulated timing of new protein production; more detailed studies are required to investigate this possibility.

6.3.2 Slave oscillators

Another interesting class of clock-regulated genes functions in 'slave oscillators': these are clock output genes that feedback to regulate the expression of a subset of genes, including themselves. Several such genes have been identified in *Arabidopsis*, including *CCR2/AtGRP7* and the transcription factor *EARLY PHYTOCHROME RESPONSIVE 1* (*EPR1*). The expression of CCR2/AtGRP7 protein is clock regulated, lagging the peak phase of transcript accumulation by about four hours (Heintzen *et al.*, 1997). Overexpression of *CCR2/AtGRP7* leads to reduced expression of the endogenous gene and a related transcript via post-transcriptional regulation (Staiger *et al.*, 2003). Overexpression of *EPR1* leads to reduced expression of the endogenous *EPR1* transcript as well as reduced and dampened expression of the unrelated gene *LIGHT-HARVESTING CHLOROPHYLL a/b PROTEIN* (*LHCB*) (Kuno *et al.*, 2003). It is unclear why slave oscillators exist, but they may serve to fine-tune the temporal expression of the subset of clock-regulated genes that they regulate. (See Chapter 3 for a more thorough discussion of slave oscillators).

6.3.3 Genes encoding other regulatory proteins

Given the large fraction of the transcriptome that is under circadian regulation, it is not surprising that many genes predicted to encode proteins with regulatory functions are also clock-regulated. These include proteins implicated in post-translational modification of proteins, regulation of protein stability, and regulation of gene expression (Harmer *et al.*, 2000; Schaffer *et al.*, 2001; Michael & McClung, 2003). This last category is of particular interest given that transcription factors are usually rather unstable and thus protein levels may often reflect transcript levels (Salghetti *et al.*, 2000; Muratani & Tansey, 2003). Approximately 13% of the detectably expressed transcriptional factors represented on the ATH1 array are under circadian regulation, higher than the rate of 10% seen when all genes on the array are considered (Covington *et al.*, 2005). Interestingly, some classes of transcription factor, such as the single Myb domain and basic helix-loop-helix factors, show an even higher incidence of circadian regulation, whereas other types of transcription factors, such as the MADS box genes, are seldom clock regulated. The significance of these observations remains to be determined.

The bulk of circadian-regulated genes are the outputs that are generally considered to be solely downstream of the clock. We will next consider this class of genes and the physiological processes in which they are involved. In discussing these genes, it will also be helpful to address which proteins and metabolites also show circadian and diurnal changes.

6.4 Regulation of plant metabolism

The circadian system acts to efficiently coordinate endogenous processes with daily (and annual) external environmental rhythms. The most conspicuous of these external rhythms is the daily light/dark cycle that is a consequence of Earth's axial rotation. Perhaps one of the primary impetuses in the evolution of clocks in land plants was to allow them to harvest the sun's energy most efficiently. It is, therefore, not surprising that many light-related processes are also under the control of the circadian clock. As mentioned above, photoreceptors and other components of light signaling pathways are influenced by the clock (see Chapters 4 and 5 for more details).

6.4.1   Photosynthesis

It has long been noted that photosynthesis, which converts light energy to fixed carbon and usable energy stores, is regulated by the circadian clock. Rhythms have been observed in both stomatal conductance and net carbon assimilation (Kerr *et al.*, 1985). The rhythm in carbon assimilation persists even when intercellular CO_2 partial pressure is held constant, indicating that non-stomatal processes also play a role (Hennessey & Field, 1991).

Plants first must absorb light energy, using chloroplast-localized antenna complexes comprised of photosynthetic pigments (carotenoids and chlorophylls) and pigment-binding proteins. Absorbed light energy is collected and channeled to the reaction centers of photosystems II and I, where it is used to produce strong oxidants and reductants that oxidize water to H^+ and O_2 and reduce $NADP^+$ to NADPH, respectively. The resulting proton gradient is used to synthesize ATP, which with NADPH drives the fixation of CO_2 to synthesize carbohydrates via the Calvin cycle. Transcripts encoding most of the components of the light harvesting machinery are clock-regulated, showing peak accumulation during the middle of the subjective day (Kloppstech, 1985; Harmer *et al.*, 2000). This co-regulation of gene expression may be important since large protein complexes must be formed to harvest light energy effectively. There is at least one report that bulk levels of chlorophyll binding proteins cycle (Adamska *et al.* (1991), see also Riesselmann & Piechulla (1992)), and it is also known that the assembly of thylakoid membranes is under circadian regulation (Beator *et al.*, 1992). However, this raises the interesting question as to how expression of components encoded by the plastid and nucleus is coordinated.

6.4.2   Partitioning of fixed carbon

The daily light/dark cycle, accompanied by fluctuations in other environmental parameters, causes significant changes in plant metabolism and physiology. Interactions might be expected between the clock and these 'indirect' metabolic changes. Indeed, many genes involved in starch metabolism are subject to circadian and/or diurnal regulation (Harmer *et al.*, 2000; Smith *et al.*, 2004); additionally, the rhythmic expression of some of these genes is altered in a photoperiod-dependent manner (Gibon *et al.* (2004b) and unpublished data). Contrary to expectations, however, the functions of these genes do not fall into two obvious groups corresponding to the synthesis and breakdown of starch (Smith *et al.*, 2004). Even more perplexing is the regulation of ADP glucose pyrophosphorylase, a key player in the regulation of starch synthesis: changes of enzyme activity are diametrically opposed to the changes of transcript levels (Gibon *et al.*, 2004a, 2004b). Despite these conundrums, sugar and starch levels are rhythmic.

In light, photosynthetic carbon fixation supports carbohydrate synthesis in 'source' leaves, and sucrose is exported to support metabolism and growth in 'sink' organs. Sucrose not immediately used is stored, usually in the form of starch in the leaves. At night, however, the plant becomes a net consumer of carbon, necessitating the remobilization of starch. Strikingly, very little starch remains at the end of the night in rapidly growing, non-stressed plants (Fondy & Geiger, 1985; Stitt *et al.*, 1987; Geiger & Servaites, 1994; Matt *et al.*, 1998). This suggests that the rate of starch accumulation in the light is coordinated with the rate of utilization during the night. This view is supported by the finding that more photoassimilate is partitioned to starch when day-length is decreased (Stitt *et al.*, 1978; Chatterton & Silvius, 1979, 1980, 1981; Matt *et al.*, 1998; Gibon *et al.*, 2004b). Analogous adjustments occur when photosynthesis is decreased by treatments such as low light, suggesting that starch synthesis responds to changes in the source-sink balance, rather than (or in addition to) photoperiod sensing (Chatterton & Silvius, 1980, 1981).

This diurnal turnover of starch acts as a buffer against fluctuations in the net carbon balance of the plant. Its vital role has been demonstrated by studies of starchless mutants. They grow like wild-type plants in continuous light or in long days, but growth is progressively impaired and even arrested as the duration of the night is increased (Caspar *et al.*, 1985; Lin *et al.*, 1988; Huber & Hanson, 1992). Such carbon depletion triggers major changes in gene expression, affecting such related metabolic processes as nitrogen assimilation (Thimm *et al.*, 2004).

6.4.3 Nitrogen assimilation

Nitrogen assimilation is one of the most energy-requiring processes in living organisms. Although molecular nitrogen comprises over 75% of the atmosphere, its exceptionally stable triple covalent bond makes it inaccessible to plants. In order for plants to incorporate nitrogen into compounds such as nucleic acids, proteins, and cofactors, they must rely on external sources such as bacteria to fix atmospheric nitrogen into useable forms such as ammonium and nitrate. The plant must spend the equivalent of 12 to 16 ATPs per nitrogen taken up, and then carefully manage the levels of the often highly toxic intermediates. These toxic products include nitrite, ammonium and NO, the latter being produced by nitrate reductase (NR) in a side reaction when nitrite is high and nitrate is low. A metabolic process with such high stakes must be both efficient and well-regulated.

Following its import via nitrate transporters, which are transcriptionally regulated by the clock (Harmer *et al.*, 2000), nitrate is converted to nitrite by NR. Transcript abundance of NR is circadian regulated, peaking prior to subjective dawn (Jones *et al.*, 1998; Lillo & Ruoff, 1989). In diurnal cycles, the steady-state level of NR transcript gradually increases during the dark period and rapidly falls once the plant is illuminated (Deng *et al.*, 1990). NR activity, on the other hand, undergoes a sharp increase upon light exposure, due in part to a light-stimulated increase in NR translation and decrease in NR degradation (Weiner & Kaiser, 1999). These complex responses of NR transcript and activity levels to diurnal cycles are due to direct modifications of NR protein (Bachmann *et al.*, 1996; Athwal & Huber, 2002) and to the indirect effects of light on the availability of the substrates and products of NR (Sivasankar & Oaks, 1995; Scheible *et al.*, 1997; Stitt *et al.*, 2002). This circuitry is very sensitive to genetic or environmental changes that affect the supply of nitrate or carbohydrate (Matt *et al.*, 1998), exemplifying how circadian and diurnal rhythms can be overridden by changes in the metabolic state of the plant.

The final steps in nitrogen assimilation are catalyzed by nitrite reductase, glutamine synthase (GS), and glutamate synthase (GOGAT) (for a thorough review, see Lam *et al.*, 1996). At least one isoform of each of these enzymes is circadian-regulated at the transcript level, all in phase with the NR transcript (Covington *et al.*, 2005). For example, Fd-GOGAT, which acts in root plastids and is involved in photorespiratory nitrogen metabolism in chloroplasts, shows clock regulation of expression (Covington *et al.*, 2005). However, unlike the activity of NR and the abundance of glutamine and total amino acids, *Arabidopsis* Fd-GOGAT activity doesn't appear to cycle in diurnal photoperiods (Gibon *et al.*, 2004a). This suggests that NR may be the key step for regulation of nitrogen assimi'ition and that adjustments to the expression and activity of the other enzymes are more important for fine-tuning.

Following assimilation into glutamine and glutamate, aminotransferases can incorporate nitrogen into other amino acids. Several aminotransferases are transcriptionally co-expressed by the clock with the nitrogen assimilation genes (Covington *et al.*, 2005). When plants are exposed to energy-limited conditions, nitrogen is channeled into the synthesis of asparagine, which is widely used for nitrogen transport and storage. Asparagine synthase expression and activity is affected by many of the same metabolic and environmental factors as plastid GS and Fd-GOGAT; however, the effects are opposite (Lam *et al.*, 1996). Furthermore, the circadian regulation of asparagine synthase is nearly anti-phasic to that of the genes involved in nitrogen assimilation (Covington *et al.*, 2005). It is believed that this opposing regulation is important for the balance of carbon and nitrogen metabolism. This sophisticated regulatory structure probably acts upon central metabolism to generate a complex system that is efficient, robust, and versatile; however, its precise operation and significance during a diurnal cycle requires further experimentation and possibly the development of new concepts (see Section 6.8).

6.4.4 *Secondary metabolic pathways*

Many clock-regulated genes encode enzymes that function in well-characterized secondary metabolic pathways. Indeed, we found that nearly all genes encoding enzymes known to function in the phenylpropanoid biosynthetic pathway exhibited coordinate circadian regulation of transcript levels. We also found that a Myb family transcription factor known to regulate the anthocyanin genes (Borevitz et al., 2000) is clock-regulated, suggesting it might mediate circadian regulation of these transcripts (Harmer et al., 2000). The anthocyanin pathway produces many secondary metabolites, including sinapate esters, lignins, flavonoids, and anthocyanins. Two products of this pathway, soluble flavonoids and a volatile scent compound, have been reported to undergo diurnal and circadian changes, respectively (Veit et al., 1996; Kolosova et al., 2001).

Another important class of plant secondary metabolites is the terpenoids. Two independent pathways provide 5-carbon subunits derived from acetyl-CoA and glycolytic intermediates for the terpenoid pathway. The plastid-localized MEP pathway is important for the formation of volatile terpenes and is under circadian regulation (Dudareva et al., 2005). The emission of a diverse collection of these volatile terpenes from various plants is rhythmic. In petunia, for example, PhCCD1 transcript is circadian regulated in leaves and corollas with peak levels during the subjective day. This gene encodes a protein that cleaves the carotenoid beta-carotene to produce beta-ionone, a terpenoid volatile that is a major contributor to the fragrance of many flowers and may attract pollinators. Beta-ionone emission is also rhythmic and has a phase similar to the transcript's abundance (Simkin et al., 2004). The role of volatile terpene emission is not restricted to reproduction. When poplar trees are infested with forest tent caterpillars, the leaves begin to release terpenoid volatiles to attract enemies of the herbivore as an indirect defense. This herbivore feeding, but not mechanical wounding, induces local and systemic diurnal terpenoid emission rhythms that are temporarily sustained after herbivore removal (Arimura et al., 2004).

Nonvolatile terpenoid compounds may also undergo circadian regulation. Genes involved in carotenoid biogenesis, for example, are clock-regulated and expressed just before dawn (Covington et al., 2005). In addition to their role as an energy donor to chlorophyll in the antenna complexes, during periods of high irradiance carotenoids also participate in nonphotochemical quenching to protect the plant from photooxidative damage that can lead to leaf bleaching and tissue necrosis if left unchecked (Havaux & Niyogi, 1999). The xanthophyll carotenoid violaxanthin is associated with the nonquenched state; however, when the light energy absorbed exceeds the rate at which photosynthesis can proceed, violaxanthin de-epoxidase converts violaxanthin to zeaxanthin, which allows quenching to occur so that much of the energy gets dissipated as heat before it reaches the reaction center. Zeaxanthin can be converted back into violaxanthin by zeaxanthin epoxidase. The transcription of the genes that encode these two enzymes is circadian-regulated with offset phases (Covington et al., 2005), such that violaxanthin might be produced in the morning and zeaxanthin when light levels increase around mid-day. It remains to be seen whether the circadian clock plays a role in the regulation of these enzymes and metabolites.

6.5   Circadian regulation of plant growth and development

Another clear effect of the daily alternation of light and dark, aided and abetted by accompanying changes in temperature, is a change in plant water status. Stomata open during the day, and in response to light (Dietrich *et al.*, 2001), to facilitate carbon dioxide uptake for photosynthesis. The accompanying increase in evaporative water loss leads to a decrease of leaf water potential, even in well-irrigated plants (McDonald & Davies, 1996). Consistent with this, a subset of genes identified as being sensitive to regulation in response to water stress do show small changes in expression during normal diurnal cycles even in the presence of adequate supplies of water (O. Bläsing, unpublished results). Changes in water potential are transmitted to the rest of the plant via the xylem and affect global metabolism; for example, post-translational regulation of sucrose phosphate synthase in response to water stress stimulates sucrose synthesis in leaves and in sink organs (Stitt, 1996; Geigenberger *et al.*, 1999).

6.5.1   Plant growth

Since expansion growth in plants is driven by changes in the osmotic potential, along with changes in cell wall extensibility, it is perhaps not surprising that aspects of plant growth show diurnal and circadian regulation. Leaf movement rhythm driven by pulvini, specialized leaf motor cells found in plants such as in legumes, are directly controlled by changes in water relations. Ion channel activity in pulvinar cells is regulated by the clock to generate changes in osmolarity that lead to movement of water into and out of specific cells on opposite sides of the petiole, resulting in rhythmic movement of leaves (Moshelion *et al.*, 2002a, 2002b). Even plants that lack pulvini can directly link turgor regulation, water movement, and the circadian clock. For example, clock-regulated hypocotyl elongation in young *Arabidopsis* seedlings is due primarily to expansion of existing cells (Dowson-Day & Millar, 1999). Leaf movement rhythms (or more accurately, cotyledon movement rhythms) in this plant, which lacks pulvini, also occur as a result of rhythmic cell expansion. Genes encoding enzymes involved in cell expansion, such as auxin efflux carriers, an expansin, an aquaporin, and cell wall hydrolases, are clock-regulated and may play a role in these clock-regulated growth processes (Harmer *et al.*, 2000). The availability of large public databases containing data from experiments in which water and salt stress and circadian responses have been investigated will make it possible to search for possible interactions between these important environmental inputs and circadian regulation of plant growth.

6.5.2   Hormone regulation

Genes implicated in both the synthesis of and response to all known plant hormones are clock-regulated at the steady-state mRNA level (Harmer *et al.*, 2000; Covington *et al.*, 2005). We focus here on two hormones that have been previously implicated in circadian regulation of plant physiology.

Ethylene

The production of the plant hormone ethylene, which is involved in numerous responses including growth regulation, is under circadian regulation in many plant species (Rikin *et al.*, 1984; Finlayson *et al.*, 1999; Thain *et al.*, 2004). Transcript levels of some of the genes involved in ethylene biosynthesis are circadian-regulated and show peak abundance during the subjective day, in phase with ethylene emission rhythms (Rikin *et al.*, 1984; Thain *et al.*, 2004). As is true of control of nitrate assimilation, other modes of regulation are superimposed upon circadian regulation of this pathway. For example, production of ethylene is positively influenced by light, and metabolic intermediates negatively feed back upon its production (Thain *et al.*, 2004). Furthermore, ethylene can elicit opposing effects on plant growth depending on light conditions (Smalle *et al.*, 1997). It is not clear what role the circadian regulation of ethylene production plays in plant physiology (Thain *et al.*, 2004).

In addition to its role in growth regulation, ethylene is involved in the response to anaerobic stress, often the result of growth in sites that become flooded. Two of the first steps in coping with growth in anaerobic conditions are an increase in ethylene synthesis caused by hypoxia (Jackson, 1982) and the resulting formation of aerenchyma (Jackson, 1985), cavities that allow for gas exchange between submerged anaerobic tissue and aerobic parts of the plant still above the water level. Some plants, such as the semi-aquatic *Rumex palustris*, have adapted to submergence in an ethylene-mediated manner. In these plants under normal conditions, ethylene levels are low and rhythmic, with peak emission during the night; however, in response to flooding ethylene levels increase and the phase shifts such that peak emission occurs during the day (Rieu *et al.*, 2005). Since ethylene exposure in the dark inhibits cell elongation while ethylene exposure in the light promotes cell elongation (Smalle *et al.*, 1997), this change in phase may allow for the appropriate growth response (i.e. elongation of vegetative structures) to hypoxic conditions.

Auxin

Another plant hormone, auxin, has also been implicated in plant growth responses. Cross-talk has been found between the auxin and ethylene signaling pathways, as is indeed probably true of all plant hormone pathways. Auxin treatment causes upregulation of ethylene biosynthetic enzymes (Zarembinski & Theologis, 1994) and ethylene may directly modulate auxin responses (Stowe-Evans *et al.*, 1998, 2001). Just as ethylene production is clock-regulated, so are levels of free auxin (Jouve *et al.*, 1999). A number of genes induced by auxin treatment show circadian rhythms in transcript abundance and the circadian clock regulates plant responses to exogenously applied auxin (Covington *et al.*, 2005). It is possible that circadian regulation of auxin signaling plays a role in modulation of vegetative growth by the circadian clock.

6.5.3   Control of flowering time

Another important physiological pathway regulated in part by the circadian clock is the transition to flowering. As described in more detail in Chapter 7, clock regulation of *CONSTANS (CO)* is central to the photoperiodic control of the vegetative to reproductive transition. Another clock-regulated gene, *FLOWERING LOCUS T (FT)*, is also involved in this developmental process. In *Arabidopsis*, coincidence between *CO* expression and the perception of light induces *FT* expression and accelerated flowering (Suarez-Lopez *et al.*, 2001; Yanovsky & Kay, 2002). This coincidence, and the resultant rapid flowering, happens only in long days; thus *Arabidopsis* is termed a long day plant. Short day plants such as rice and *Pharbitis nil* flower more rapidly when the days are short. Homologs of *CO* are also clock-regulated in these plants, with a similar phase of expression as their *Arabidopsis* counterpart (Liu *et al.*, 2001a; Izawa *et al.*, 2002). However, at least in rice, coincidence between *CO* and light perception inhibits the expression of *FT* homologs and flowering is delayed (Izawa *et al.*, 2002). Thus long day plants and short day plants make use of the same clock outputs to regulate flowering time, but have altered their mode of regulation to suit environmental constraints. However, this is clearly not the only way plant clock output pathways can regulate flowering time. Recently, a gene with no apparent *Arabidopsis* homolog was shown to be clock-regulated and to function in the photoperiodic control of flowering in rice (Doi *et al.*, 2004). Doubtless the investigation of clock output pathways in varied plant species will lead to the discovery of many more such interesting parallel and divergent pathways.

6.6   How is clock-regulation of gene expression achieved?

An important question raised by the identification of the large number of clock-regulated genes is their mode of regulation. How are their circadian patterns of expression, with peak expression at diverse phases (Fig. 6.2), generated?

6.6.1   The evening element (EE)

One approach to this problem is to look for promoter motifs that are over-represented in the promoters of cycling genes relative to the rest of genome. This method led to the identification of an invariant nine-nucleotide element present in the promoters of many evening-phased, clock-regulated genes. This motif was thus dubbed the evening element (EE). Consistent with its over-representation in clock-regulated genes, a wild-type EE was shown to be essential for the ability of a fragment of the *CCR2* promoter to confer rhythmicity on a reporter gene (Harmer *et al.*, 2000).

CT · Individual Genes

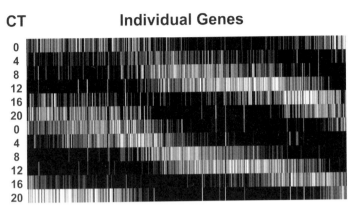

Figure 6.2 Heatmap indicating times of peak expression for clock-regulated genes. Expression profiles of 453 clock-regulated genes (Harmer *et al.*, 2000) are sorted by phase. Circadian time (CT), with CT0 equivalent to subjective dawn, is indicated on the *y*-axis. Individual genes are graphed on the *x*-axis, with white boxes indicating times of peak expression and black boxes indicating times of trough expression. Note the near-continuous distribution of phases.

In more recent work, we have found that multimerized EE are sufficient to confer evening-phased rhythms (Harmer & Kay, 2005). Circadian microarray studies using arrays representing approximately 22 000 *Arabidopsis* genes have revealed that over one third of genes with peak expression in the late day or early night have an EE within their promoters (Covington *et al.*, 2005), suggesting that this motif plays an important role in regulation of evening-phased gene expression. This is confirmed by permutation testing, which has shown that the EE is statistically over-represented in the promoters of evening-phased genes and under-represented in the promoters of day-phased genes. Notably, the EE is also present in the promoters of evening-phased genes that function close to the circadian oscillator (*TOC1*, *GI*, and *ELF4*), suggesting that it may play a role in the functioning of the central oscillator itself. Indeed, it has been shown that CCA1 and LHY bind directly to the EE in the *TOC1* promoter and it is likely to be important for the reciprocal regulation between *CCA1/LHY* and *TOC1* (Alabadi *et al.*, 2001). (See Chapter 1 for more discussion of this point.)

6.6.2   Other circadian associated motifs

Other motifs have also been reported to function in circadian regulation of gene expression. A motif differing from the EE by only one nucleotide, the CCA1 binding site (CBS), has been suggested to be involved in the generation of day-phased rhythms (Michael & McClung, 2002). Consistent with this, we find the CBS to be statistically over-represented in the promoters of cycling genes that have peak expression between CT2 and CT6 (Covington *et al.*, 2005). However, multimerized CBS motifs confer the same evening phase as mutimerized EE (Harmer & Kay,

2005), suggesting that either tandemly repeated CBS motifs function differently from a single CBS or that CBS function can be modified by other regulatory motifs found in native promoters. Another motif observed more frequently in the promoters of clock-regulated genes than expected by chance is the G-box (Michael & McClung, 2003; Hudson & Quail, 2003). The G-box was originally identified as a motif found in the promoters of many light-regulated genes (Giuliano *et al.*, 1988). Since many clock-regulated genes are also light-induced (Harmer *et al.*, 2000), this raises the question as to whether the G-box is found in many clock regulated promoters simply because they are also light regulated, or if it plays an independent role in circadian regulation as well.

How many clock motifs are there in plants? Is there a separate motif for every phase of gene expression? This seems unlikely given the range of phases of gene expression that are observed. Another possibility is that these diverse phases are produced through combinatorial regulation, a mechanism that has indeed been demonstrated in mouse. Two promoter elements, each separately conferring nearly opposite phases of gene expression, produce peak expression at an intermediate phase when combined in the same promoter (Etchegaray *et al.*, 2003; Ueda *et al.*, 2005). It will be exciting to determine whether this model of regulation holds true in plants as well.

6.7   Cross-species comparisons

In recent years, genomic techniques have been used to assess the complement of circadian-regulated genes in a variety of model organisms. Numerous circadian microarray studies performed in plants, animals, and fungi have concluded that a significant fraction of each transcriptome (from 1% to 19% of expressed genes) is clock-regulated at the steady-state mRNA level. However, the fraction of genes that are clock-regulated at the level of transcription may be considerably higher (Liu *et al.*, 1995; Michael & McClung, 2003). Perhaps not surprisingly given the large fraction of the genome under circadian regulation, the protein products of clock-regulated genes perform a variety of cellular functions, from primary metabolism to modulation of behavioral rhythms.

6.7.1   Are similar genes clock regulated in all organisms?

As discussed in two excellent reviews (Panda *et al.*, 2003; Duffield, 2003), there is limited overlap in genes classified as cycling even when the studies being compared using the same species and similar environmental conditions. Reasons for this may include biological differences, technical differences, and differences in low-level and high-level analysis. In fact, correlation between microarray data generated using different array platforms, even when done by the same group, is limited at best (Tan *et al.*, 2003). Therefore, it is difficult to compare individual genes found to be clock-regulated in different organisms and in different studies.

However, a comparison of the types of genes under clock regulation may be informative. Genes that function in signal transduction, transcriptional regulation, protein folding, and modulation of protein stability have been identified as cycling in virtually all published microarray studies, encompassing plants, animals, and fungi. Also prominent are genes whose products may be involved in detoxification or other stress responses, including catalases, cytochrome P450s, glutathione transferases, and thioredoxins (Harmer *et al.*, 2000; Schaffer *et al.*, 2001; Claridge-Chang *et al.*, 2001; McDonald & Rosbash, 2001; Akhtar *et al.*, 2002; Ceriani *et al.*, 2002; Panda *et al.*, 2002; Storch *et al.*, 2002; Correa *et al.*, 2003). Clock regulation of genes implicated in stress responses may, in fact, be a general theme, especially in plants. Two-thirds of genes regulated by the circadian clock in *Arabidopsis* are also induced by cold, osmotic, or salt stress (Kreps *et al.*, 2002). Does this commonality reflect a shared evolutionary relationship? Not necessarily. Since substantial fractions of most genomes surveyed seem to be clock-regulated, at least at the transcriptional level, this extensive overlap in the types of genes that are clock-regulated in diverse species could occur simply by chance.

6.7.2 Rhythmic gene expression in photosynthetic species

Cyanobacteria

The first genome-wide glimpse of the effect of circadian rhythms on gene expression was provided by the cyanobacterium *Synechococcus elongates*. This photosynthetic prokaryote does not share central clock genes with eukaryotes, and in fact appears to use a fundamentally different mechanism to drive rhythms (Tomita *et al.*, 2004). Promoter activity of almost all genes appears to be clock-regulated (Liu *et al.*, 1995). However, it is not yet clear what fraction of cyanobacterial transcripts, much less proteins, show circadian oscillations. It is therefore impossible to speculate whether all or a subset of biological processes are clock-regulated in these organisms. In fact, no specific biochemical pathway has as yet been shown to be under circadian regulation in cyanobacteria, despite the clear biological advantage these rhythms confer (Ouyang *et al.*, 1998, Woelfle *et al.*, 2004).

Dinoflagellates

Recently, a circadian microarray experiment was performed in a eukaryotic relative of green plants, the dinoflagellate *Pyrocystis lumula* (Okamoto & Hastings, 2003). In this study, estimated to cover roughly 50% of the genome, expression of 3% of the transcriptome was classified as circadian. Despite the considerable evolutionary distance between dinoflagellates and land plants (estimated to have diverged 1,500 million years ago (Yoon *et al.*, 2004; Bhattacharya *et al.*, 2004), many of these clock-controlled genes fall into categories with significant circadian regulation in *Arabidopsis*, such as light perception, light harvesting, and intracellular transport. However, there are significant differences in clock regulation of gene expression in these systems as well. Roughly 90% of the dinoflagellate circadian regulated transcripts showed peak expression in the late night or early day (compared to a roughly even distribution in *Arabidopsis*), with the most variable transcript showing only a 2.4-fold peak-to-trough ratio (compare this to the 20-fold or greater daily variation seen in *CCA1* and *LHY* levels in *Arabidopsis*) (Harmer *et al.*, 2000; Schaffer *et al.*, 2001). Although post-transcriptional regulation plays an important role in rhythms in dinoflagellates (Morse *et al.*, 1989; Mittag *et al.*, 1994), the importance of transcriptional regulation has yet to be established. Nothing is known about the molecular nature of the central clock in dinoflagellates, and none of the cycling transcripts identified by Okamoto and Hastings has obvious homology to plant clock-associated genes. Therefore, it is not clear if the similarities between the types of genes that are clock-regulated in land plants and dinoflagellates are due to common evolutionary descent or convergent evolution.

Plants

Although most recent work on circadian regulation of gene expression in plants has been carried out in *Arabiodopsis*, some comparisons can be made to other species. Several genes that are clock-regulated in angiosperms have also been found to be clock-regulated in bryophytes (Ichikawa *et al.*, 2004; Shimizu *et al.*, 2004; Aoki *et al.*, 2004) and in dinoflagellates (described above). It is not yet clear if these highly divergent relatives of flowering plants have homologs of the known clock-associated genes *CCA1*, *LHY*, and *TOC1* and if so whether they have conserved functions. Studies of the monocots *Lemna gibba* (duckweed) and rice have found circadian-regulated expression of homologs of genes essential for normal clock function in *Arabidopsis*, suggesting that the central clocks of monocots and dicots may be fundamentally similar (Hayama *et al.*, 2002; Murakami *et al.*, 2003; Oyama, 2005). Given the conservation of clock genes between *Drosophila* and mammals (Young & Kay, 2001), it would not be surprising to find that mosses and vascular plants share similar clock mechanisms since these plants are estimated to have diverged more recently than insects and mammals.

Although studies in *Arabidopsis* may help reveal the workings of a central clock mechanism that is universal to land plants, it is necessary to examine clock-regulated processes in other species to learn how the circadian clock helps plants adapt to diverse environmental niches. A particularly interesting example of such specialization of the circadian system is found in Crassulacean acid metabolism (CAM) plants. These remarkable plants, specialized to live in arid environments, have temporally separated CO_2 uptake and photosynthesis. Homologs of *CCA1/LHY*, *TOC1*, and *ELF3* have been found to be clock-regulated in the facultative CAM plant *Mesembryanthemum crystallinum* (Boxall *et al.*, 2005). Furthermore, most genes implicated in CAM metabolism seem to be under circadian regulation (Boxall *et al.*, 2004). It will be fascinating to learn how circadian regulation of sugar metabolism and other aspects of physiology differ in CAM and C3 plants such as *Arabidopsis*.

(See Chapter 9 for a more thorough discussion of the role of the circadian clock in CAM metabolism.)

Other processes under circadian control in some species but not *Arabidopsis* may also involve circadian regulation of gene expression. For example, many cross-pollinated plants such as night-blooming jasmine (*Cestrum nocturnum*) have obvious circadian rhythms in flower opening and scent emission (Overland, 1960), presumably timed to attract appropriate pollinators. The chemical nature of these clock-regulated volatiles and their modes of regulation have been studied in a variety of fragrant species. Benzenoids are important scent compounds released in a circadian fashion by snapdragon, jasmine (*Stephanotis floribunda*) and petunia flowers (Kolosova *et al.*, 2001; Pott *et al.*, 2003; Verdonk *et al.*, 2003). Transcripts encoding the enzymes that catalyze the final steps in the production of benzenoids are clock-regulated in snapdragon and *Stephanotis*, as are the activities of the enzymes themselves (Kolosova *et al.*, 2001; Pott *et al.*, 2003). The benzenoids are synthesized from benzoic acid, which is made as part of the phenylpropanoid pathway. As noted above, most or all components of the phenylpropanoid pathway are clock-regulated at the transcriptional level in *Arabidopsis*, as are the activities of at least some of these enzymes in other plant species (Podstolsky & Brown, 1974; McClure, 1974; Gordon & Koukkari, 1978; Knypl *et al.*, 1986; Peter *et al.*, 1991; Harmer *et al.*, 2000; Kolosova *et al.*, 2001). Thus emission of benzenoids is a species-specific physiological rhythm whose rhythmicity is based at least in part upon a more general pathway that is clock-regulated in many plants. It will be interesting to learn if other specialized physiological clock outputs show such intersections with more widespread rhythms.

6.8 Comparing gene expression in constant and diurnal conditions

The availability of expression arrays now makes it possible to compare circadian and diurnal regulation of gene expression on a global scale. The following section presents some general conclusions that were obtained by comparing the changes of circadian-regulated genes monitored by the 8K Affymetrix array (Harmer *et al.*, 2000) with a recently completed study of the diurnal changes of gene expression.

6.8.1 Circadian rhythms vs. diurnal responses

To investigate patterns of gene expression when plants are grown in diurnal conditions, we grew plants in 12 h light/12 h dark, harvested samples at six time points, and analyzed gene expression in three true biological replicates using the 22k ATH1 Affymetrix array (O.E. Bläsing, unpublished results; data available in the AtGenExpress database). Harmer *et al.* (2000) identified 453 genes as subject to circadian regulation, using the 8K Affymetrix array, representing 6% of the genes on the array. Of these, 369 were present on the 22K ATH1 array and all of these showed a significant diurnal change. During a diurnal cycle, 30–50% of the genes

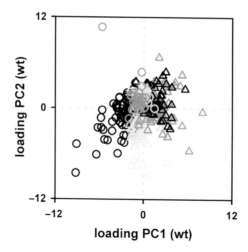

Figure 6.3 Principle Component Analysis (PCA) of the diurnal responses of circadian regulated genes from Arabidopsis, as identified by Harmer *et al.* (2000). Clock-regulated genes are separated by their partitioning in the first (*x*-axis) and second (*y*-axis) components after PCA analysis of 13 690 genes that were detected by microarray analysis of Col-0 wild type (wt) plants, which were harvested at 4 h-intervals in a 12 h light/12 h dark diurnal cycle. Genes are shaded in grey according to their phase of peak expression in circadian conditions (circles indicate plants harvested during the subjective dark period, triangles indicate plants harvested during the subjective light period; duration is indicated by color: 4 h in each condition is light grey, 8 h is dark grey and 12 h is black).

called present on the array showed diurnal changes, depending on the stringency of the statistical procedures applied.

We first studied whether the expression patterns of the 369 circadian-regulated genes were consistent with circadian regulation playing a major role in the diurnal cycle (O.E. Bläsing *et al.*, unpublished results). To do this, a principle components analysis (PCA) was carried out on the entire set of 13 700 genes called 'present' on the ATH1 array in our triplicated study of the diurnal cycle. This separated the six time points well, and the first two components captured more than 60% of all the variation (data not shown). The partitioning of individual genes into the first two components was then inspected. Figure 6.3 shows the separation of the 369 circadian-regulated genes identified by Harmer *et al.* (2000). Genes that peak at different circadian times are shown as different symbols and shades. At a global scale, many genes that are circadian-regulated have moderate to high weightings in both of the principle components, indicating that they vary under diurnal conditions as well. Even more importantly, genes that peak at different times in the circadian cycle have different weightings in the two components, causing them to cluster in different places on the two-dimensional plot. This suggests that, in general, genes with similar phases of expression in constant conditions are co-regulated in light/dark cycles.

We next examined whether environmental and metabolic factors might also play roles in the regulation of gene expression in diurnal cycles. We extracted lists of genes that are induced or repressed by sugars, light, nitrogen or water stress from published and unpublished studies and examined their groupings on the diurnal PCA described above. This analysis suggested that diurnal changes of sugars play a major role in the diurnal regulation of gene expression. Surprisingly, light seemed to have a somewhat less significant role, and nitrogen and water stress do not appear to make a major contribution under these growth conditions (O.E. Bläsing & M. Stitt, unpublished).

6.8.2 Interactions between sugar and circadian regulation

To investigate how one metabolic signal interacts with circadian regulation in a diurnal cycle, we examined diurnal expression patterns in the starchless *pgm* mutant. Starchless mutants undergo much larger diurnal changes of sugars than wild-type plants, but the basic patterns of sugar accumulation are similar except that sugars fall more rapidly during the night. PCA of diurnal regulation of gene expression in *pgm* shows a less obvious separation between circadian-regulated genes than in wild-type plants (data not shown), indicating there may indeed be interactions between regulation of gene expression by the circadian clock and sugars. Consistent with this, about 30% of the clock-regulated genes identified by Harmer *et al.* (2000) showed increased amplitude and another 5% showed a shift of phase in *pgm* mutants relative to wild-type plants (data not shown).

To further explore interactions between circadian regulation and regulation by sugars, the circadian and diurnal changes of gene expression in wild-type plants were inspected more closely. Figure 6.4 examines the relationship between the peak circadian phases of the 369 genes identified by Harmer *et al.* (2000) with their peak phases during a diurnal cycle. Each of the six panels shows genes that peak at a particular time in constant light (the circadian peak). The *x*-axis indicates whether that peak is identical to the time of peak expression in a diurnal cycle, or whether it is shifted forwards or backwards. While most genes retain the same approximate time of peak expression in both circadian and diurnal conditions, there are obvious exceptions. For example, some genes that peak 8 h after lights on in circadian conditions show a phase shift that can be as large as 8 or 12 h in diurnal conditions. Since the plants assayed by Harmer *et al.* (2000) were allowed to free-run in constant conditions for 24 h before sampling began, some of these apparent differences in phase may be due to phase drift. Other potentially confounding factors may include experimental noise and differing mathematical procedures. However, the magnitude of some phase differences (up to 8–12 h) suggests additional regulatory mechanisms may be at work in diurnal conditions. To explore one such possible mechanism, we have separated the genes on the *y*-axis according to their responsiveness to glucose. (This data is taken from experiments in which glucose was added to seedlings that had been grown in continuous low light in the absence of added external sugar for two days (W. R. Scheible, unpublished data)).

While a detailed analysis would be beyond the scope of this chapter, some general conclusions can be made. First, a very large proportion of the circadian-regulated

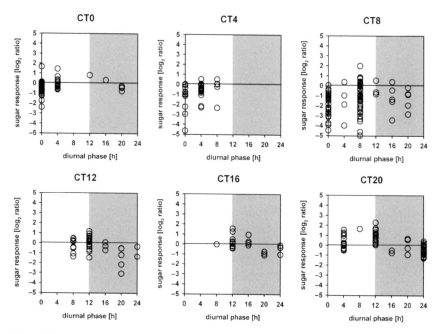

Figure 6.4 Phase shift and sugar response of circadian regulated genes in a diurnal cycle. Genes were separated into six groups according to their estimated peak phase of expression in constant conditions (CT0 to CT20). Each sub-panel shows the difference between the timing of the circadian and diurnal peaks on an absolute timescale (x-axis). The sugar response (y-axis) shows induction (positive values) or repression (negative values) of each gene to glucose addition in a seedling culture system after 3 days of sugar depletion.

genes are also subject to induction or repression by sugars, suggesting there is a strong interaction between circadian and sugar regulation of gene expression. Second, the changes in the timing of expression between a circadian and a diurnal cycle in some cases may be due to sugar regulation overriding circadian regulation. Many of the genes that peak at the end of the subjective day in constant conditions but in the later part or the end of the night in a diurnal cycle are sugar-represssed. Sugars are high during the day and decrease to a minimum at the end of the night, suggesting that in these cases metabolic regulation counteracts the increase in expression that would otherwise be driven by circadian regulation at this time. Furthermore, many genes that show peak expression during the subjective night in constant conditions and that are also induced by sugars show peak transcript levels during the light period in a light/dark cycle. The possible physiological implications of such antagonistic interactions between circadian regulation and changes of sugars will be discussed below.

The availability of large databases of expression arrays will make it possible in the very near future to carry out these types of comparisons for a wide range of environmental and physiological inputs that vary during the diurnal cycle. This will help us better understand how the circadian system interacts with changes that occur in natural day/night cycles.

6.9 Why have a clock rather than relying on driven rhythms?

This is a far-reaching question, which empirically could be answered by examining genotypes that are defective in circadian signaling for deficiencies in their physiological responses. It has indeed been shown that arrhythmic *Arabidopsis* plants have a low-viability phenotype (Green *et al.*, 2002). The ability to detect physiological deficits due to disrupted clock function will be increased in the future by the ability to monitor a wide range of transcripts, proteins and metabolites. This will allow the detection of small changes in metabolism or signaling pathways produced when interactions between the circadian system and environmental or physiological signaling are altered. But until such studies can be performed, we can begin to address this question by integrating our current understanding of circadian regulation with the known roles of clock-regulated genes in different metabolic, cellular and physiological processes.

6.9.1 *Might anticipation be key?*

One frequently cited role for circadian regulation is that it allows anticipation of changes that occur on a regular basis in diurnal cycles. This is consistent with the regulation of many genes involved in photosynthesis and nutrient assimilation that peak in the last part of the night, not long before their gene products are thought to be required. Circadian regulation in these cases would not only anticipate but might also 'prime' the plant to respond particularly strongly to daily changes in environmental and physiological inputs, reinforcing these cues.

However, there are also cases where circadian regulation and physiological inputs appear to oppose one another. Examples discussed above include the subset of sugar-induced genes whose circadian peak is at the end of the night, and the subset of sugar-repressed genes whose circadian peak is at the end of the day. In a light-dark cycle, the transcripts for these genes peak at the time predicted based on the diurnal changes of sugar levels rather than those found in a free-running circadian rhythm. Although further research is needed to confirm such differences and investigate whether they have any physiological role, one possible reason for such antagonistic regulation might be to provide a temporal dimension to the sensing of dynamic physiological signals, facilitating more refined sensing of different physiological states. For example, the changes of sugars described so far in this chapter are typical for the source levels of well-fertilized and vigorously growing plants. The rise of sugars during the day is depressed in conditions where photosynthesis is slow, whereas carbohydrates are high throughout the day and night in conditions where growth is slow (Stitt & Krapp, 1999). It will be interesting to learn whether in these non-optimal growth conditions the transcripts for the above sets of genes shift to match the free-running circadian pattern.

6.9.2   Do transcript levels correlate with protein levels?

An important question in assessing the physiological roles of circadian and diurnal regulation is whether the changes of transcript levels lead within a relevant time frame to a change in the level of the encoded protein. In some cases, circadian regulation of transcript abundance is a reliable indicator of rhythmic levels of the encoded proteins. Examples include the genes that encode clock components and important members of clock-driven signaling pathways such as *CO* (Valverde *et al.*, 2004). However, this is clearly not the case for all genes, including genes that encode enzymes acting in primary metabolism (Gibon *et al.*, 2004a). This introduces the question as to why genes encoding stable proteins show circadian or diurnal regulation. A tentative answer to the latter part of the question might be as follows: it is important during development and for mid- to long-term adaptation to changing environmental conditions that expression of these genes responds to signals like light, and sugar and nitrate availability. This will require that the genes contain elements in their promoters that respond to physiological inputs that change dramatically during the diurnal cycle and will automatically lead to changes of transcription. However, for genes such as those characterized by Gibon *et al.* (2004a), this will not lead to an immediate change in protein levels.

Slow turnover of some types of protein might be advantageous, ensuring that protein levels would be determined by environmental conditions experienced over several days rather than due to transient changes in local conditions. This can be illustrated by considering how glutamate dehydrogenase (GDH) activity and transcripts for *GDH1* and *GDH2* respond in diurnal cycles and in extended darkness (Gibon *et al.*, 2004a). In a diurnal cycle, *GDH* transcripts are low, as is GDH activity. An extended night leads within 2–4 h to a dramatic increase of *GDH* transcripts, whereas GDH activity remains initially unaltered and rises slowly by two- to three-fold over the next 4–7 days. Some enzymes that are required for photosynthesis show a reversed but analogous response, with a rapid decrease of transcripts but a slow decrease of enzyme activity upon transfer to darkness. The relatively slow turnover of these proteins might speculatively be seen as an adaptation, allowing gene expression to respond to a large number of factors that change both on a long-term and on a daily basis but with significant changes in the level of protein seen only in response to long term changes in local conditions. Such buffering might be especially important for enzymes involved in key processes such as photosynthesis and central metabolism.

6.9.3   Why are the transcripts of stable proteins under circadian regulation?

Another important factor that should be considered is the impact of repeated transient changes, as illustrated by regulation of *GDH* expression and enzyme activity in *pgm* mutants (Gibon *et al.*, 2004a). These mutants display a transient, large peak of

GDH expression in the second part of every night, triggered by the nightly depletion of sugars in this starchless mutant. GDH activity in *pgm* plants is stable and high throughout the diurnal cycle, and, notably, is similar to that seen in wild-type plants that have been in the dark for seven days. Therefore although a single transient change in transcript levels may have little effect on protein levels, repetition of such a change over several days may lead to a significant increase in protein levels. Perhaps circadian regulation of gene expression for such stable proteins is not primarily important for the regulation of events on the same day but, rather, has its effects by interacting with environmental and physiological signals that are repeated on a daily basis.

Such an interaction would result in daily signals whose strength is dependent upon circadian, environmental, and physiological inputs. These repeated signals could then be integrated over several days to cause the appropriate change in protein levels and enzyme activity. In a perhaps analogous fashion, many plant species integrate information flow between the environment and the circadian clock over multiple days to control the transition to flowering (Yanovsky & Kay, 2003). Clearly, if changes of transcripts do not lead to changes of the encoded protein within a circadian time frame, it is difficult to see how circadian regulation of the expression of these genes can be seen as 'anticipation', at least in a simple sense. The concept probably needs to be extended to include the interaction between circadian regulation and the probable – and shifting – timing of other environmental and physiological inputs that regulate the expression of the gene.

6.10 Future prospects

Advances in proteomics will allow us to determine whether most enzymes show the same damped relationship between transcript and protein levels as the central metabolic enzymes discussed in this last section. New techniques will also be required to determine the relationship between transcript, protein, and protein activity at different stages of development and in different tissues. It may be that some pathways or processes are under circadian control in specific tissues, but that this regulation cannot be measured when the plant is considered as a whole. Indeed, this type of tissue-specific regulation of clock outputs has previously been observed in plants (Dudareva *et al.*, 2000; Kolosova *et al.*, 2001; Thain *et al.*, 2002) and is the rule rather than the exception in mammals (Panda *et al.*, 2002; Storch *et al.*, 2002). The study of the diverse outputs of the plant circadian clock promises to yield a better understanding of the many ways plants adapt to their ever-changing environment and the role of clock-regulated gene expression in these processes.

Acknowledgements

We are grateful to the following agencies for funding our research: the National Institutes of Health (5R01GM069418-02 to SLH) and the National Research Initiative of the USDA (2004-35100-14903 to MFC).

References

Adamska, I., Scheel, B. & Kloppstech, K. (1991) Circadian oscillations of nuclear-encoded chloroplast proteins in pea (*Pisum sativum*). *Plant Mol. Biol.*, **17**, 1055–65.

Akhtar, R.A., Reddy, A.B., Maywood, E.S., Clayton, J.D., King, V.M., Smith, A.G., Gant, T.W., Hastings, M.H. & Kyriacou, C.P. (2002) Circadian cycling of the mouse liver transcriptome, as revealed by cDNA microarray, is driven by the suprachiasmatic nucleus. *Curr. Biol.*, **12**, 540–50.

Alabadi, D., Oyama, T., Yanovsky, M.J., Harmon, F.G., Mas, P. & Kay, S.A. (2001) Reciprocal regulation between TOC1 and LHY/CCA1 within the *Arabidopsis* circadian clock. *Science*, **293**, 880–3.

Aoki, S., Kato, S., Ichikawa, K. & Shimizu, M. (2004) Circadian expression of the PpLhcb2 gene encoding a major light-harvesting chlorophyll a/b-binding protein in the moss *Physcomitrella patens*. *Plant Cell Physiol.*, **45**, 68–76.

Arimura, G., Huber, D.P. & Bohlmann, J. (2004) Forest tent caterpillars (*Malacosoma disstria*) induce local and systemic diurnal emissions of terpenoid volatiles in hybrid poplar (*Populus trichocarpa* x *deltoides*): cDNA cloning, functional characterization, and patterns of gene expression of (-)-germacrene D synthase, PtdTPS1. *Plant J.*, **37**, 603–16.

Athwal, G.S. & Huber, S.C. (2002) Divalent cations and polyamines bind to loop 8 of 14-3-3 proteins, modulating their interaction with phosphorylated nitrate reductase. *Plant J.*, **29**, 119–29.

Bachmann, M., Huber, J.L., Liao, P.C., Gage, D.A. & Huber, S.C. (1996) The inhibitor protein of phosphorylated nitrate reductase from spinach (*Spinacia oleracea*) leaves is a 14-3-3 protein. *FEBS Lett.*, **387**, 127–31.

Beator, J., Potter, E. & Kloppstech, K. (1992) The effect of heat shock on morphogenesis in barley: coordinated circadian regulation of mRNA levels for light-regulated genes and of the capacity for accumulation of chlorophyll protein complexes. *Plant Physiol.*, **100**, 1780–1786.

Bhattacharya, D., Yoon, H.S. & Hackett, J.D. (2004) Photosynthetic eukaryotes unite: endosymbiosis connects the dots. *Bioessays*, **26**, 50–60.

Bognar, L.K., Hall, A., Adam, E., Thain, S.C., Nagy, F. & Millar, A.J. (1999) The circadian clock controls the expression pattern of the circadian input photoreceptor, phytochrome B. *Proc. Natl. Acad. Sci. USA*, **96**, 14652–7.

Borevitz, J.O., Xia, Y., Blount, J., Dixon, R.A. & Lamb, C. (2000) Activation tagging identifies a conserved MYB regulator of phenylpropanoid biosynthesis. *Plant Cell*, **12**, 2383–2394.

Boxall, S.F., Foster, J.M., Bohnert, H.J., Cushman, J.C., Nimmo, H.G. & Hartwell, J. (2004) In *13th International Congress of Photosynthesis*, Vol. Abstract #165 Montreal, Canada.

Boxall, S.F., Foster, J.M., Bohnert, H.J., Cushman, J.C., Nimmo, H.G. & Hartwell, J. (2005) Conservation and divergence of the central circadian clock in the stress-inducible CAM plant *Mesembryanthemum crystallinum*: clock operation in a CAM halophyte reveals clock compensation against abiotic stress. *Plant Physiol.*, **137**:969–82.

Carpenter, C.D., Kreps, J.A. & Simon, A.E. (1994) Genes encoding glycine-rich *Arabidopsis thaliana* proteins with RNA-binding motifs are influenced by cold treatment and an endogenous circadian rhythm. *Plant Physiol.*, **104**, 1015–25.

Caspar, T., Huber, S.C. & Somerville, C.R. (1985) Alterations in growth, photosynthesis and respiration in a starch deficient mutant of *Arabidopsis thaliana* (L.) Heynh deficient in chloroplast phospho-glucomutase. *Plant Physiol.*, **79**, 11–17.

Ceriani, M.F., Hogenesch, J.B., Yanovsky, M., Panda, S., Straume, M. & Kay, S.A. (2002) Genome-wide expression analysis in Drosophila reveals genes controlling circadian behavior. *J. Neurosci.*, **22**, 9305–19.

Chatterton, N.J. & Silvius, J.E. (1979) Photosynthate partitioning into starch in soybean leaves: effects of photoperiod versus photosynthetic period duration. *Plant Physiol.*, **64**, 749–53.

Chatterton, N.J. & Silvius, J.E. (1980) Photosynthate partitioning as affected by daily photosynthetic period duration in six species. *Physiol. Plant.*, **49**, 141–144.

Chatterton, N.J. & Silvius, J.E. (1981) Photosynthate partitioning into starch in soybean leaves: irradiance level and daily photsynthetic period duration effects. *Plant Physiol.*, **67**, 257–60.

Chelius, D., Zhang, T., Wang, G. & Shen, R.F. (2003) Global protein identification and quantification technology using two-dimensional liquid chromatography nanospray mass spectrometry. *Anal. Chem.*, **75**, 6658–65.

Claridge-Chang, A., Wijnen, H., Naef, F., Boothroyd, C., Rajewsky, N. & Young, M.W. (2001) Circadian regulation of gene expression systems in the *Drosophila* head. *Neuron*, **32**, 657–71.

Correa, A., Lewis, Z.A., Greene, A.V., March, I.J., Gomer, R.H. & Bell-Pedersen, D. (2003) Multiple oscillators regulate circadian gene expression in Neurospora. *Proc. Natl. Acad. Sci. USA*, **100**, 13597–602.

Covington, M.F., Kay, S., Maloof, J.N., Straume, M. & Harmer, S.L. (2005), In preparation.

Czechowski, T., Bari, R.P., Stitt, M., Scheible, W.R. & Udvardi, M.K. (2004) Real-time RT-PCR profiling of over 1400 *Arabidopsis* transcription factors: unprecedented sensitivity reveals novel root- and shoot-specific genes. *Plant J.*, **38**, 366–79.

Daniel, X., Sugano, S. & Tobin, E.M. (2004) CK2 phosphorylation of CCA1 is necessary for its circadian oscillator function in *Arabidopsis*. *Proc. Natl. Acad. Sci. USA*, **101**, 3292–3297.

Deng, M.D., Moureaux, T., Leydecker, M.T. & Caboche, M. (1990) Nitrate-reductase expression is under the control of a circadian rhythm and is light inducible in Nicotiana tabacum leaves. *Planta*, **180**, 257–61.

Dietrich, P., Sanders, D. & Hedrich, R. (2001) The role of ion channels in light-dependent stomatal opening. *J. Exp. Bot.*, **52**, 1959–67.

Doi, K., Izawa, T., Fuse, T., Yamanouchi, U., Kubo, T., Shimatani, Z., Yano, M. & Yoshimura, A. (2004) Ehd1, a B-type response regulator in rice, confers short-day promotion of flowering and controls FT-like gene expression independently of Hd1. *Genes Dev.*, **18**, 926–36.

Dowson-Day, M.J. & Millar, A.J. (1999) Circadian dysfunction causes aberrant hypocotyl elongation patterns in *Arabidopsis*. *Plant J.*, **17**, 63–71.

Dudareva, N., Andersson, S., Orlova, I., Gatto, N., Reichelt, M., Rhodes, D., Boland, W. & Gershenzon, J. (2005) The nonmevalonate pathway supports both monoterpene and sesquiterpene formation in snapdragon flowers. *Proc. Natl. Acad. Sci. USA*, **102**, 933–938.

Dudareva, N., Murfitt, L.M., Mann, C.J., Gorenstein, N., Kolosova, N., Kish, C.M., Bonham, C. & Wood, K. (2000) Developmental regulation of methyl benzoate biosynthesis and emission in snapdragon flowers. *Plant Cell*, **12**, 949–61.

Duffield, G.E. (2003) DNA microarray analyses of circadian timing: the genomic basis of biological time. *J. Neuroendocrinol.*, **15**, 991–1002.

Eriksson, M.E., Hanano, S., Southern, M.M., Hall, A. & Millar, A.J. (2003) Response regulator homologs have complementary, light-dependent functions in the *Arabidopsis* circadian clock. *Planta*, **218**, 159–62.

Etchegaray, J.P., Lee, C., Wade, P.A. & Reppert, S.M. (2003) Rhythmic histone acetylation underlies transcription in the mammalian circadian clock. *Nature*, **421**, 177–82.

Fernie, A.R., Geigenberger, P. & Stitt, M. (2005) Flux an important, but neglected, component of functional genomics. *Curr. Opin. in Plant Bio.*, **8**:174–82.

Fiehn, O. (2002) Metabolomics–the link between genotypes and phenotypes. *Plant Mol. Biol.*, **48**, 155–71.

Finlayson, S.A., Lee, I.J., Mullet, J.E. & Morgan, P.W. (1999) The mechanism of rhythmic ethylene production in sorghum. The role of phytochrome B and simulated shading. *Plant Physiol.*, **119**, 1083–9.

Fondy, B.R. & Geiger, D.R. (1985) Diurnal changes of allocation of newly fixed carbon in exporting sugar beet leaves. *Plant Physiol.*, **78**, 753–57.

Geigenberger, P., Reimholz, R., Deiting, U., Sonnewald, U. & Stitt, M. (1999) Decreased expression of sucrose phosphate synthase strongly inhibits the water stress-induced synthesis of sucrose in growing potato tubers. *Plant J.*, **19**, 119–129.

Geiger, D.R. & Servaites, J.C. (1994) Diurnal regulation of photosynthetic carbon metabolism in C3 plants. *Annu. Rev. Plant Biol.*, **45**, 235–56.

Gibon, Y., Blaesing, O.E., Hannemann, J., Carillo, P., Hohne, M., Hendriks, J.H., Palacios, N., Cross, J., Selbig, J. & Stitt, M. (2004a) A robot-based platform to measure multiple enzyme activities in *Arabidopsis* using a set of cycling assays: comparison of changes of enzyme activities and transcript levels during diurnal cycles and in prolonged darkness. *Plant Cell*, **16**, 3304–25.

Gibon, Y., Blasing, O.E., Palacios-Rojas, N., Pankovic, D., Hendriks, J.H., Fisahn, J., Hohne, M., Gunther, M. & Stitt, M. (2004b) Adjustment of diurnal starch turnover to short days: depletion of sugar during the night leads to a temporary inhibition of carbohydrate utilization, accumulation of sugars and post-translational activation of ADP-glucose pyrophosphorylase in the following light period. *Plant J.*, **39**, 847–62.

Giuliano, G., Hoffman, N.E., Ko, K., Scolnik, P.A. & Cashmore, A.R. (1988) A light-entrained circadian clock controls transcription of several plant genes. *Embo J.*, **7**, 3635–42.

Gordon, W.R. & Koukkari, W.L. (1978) Circadian rhythmicity in the activities of phenylalanine ammonia lyase from *Lemna perpusilla* and *Spirodela polyrhiza*. *Plant Physiol.*, **62**, 612–615.

Green, R.M., Tingay, S., Wang, Z.Y. & Tobin, E.M. (2002) Circadian rhythms confer a higher level of fitness to *Arabidopsis* plants. *Plant Physiol.*, **129**, 576–84.

Gutierrez, R.A., Ewing, R.M., Cherry, J.M. & Green, P.J. (2002) Identification of unstable transcripts in *Arabidopsis* by cDNA microarray analysis: rapid decay is associated with a group of touch- and specific clock-controlled genes. *Proc. Natl. Acad. Sci. USA*, **99**, 11513–8.

Halberg, F., Halberg, E., Barnum, C. & Bittner, J. (1959) In *Photoperiodism and related phenomena in plants and animals*, Vol. 55 (ed R.B. Withrow), pp. 803–78. American Association for the Advancement of Science, Washington.

Harmer, S.L., Hogenesch, J.B., Straume, M., Chang, H.S., Han, B., Zhu, T., Wang, X., Kreps, J.A. & Kay, S.A. (2000) Orchestrated transcription of key pathways in *Arabidopsis* by the circadian clock. *Science*, **290**, 2110–3.

Harmer, S.L. & Kay, S. (2005) Positive and negative factors confer phase-specific circadian regulation of transcription in *Arabidopsis*. *Plant Cell*, **17**, 1926–40.

Havaux, M. & Niyogi, K.K. (1999) The violaxanthin cycle protects plants from photooxidative damage by more than one mechanism. *Proc. Natl. Acad. Sci. USA*, **96**, 8762–7.

Hayama, R., Izawa, T. & Shimamoto, K. (2002) Isolation of rice genes possibly involved in the photoperiodic control of flowering by a fluorescent differential display method. *Plant Cell Physiol.*, **43**, 494–504.

Heintzen, C., Fischer, R., Melzer, S., Kappeler, K., Apel, K. & Staiger, D. (1994a) Circadian oscillations of a transcript encoding a germin-like protein that is associated with cell walls in young leaves of the long-day plant *Sinapis alba* L. *Plant Physiol.*, **106**, 905–15.

Heintzen, C., Melzer, S., Fischer, R., Kappeler, S., Apel, K. & Staiger, D. (1994b) A light- and temperature-entrained circadian clock controls expression of transcripts encoding nuclear proteins with homology to RNA-binding proteins in meristematic tissue. *Plant J.*, **5**, 799–813.

Heintzen, C., Nater, M., Apel, K. & Staiger, D. (1997) AtGRP7, a nuclear RNA-binding protein as a component of a circadian-regulated negative feedback loop in *Arabidopsis thaliana*. *Proc. Natl. Acad. Sci. USA*, **94**, 8515–20.

Hennessey, T.L. & Field, C.B. (1991) Circadian rhythms in photosynthesis: Oscillations in carbon assimilation and stomatal conductance under constant conditions. *Plant Physiol.*, **96**, 831–36.

Huber, S.C. & Hanson, K.R. (1992) Carbon partitioning and growth of a starchless mutant of *Nicotiana sylvestris*. *Plant Physiol.*, **99**, 1449–1454.

Hudson, M.E. & Quail, P.H. (2003) Identification of promoter motifs involved in the network of phytochrome A-regulated gene expression by combined analysis of genomic sequence and microarray data. *Plant Physiol.*, **133**, 1605–16.

Ichikawa, K., Sugita, M., Imaizumi, T., Wada, M. & Aoki, S. (2004) Differential expression on a daily basis of plastid sigma factor genes from the moss physcomitrella patens. Regulatory interactions among PpSig5, the circadian clock, and blue light signaling mediated by cryptochromes. *Plant Physiol.*, **136**, 4285–98.

Izawa, T., Oikawa, T., Sugiyama, N., Tanisaka, T., Yano, M. & Shimamoto, K. (2002) Phytochrome mediates the external light signal to repress FT orthologs in photoperiodic flowering of rice. *Genes Dev.*, **16**, 2006–20.

Jackson, M.B. (1982) In *Plant Growth Substances* (ed P.F. Wareing), pp. 291–301. Academic Press, London.

Jackson, M.B. (1985) Ethylene and responses of plants to soil waterlogging and submergence. *Ann. Rev. Plant Physiol.*, **36**, 145–74.

Jones, T.L., Tucker, D.E. & Ort, D.R. (1998) Chilling delays circadian pattern of sucrose phosphate synthase and nitrate reductase activity in tomato. *Plant Physiol.*, **118**, 149–58.

Jouve, L., Gaspar, T., Kevers, C., Greppin, H. & Degli Agosti, R. (1999) Involvement of indole-3-acetic acid in the circadian growth of the first internode of *Arabidopsis*. *Planta*, **209**, 136–42.

Kerr, P.S., Rufty, T.W. & Huber, S.C. (1985) Endogenous rhythms in photosynthesis, sucrose phosphate synthase activity, and stomatal resistance in leaves of soybean (*Glycine max* [L.] Merr.). *Plant Physiol.*, **77**, 275–80.

Kim, J.Y., Song, H.R., Taylor, B.L. & Carre, I.A. (2003a) Light-regulated translation mediates gated induction of the *Arabidopsis* clock protein LHY. *Embo J.*, **22**, 935–44.

Kim, W.Y., Geng, R. & Somers, D.E. (2003b) Circadian phase-specific degradation of the F-box protein ZTL is mediated by the proteasome. *Proc. Natl. Acad. Sci. USA*, **100**, 4933–38.

Kloppstech, K. (1985) Diurnal and circadian rhythmicity in the expression of light-induced plant nuclear messenger RNAs. *Planta*, **165**, 502–506.

Knypl, J.S., Janas, K.M. & Wolska, M. (1986) Rhythmicity of L-phenylalanine ammonia-lyase activity in *Spirodela oligorhiza*. Effects of darkening, abscisic acid, and 1-amino-2-phenyl-ethylphosphoric acid. *Physiol Plant.*, **66**, 543–549.

Kolosova, N., Gorenstein, N., Kish, C.M. & Dudareva, N. (2001) Regulation of circadian methyl benzoate emission in diurnally and nocturnally emitting plants. *Plant Cell*, **13**, 2333–47.

Kopka, J., Fernie, A., Weckwerth, W., Gibon, Y. & Stitt, M. (2004) Metabolite profiling in plant biology: platforms and destinations. *Genome Biol.*, **5**, 109.

Kreps, J.A., Muramatsu, T., Furuya, M. & Kay, S.A. (2000) Fluorescent differential display identifies circadian clock-regulated genes in *Arabidopsis thaliana*. *J. Biol. Rhythms*, **15**, 208–17.

Kreps, J.A., Wu, Y., Chang, H.S., Zhu, T., Wang, X. & Harper, J.F. (2002) Transcriptome changes for *Arabidopsis* in response to salt, osmotic, and cold stress. *Plant Physiol.*, **130**, 2129–41.

Kuno, N., Moller, S.G., Shinomura, T., Xu, X., Chua, N.H. & Furuya, M. (2003) The novel MYB protein EARLY-PHYTOCHROME-RESPONSIVE1 is a component of a slave circadian oscillator in *Arabidopsis*. *Plant Cell*, **15**, 2476–88.

Lam, H.M., Coschigano, K.T., Oliveira, I.C., Melo-Oliveira, R. & Coruzzi, G.M. (1996) The molecular-genetics of nitrogen assimilation into amino acids in higher plants. *Annu. Rev. Plant. Physiol. Plant Mol. Biol.*, **47**, 569–93.

Lillo, C. & Ruoff, P. (1989) An unusually rapid light-induced nitrate reductase mRNA pulse and circadian oscillations. *Naturwissenschaften*, **76**, 526–8.

Lin, T.P., Caspar, T., Somerville, C.R. & Preiss, J. (1988) A starch deficient mutant of *Arabidopsis thaliana* with low ADP-glucose pyrophosphorylase activity lacks one of the two subunits of the enzyme. *Plant Physiol.*, **88**, 1175–81.

Liu, J., Yu, J., McIntosh, L., Kende, H. & Zeevaart, J.A. (2001a) Isolation of a CONSTANS ortholog from Pharbitis nil and its role in flowering. *Plant Physiol.*, **125**, 1821–30.

Liu, X.L., Covington, M.F., Fankhauser, C., Chory, J. & Wagner, D.R. (2001b) ELF3 encodes a circadian clock-regulated nuclear protein that functions in an *Arabidopsis* PHYB signal transduction pathway. *Plant Cell*, **13**, 1293–304.

Liu, Y., Tsinoremas, N.F., Johnson, C.H., Lebedeva, N.V., Golden, S.S., Ishiura, M. & Kondo, T. (1995) Circadian orchestration of gene expression in cyanobacteria. *Genes Dev.*, **9**, 1469–78.

Mas, P., Kim, W.Y., Somers, D.E. & Kay, S.A. (2003) Targeted degradation of TOC1 by ZTL modulates circadian function in *Arabidopsis thaliana*. *Nature*, **426**, 567–70.

Matt, P., Schurr, U., Klein, D., Krapp, A. & Stitt, M. (1998) Growth of tobacco in short-day conditions leads to high starch, low sugars, altered diurnal changes in the Nia transcript and low nitrate reductase activity, and inhibition of amino acid synthesis. *Planta*, **207**, 27–41.

McClure, J.W. (1974) Phytochrome control of oscillating levels of phenylalanine ammonia-lyase in *Hordeum vulgare* shoots. *Phytochemistry*, **13**, 1065–69.

McDonald, A.J.S. & Davies, W.J. (1996) Keeping in touch: responses of whole plant to deficits in water and nutrient supply. *Adv. Botan. Res.*, **22**, 229–300.

McDonald, M.J. & Rosbash, M. (2001) Microarray analysis and organization of circadian gene expression in *Drosophila*. *Cell*, **107**, 567–78.

Michael, T.P. & McClung, C.R. (2002) Phase-specific circadian clock regulatory elements in *Arabidopsis*. *Plant Physiol.*, **130**, 627–38.

Michael, T.P. & McClung, C.R. (2003) Enhancer trapping reveals widespread circadian clock transcriptional control in *Arabidopsis*. *Plant Physiol.*, **132**, 629–39.

Michael, T.P., Salome, P.A. & McClung, C.R. (2003) Two *Arabidopsis* circadian oscillators can be distinguished by differential temperature sensitivity. *Proc. Natl. Acad. Sci. USA*, **100**, 6878–83.

Millar, A.J., Carre, I.A., Strayer, C.A., Chua, N.H. & Kay, S.A. (1995) Circadian clock mutants in *Arabidopsis* identified by luciferase imaging. *Science*, **267**, 1161–63.

Millar, A.J., Short, S.R., Chua, N.H. & Kay, S.A. (1992) A novel circadian phenotype based on firefly luciferase expression in transgenic plants. *Plant Cell*, **4**, 1075–87.

Mittag, M., Lee, D.H. & Hastings, J.W. (1994) Circadian expression of the luciferin-binding protein correlates with the binding of a protein to the 3' untranslated region of its mRNA. *Proc. Natl. Acad. Sci. USA*, **91**, 5257–61.

Morse, D., Milos, P.M., Roux, E. & Hastings, J.W. (1989) Circadian regulation of bioluminescence in Gonyaulax involves translational control. *Proc. Natl. Acad. Sci. USA*, **86**, 172–76.

Moshelion, M., Becker, D., Biela, A., Uehlein, N., Hedrich, R., Otto, B., Levi, H., Moran, N. & Kaldenhoff, R. (2002a) Plasma membrane aquaporins in the motor cells of *Samanea saman*: diurnal and circadian regulation. *Plant Cell*, **14**, 727–39.

Moshelion, M., Becker, D., Czempinski, K., Mueller-Roeber, B., Attali, B., Hedrich, R. & Moran, N. (2002b) Diurnal and circadian regulation of putative potassium channels in a leaf moving organ. *Plant Physiol.*, **128**, 634–42.

Murakami, M., Ashikari, M., Miura, K., Yamashino, T. & Mizuno, T. (2003) The evolutionarily conserved OsPRR quintet: rice pseudo-response regulators implicated in circadian rhythm. *Plant Cell Physiol.*, **44**, 1229–36.

Muratani, M. & Tansey, W.P. (2003) How the ubiquitin-proteasome system controls transcription. *Nat. Rev. Mol. Cell Biol.*, **4**, 192–201.

Okamoto, K. & Hastings, J.W. (2003) Novel dinoflagellate clock-related genes identified through microarray analysis. *J. Phycol.*, **39**, 519–526.

Onai, K., Okamoto, K., Nishimoto, H., Morioka, C., Hirano, M., Kami-Ike, N. & Ishiura, M. (2004) Large-scale screening of *Arabidopsis* circadian clock mutants by a high-throughput real-time bioluminescence monitoring system. *Plant J.*, **40**, 1–11.

Ouyang, Y., Andersson, C.R., Kondo, T., Golden, S.S. & Johnson, C.H. (1998) Resonating circadian clocks enhance fitness in cyanobacteria. *Proc. Natl. Acad. Sci. USA*, **95**, 8660–64.

Overland, L. (1960) Endogenous rhythm in opening and odor of flowers of *Cestrum nocturnum*. *Amer. J. Bot.*, **47**, 378–382.

Oyama, T. (2005). In preparation.

Panda, S., Antoch, M.P., Miller, B.H., Su, A.I., Schook, A.B., Straume, M., Schultz, P.G., Kay, S.A., Takahashi, J.S. & Hogenesch, J.B. (2002) Coordinated transcription of key pathways in the mouse by the circadian clock. *Cell*, **109**, 307–20.

Panda, S., Sato, T.K., Hampton, G.M. & Hogenesch, J.B. (2003) An array of insights: application of DNA chip technology in the study of cell biology. *Trends Cell Biol.*, **13**, 151–6.

Peter, H.-J., Kruger-Alef, C., Knogge, W., Brinkmann, K. & Weissenbock, G. (1991) Diurnal periodicity of chalcone-synthase activity during the development of oat primary leaves. *Planta*, **183**, 409–415.

Plautz, J.D., Straume, M., Stanewsky, R., Jamison, C.F., Brandes, C., Dowse, H.B., Hall, J.C. & Kay, S.A. (1997) Quantitative analysis of *Drosophila* period gene transcription in living animals. *J. Biol. Rhythms*, **12**, 204–17.

Podstolsky, A.J. & Brown, G.N. (1974) L-phenylalanine ammonia-lyase activity in Robinia pseudoacacia seedlings: Cyclic phenomenon activity during continuous light. *Plant Physiology*, **54**, 41–43.

Pott, M.B., Effmert, U. & Piechulla, B. (2003) Transcriptional and post-translational regulation of S-adenosyl-L-methionine: salicylic acid carboxyl methyltransferase (SAMT) during *Stephanotis floribunda* flower development. *J. Plant Physiol.*, **160**, 635–43.

Riesselmann, S. & Piechulla, B. (1992) Diurnal and circadian light-harvesting complex and quinone B-binding protein synthesis in leaves of tomato (*Lycopersicon esculentum*). *Plant Physiol.*, **100**, 1840–45.

Rieu, I., Cristescu, S.M., Harren, F.J., Huibers, W., Voesenek, L.A., Mariani, C. & Vriezen, W.H. (2005) RP-ACS1, a flooding-induced 1-aminocyclopropane-1-carboxylate synthase gene of *Rumex palustris*, is involved in rhythmic ethylene production. *J. Exp. Bot.*, **56**, 841–49.

Rikin, A., Chalutz, E. & Anderson, J.D. (1984) Rhythmicity in ethylene production in cotton seedlings. *Plant Physiol.*, **75**, 493–95.

Sage-Ono, K., Ono, M., Harada, H. & Kamada, H. (1998) Accumulation of a clock-regulated transcript during flower-inductive darkness in pharbitis nil. *Plant Physiol.*, **116**, 1479–85.

Salghetti, S.E., Muratani, M., Wijnen, H., Futcher, B. & Tansey, W.P. (2000) Functional overlap of sequences that activate transcription and signal ubiquitin-mediated proteolysis. *Proc. Natl. Acad. Sci. USA*, **97**, 3118–23.

Sato, T.K., Panda, S., Kay, S.A. & Hogenesch, J.B. (2003) DNA arrays: applications and implications for circadian biology. *J. Biol. Rhythms*, **18**, 96–105.

Schaffer, R., Landgraf, J., Accerbi, M., Simon, V., Larson, M. & Wisman, E. (2001) Microarray analysis of diurnal and circadian-regulated genes in *Arabidopsis*. *Plant Cell*, **13**, 113–23.

Schaffer, R., Ramsay, N., Samach, A., Corden, S., Putterill, J., Carre, I.A. & Coupland, G. (1998) The late elongated hypocotyl mutation of *Arabidopsis* disrupts circadian rhythms and the photoperiodic control of flowering. *Cell*, **93**, 1219–29.

Scheible, W.R., Gonzalez-Fontes, A., Lauerer, M., Muller-Rober, B., Caboche, M. & Stitt, M. (1997) Nitrate acts as a signal to induce organic acid metabolism and repress starch metabolism in tobacco. *Plant Cell*, **9**, 783–798.

Scheible, W.R., Morcuende, R., Czechowski, T., Fritz, C., Osuna, D., Palacios-Rojas, N., Schindelasch, D., Thimm, O., Udvardi, M.K. & Stitt, M. (2004) Genome-wide reprogramming of primary and secondary metabolism, protein synthesis, cellular growth processes, and the regulatory infrastructure of *Arabidopsis* in response to nitrogen. *Plant Physiol.*, **136**, 2483–99.

Shimizu, M., Ichikawa, K. & Aoki, S. (2004) Photoperiod-regulated expression of the PpCOL1 gene encoding a homolog of CO/COL proteins in the moss *Physcomitrella patens*. *Biochem. Biophys. Res. Commun.*, **324**, 1296–301.

Simkin, A.J., Underwood, B.A., Auldridge, M., Loucas, H.M., Shibuya, K., Schmelz, E., Clark, D.G. & Klee, H.J. (2004) Circadian regulation of the PhCCD1 carotenoid cleavage dioxygenase controls emission of beta-ionone, a fragrance volatile of petunia flowers. *Plant Physiol.*, **136**, 3504–14.

Sivasankar, S. & Oaks, A. (1995) Regulation of nitrate reductase during early seedling growth (A role for asparagine and glutamine). *Plant Physiol.*, **107**, 1225–1231.

Smalle, J., Haegman, M., Kurepa, J., Van Montagu, M. & Van der Straeten, D. (1997) Ethylene can stimulate *Arabidopsis* hypocotyl elongation in the light. *Proc. Natl. Acad. Sci. USA*, **94**, 2756–2761.

Smith, S.M., Fulton, D.C., Chia, T., Thorneycroft, D., Chapple, A., Dunstan, H., Hylton, C., Zeeman, S.C. & Smith, A.M. (2004) Diurnal changes in the transcriptome encoding enzymes of starch metabolism provide evidence for both transcriptional and posttranscriptional regulation of starch metabolism in *Arabidopsis* leaves. *Plant Physiol.*, **136**, 2687–99.

Somers, D.E., Devlin, P.F. & Kay, S.A. (1998) Phytochromes and cryptochromes in the entrainment of the *Arabidopsis* circadian clock. *Science*, **282**, 1488–90.

Staiger, D., Zecca, L., Wieczorek Kirk, D.A., Apel, K. & Eckstein, L. (2003) The circadian clock regulated RNA-binding protein AtGRP7 autoregulates its expression by influencing alternative splicing of its own pre-mRNA. *Plant J.*, **33**, 361–71.

Stitt, M. (1996) In *Environmental Stress and Photosynthesis*, Vol. 3 (ed N. Baker), Academic Press, New York.

Stitt, M., Bulpin, P.V. & Rees, T. (1978) Pathway of starch breakdown in photosynthetic tissues of *Pisum sativum. Biochim. Biophys. Acta*, **544**, 200–14.

Stitt, M., Huber, S. & Kerr, P. (1987) In *The Biochemistry of Plants*, Vol. 10 (eds M.D. Hatch & N.K. Boardman), pp. 327–409, Academic Press, New York.

Stitt, M. & Krapp, A. (1999) The interaction between elevated carbon dioxide and nitrogen nutrition: the physiological and molecular background. *Plant Cell Environ.*, **22**, 583–621.

Stitt, M., Muller, C., Matt, P., Gibon, Y., Carillo, P., Morcuende, R., Scheible, W.R. & Krapp, A. (2002) Steps towards an integrated view of nitrogen metabolism. *J. Exp. Bot.*, **53**, 959–70.

Storch, K.F., Lipan, O., Leykin, I., Viswanathan, N., Davis, F.C., Wong, W.H. & Weitz, C.J. (2002) Extensive and divergent circadian gene expression in liver and heart. *Nature*, **417**, 78–83.

Stowe-Evans, E.L., Harper, R.M., Motchoulski, A.V. & Liscum, E. (1998) NPH4, a conditional modulator of auxin-dependent differential growth responses in *Arabidopsis. Plant Physiol.*, **118**, 1265–75.

Stowe-Evans, E.L., Luesse, D.R. & Liscum, E. (2001) The enhancement of phototropin-induced phototropic curvature in *Arabidopsis* occurs via a photoreversible phytochrome A-dependent modulation of auxin responsiveness. *Plant Physiol.*, **126**, 826–34.

Strayer, C., Oyama, T., Schultz, T.F., Raman, R., Somers, D.E., Mas, P., Panda, S., Kreps, J.A. & Kay, S.A. (2000) Cloning of the *Arabidopsis* clock gene TOC1, an autoregulatory response regulator homolog. *Science*, **289**, 768–71.

Suarez-Lopez, P., Wheatley, K., Robson, F., Onouchi, H., Valverde, F. & Coupland, G. (2001) CONSTANS mediates between the circadian clock and the control of flowering in *Arabidopsis. Nature*, **410**, 1116–20.

Tan, P.K., Downey, T.J., Spitznagel, E.L., Jr., Xu, P., Fu, D., Dimitrov, D.S., Lempicki, R.A., Raaka, B.M. & Cam, M.C. (2003) Evaluation of gene expression measurements from commercial microarray platforms. *Nucleic Acids Res.*, **31**, 5676–84.

Taybi, T., Patil, S., Chollet, R. & Cushman, J.C. (2000) A minimal serine/threonine protein kinase circadianly regulates phosphoenolpyruvate carboxylase activity in crassulacean acid metabolism-induced leaves of the common ice plant. *Plant Physiol.*, **123**, 1471–82.

Thain, S.C., Murtas, G., Lynn, J.R., McGrath, R.B. & Millar, A.J. (2002) The circadian clock that controls gene expression in *Arabidopsis* is tissue specific. *Plant Physiol.*, **130**, 102–10.

Thain, S.C., Vandenbussche, F., Laarhoven, L.J., Dowson-Day, M.J., Wang, Z.Y., Tobin, E.M., Harren, F.J., Millar, A.J. & Van Der Straeten, D. (2004) Circadian rhythms of ethylene emission in *Arabidopsis. Plant Physiol.*, **136**, 3751–61.

Thimm, O., Blasing, O., Gibon, Y., Nagel, A., Meyer, S., Kruger, P., Selbig, J., Muller, L.A., Rhee, S.Y. & Stitt, M. (2004) MAPMAN: a user-driven tool to display genomics data sets onto diagrams of metabolic pathways and other biological processes. *Plant J.*, **37**, 914–39.

Tomita, J., Nakajima, M., Kondo, T. & Iwasaki, H. (2004) No transcription-translation feedback in circadian rhythm of KaiC phosphorylation. *Science*, **307**, 251–54.

Toth, R., Kevei, E., Hall, A., Millar, A.J., Nagy, F. & Kozma-Bognar, L. (2001) Circadian clock-regulated expression of phytochrome and cryptochrome genes in *Arabidopsis. Plant Physiol.*, **127**, 1607–16.

Ueda, H.R., Hayashi, S., Chen, W., Sano, M., Machida, M., Shigeyoshi, Y., Iino, M. & Hashimoto, S. (2005) System-level identification of transcriptional circuits underlying mammalian circadian clocks. *Nat. Genet.*, **37**, 187–92.

Valverde, F., Mouradov, A., Soppe, W., Ravenscroft, D., Samach, A. & Coupland, G. (2004) Photoreceptor regulation of CONSTANS protein in photoperiodic flowering. *Science*, **303**, 1003–6.

Veit, M., Bilger, W., Muhlbauer, T., Brummet, W. & Winter, K. (1996) Diurnal changes in flavonoids. *J. Plant Physiol.*, **148**, 478–82.

Verdonk, J.C., Ric de Vos, C.H., Verhoeven, H.A., Haring, M.A., van Tunen, A.J. & Schuurink, R.C. (2003) Regulation of floral scent production in petunia revealed by targeted metabolomics. *Phytochemistry*, **62**, 997–1008.

Wang, Z.Y. & Tobin, E.M. (1998) Constitutive expression of the CIRCADIAN CLOCK ASSOCIATED 1 (CCA1) gene disrupts circadian rhythms and suppresses its own expression. *Cell*, **93**, 1207–17.

Weckwerth, W., Wenzel, K. & Fiehn, O. (2004) Process for the integrated extraction, identification and quantification of metabolites, proteins and RNA to reveal their co-regulation in biochemical networks. *Proteomics*, **4**, 78–83.

Weiner, H. & Kaiser, W.M. (1999) 14-3-3 proteins control proteolysis of nitrate reductase in spinach leaves. *FEBS Lett.*, **455**, 75–78.

Woelfle, M.A., Ouyang, Y., Phanvijhitsiri, K. & Johnson, C.H. (2004) The adaptive value of circadian clocks: an experimental assessment in cyanobacteria. *Curr. Biol.*, **14**, 1481–6.

Yanovsky, M.J. & Kay, S.A. (2002) Molecular basis of seasonal time measurement in *Arabidopsis. Nature*, **419**, 308–12.

Yanovsky, M.J. & Kay, S.A. (2003) Living by the calendar: how plants know when to flower. *Nat. Rev. Mol. Cell Biol.*, **4**, 265–75.

Yoon, H.S., Hackett, J.D., Ciniglia, C., Pinto, G. & Bhattacharya, D. (2004) A molecular timeline for the origin of photosynthetic eukaryotes. *Mol. Biol. Evol.*, **21**, 809–18.

Young, M.W. & Kay, S.A. (2001) Time zones: a comparative genetics of circadian clocks. *Nat. Rev. Genet.*, **2**, 702–15.

Zarembinski, T.I. & Theologis, A. (1994) Ethylene biosynthesis and action: a case of conservation. *Plant Mol. Biol.*, **26**, 1579–97.

Zhang, F., Bartels, M.J. & Stott, W.T. (2004) Quantitation of human glutathione S-transferases in complex matrices by liquid chromatography/tandem mass spectrometry with signature peptides. *Rapid Commun. Mass. Spectrom*, **18**, 491–8.

7 Photoperiodic responses and the regulation of flowering

Isabelle Carré, George Coupland and Joanna Putterill

7.1 Introduction to photoperiodism

Photoperiodism is defined as a response to the length of day or night that enables adaptation of an organism to seasonal changes, and it was first described in detail by Garner and Allard (1920). They observed that Maryland Mammoth tobacco plants would not flower when exposed to the long summer days experienced around Washington DC, but that flowering occurred rapidly if the plants were placed in a dark chamber for most of the day so that they were only exposed to daylight for 7 h. Similar responses to day length were shown for a wide range of species including soybean, cabbage and lettuce, and the authors commented that 'the relative length of the day is really a dominating factor in plant reproduction processes' and that 'this seems not to have been suspected by previous workers in this field.'

The discovery of photoperiodism led to the classification of plants into three major groups based on their responses to day length (Thomas & Vince-Prue, 1997). Those species that flower if the duration of daylight falls below a critical day length were called short-day plants. Rice, maize, chrysanthemum and soybean are examples of species that show this response. In contrast, flowering of species such as Arabidopsis, barley, wheat and pea is accelerated under day lengths longer than a critical day length, and these are classified as long-day plants. Flowering of a third class of plants is not affected by day length, and these are therefore called day neutral. Both long and short day response types appear in different Angiosperm families, suggesting that the distinction between these responses has evolved more than once.

The critical day length that induces flowering can differ between accessions of the same species. For example, accessions of the short-day plant *Xanthium strumarium* were collected throughout the United States and their response to photoperiod studied in the laboratory (Ray & Alexander, 1966). Those accessions from the extreme north flowered in response to longer day lengths than those from the south. This was proposed to represent an adaptation ensuring that northern varieties flowered soon after the summer solstice and had therefore completed seed maturation before the onset of extreme winter conditions. In contrast, in the south winter conditions occurred later in the year and therefore flowering of accessions from this region could be delayed without endangering seed maturation. Recently, a north–south cline in the flowering times of accessions of Arabidopsis was reported, with accessions from higher latitude flowering later (Stinchcombe *et al.*, 2004). However, this cline was only observed for accessions carrying an active *FRIGIDA* (*FRI*) gene, and therefore is probably associated with the response of flowering to prolonged cold, a process called vernalization, rather than to photoperiod (Johanson *et al.*, 2000). Similarly, natural genetic variation in circadian rhythms between accessions of Arabidopsis was shown to correlate with the day length at the latitude at which the accession originated (Michael *et al.*, 2003). However, no correlation between this variation and flowering time was described, although circadian rhythms are an important part of the mechanism by which photoperiodic flowering is controlled (see below).

7.2   Models for the measurement of day length

7.2.1   The hourglass timer

Several models have been suggested to explain how plants and animals measure day length (Fig. 7.1). The simplest of these models is a timer of the hourglass category. In this type of model, light induces changes in the level or the activity of a regulatory molecule and a response is triggered when levels of this molecule rise above (or drop below) a certain threshold (Fig. 7.1A). Thus, Borthwick and colleagues (Borthwick *et al.*, 1952) suggested that the red-light photoreceptor, phytochrome, might allow measurement of night-length. Plants accumulate high amounts of the far-red absorbing form of phytochrome (Pfr) during the light period. This pool of active phytochrome reverts slowly to its inactive, red-absorbing form (Pr) during the night period (for a detailed description of phytochromes, see Chapter 5). It was therefore proposed that a floral response might be triggered when the duration of darkness is sufficient to allow levels of active phytochrome to drop below a critical threshold. This hypothesis was supported by the observation, that, in the short-day plant *Xanthium*, the effect of inductive long-nights was suppressed by single, 2-min night-breaks of red light (R), which presumably converted all of the inactive Pr to the inhibitory Pfr form. Furthermore, the effect of red night-breaks was reversed by immediate, 3-min exposure to far-red light (FR), converting the inactive Pr back to the inhibitory Pfr form. The response was repeatedly reversible and the floral response reflected whether R or FR light was given as the final exposure.

7.2.2   The circadian clock

As an alternative to the hourglass model, the circadian clock was proposed to provide the time measurement system for photoperiodism (Bünning, 1936; Pittendrigh & Minis, 1964). The involvement of the circadian clock was convincingly demonstrated in a set of experiments in which the short-day plant *Chenopodium rubrum* was grown in continuous light and then transferred to an extended, 72 h dark period that induced flowering. This dark interval was then systematically scanned with brief night-breaks (Cumming *et al.*, 1965). These experiments indicated that plants exhibit rhythmic changes in responsiveness to short night-breaks, which could not easily be accounted for by the hourglass model. Sensitivity to night-breaks increased

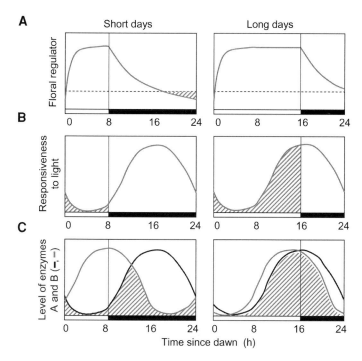

Figure 7.1 Three possible mechanisms for photoperiodic time measurement. Hatched areas under the curves indicate perceived photoperiodic signal. (A) The sand-timer model relies on the light-driven accumulation and dark-induced degradation of a regulatory molecule. A response is triggered when levels of this regulatory molecule fall below a critical threshold. (B) The external coincidence model is based on a circadian rhythm of responsiveness to light. A response is triggered when light overlaps with the sensitive phase of the cycle. (C) The internal coincidence model predicts that changes of photoperiod alter the phase-relationship of two circadian rhythms. These two rhythms may drive expression of two enzymes within a signaling pathway, for example. Activity of the pathway depends on the degree of overlap between the two rhythms.

gradually until the middle of the night, reaching a minimum around the next subjective dawn. Further peaks of sensitivity followed in the subsequent subjective night and at approximately 24 h intervals. Similar results were obtained for short and long-day plants, although long-day plants were generally much less sensitive to night-breaks than short-day plants and the floral response to night breaks was opposite (reviewed in Thomas & Vince-Prue, 1997). These data provided convincing experimental evidence that a circadian clock was associated with photoperiodic time perception.

The link between circadian rhythms and photoperiodic timing, initially suggested by the results of night-break experiments, was substantiated by the analysis of flowering-time defects in Arabidopsis plants exhibiting impaired circadian rhythms. There is a growing body of evidence suggesting that abnormal circadian timing causes aberrant perception of photoperiod and altered flowering times. For example, loss of function of the putative oscillator components *LHY*, *CCA1* or *TOC1* shortened circadian period and caused plants to flower earlier than wild-type under short-day conditions (Somers *et al.*, 1998; Mizoguchi *et al.*, 2002). Transgenic plants carrying over expressed copies of the *LHY* or *CCA1* genes were incapable of free-running rhythmicity in constant light and exhibited late-flowering phenotypes under short or long-day conditions (Schaffer *et al.*, 1998; Wang and Tobin, 1998). The arrhythmic mutant *elf3* was also insensitive to day length but flowered as early, or earlier, than wild-type plants grown under inductive, long-day conditions (Hicks *et al.*, 1996). It is now generally accepted that the circadian clock constitutes an integral component of the photoperiodic timing mechanism.

7.3   Internal and external coincidence models

Two models have been proposed by which a circadian clock might mediate perception of day length (Pittendrigh & Minis 1964). The external coincidence model suggests that photoperiodic responses might be triggered when an external signal (light) coincides with the sensitive phase of an endogenous, circadian rhythm of photoresponsiveness. The phase of this rhythm relative to dawn and dusk would in turn determine whether a response is obtained under any given photoperiod. In contrast, the internal coincidence model predicts that the effect of changing photoperiod is to alter the phase-relationship between two endogenous circadian rhythms, so that a response can only be triggered when the two rhythms are in phase with each other. For example, two enzymes or signaling molecules that function as part of the same pathway might be controlled by distinct circadian rhythms. Photoperiods that allow both of these rhythms to come in phase with each other would then allow the pathway to become active and result in a floral response. Whether the internal or external coincidence hypothesis turns out to be correct, a common prediction from these models is that the expression or activity of some floral regulators must be under circadian control. Genetic analysis of the photoperiodic response pathway in Arabidopsis identified a number of regulatory proteins that fulfil this prediction.

7.4   Arabidopsis molecular genetics identifies a regulatory pathway that controls flowering in long photoperiods

The flowering time of Arabidopsis is strongly photoperiodic with many varieties flowering in spring and early summer as the days lengthen. Similarly, in the laboratory, flowering is promoted in long days (LD; 16 h light, 8 h dark), and delayed but not abolished in short days (SD; 8 h light, 16 h dark). The genes underpinning this photoperiodic response were found by analyzing mutants that were late flowering in LD, but flowered at a similar time to wild type in SD. Characterization of these mutants began over 40 years ago with the work of Redei, later Koornneef and colleagues systematically examined late-flowering mutants using genetic and physiological tests and described an epistatic group that were unable to respond to LDs (Redei, 1962; Koornneef et al., 1991, 1998). The corresponding genes comprise the photoperiodic flowering pathway, which promotes flowering specifically in response to LDs (Table 7.1; Searle & Coupland, 2004). Subsequently, molecular and genetic analyses of these and other flowering-time mutants, including some that flower early under SD, have defined a regulatory hierarchy based on the order of function of the genes (Fig. 7.2).

7.4.1   The GI-CO-FT regulatory hierarchy

Three of the genes identified by characterization of the late-flowering mutants, form a regulatory hierarchy that functions in the order *GI-CO-FT*. These genes are rhythmically expressed during the daily cycle and thus form an output pathway of the circadian clock that functions to promote flowering of Arabidopsis under LDs (Park et al., 1999; Fowler et al., 1999; Suarez-Lopez et al., 2001). *GI* is positioned upstream of *CO* in the hierarchy, because *gi* mutations strongly reduce *CO* mRNA abundance, while overexpression of *CO* from the constitutive, strong CaMV 35S promoter (*35S*) rescues the late flowering of *gi* mutants (Suarez-Lopez et al., 2001). These results are consistent with the late-flowering phenotype of *gi* mutants being largely due to reduced *CO* expression. However, how directly GI regulates *CO* is not yet clear, particularly because *GI* feeds back to influence clock function, probably via a light-input pathway, to promote robust rhythmicity (Fowler et al., 1999; Park et al., 1999; Huq et al., 2000). Thus, GI could regulate *CO* expression indirectly via its general effect on rhythms. In addition, the biochemical role of GI remains to be demonstrated. GI is a novel, large, nuclear protein that is highly conserved in seed plants, including the monocot rice and the gymnosperm, loblolly pine, but not found so far in other organisms (Fowler et al., 1999; Park et al., 1999; Hayama et al., 2002). It has no domains of known biochemical function, apart from a nuclear localization region in the middle of the protein (Huq et al., 2000) and a carboxyl-terminal portion which interacts with SPINDLY (SPY) in yeast and *E. coli* (Tseng et al., 2004). SPY is a regulator of gibberellin-hormone signaling and is proposed to negatively regulate GI activity, possibly via O-N-acetyl-glucosamine modification of target proteins.

   FT, the third gene in the hierarchy, is a direct target of CO. Experiments with a glucocorticoid-inducible form of CO expressed from the 35*S* promoter, show that *FT* transcript is rapidly upregulated after hormone induction, in the absence of further translation (Samach et al., 2000). This increase in *FT* is important as overexpression of *CO*, or additional copies of it, strongly accelerate flowering in a *FT*-dependent manner and *FT* expression is reduced in *co* mutants (Putterill et al., 1995; Onouchi et al., 2000; Samach et al., 2000). *CO* encodes a nuclear protein with two zinc fingers at the amino terminus and a conserved carboxyl-terminal domain, the CCT domain,

Table 7.1 Arabidopsis and rice genes involved in photoperiodic flowering

Abbreviation	Gene name	Predicted gene product	Function
Arabidopsis			
CCA1	CIRCADIAN CLOCK ASSOCIATED 1	Myb domain transcription factor	Circadian clock, floral repressor
CO	CONSTANS	Nuclear protein with two B-box type zinc fingers and plant-specific CCT domain	Floral promoter
CRY2	CRYPTOCHROME 2	Blue/UV light photoreceptor	Floral promoter
ELF3	EARLY FLOWERING 3	Novel nuclear protein	Circadian clock, floral repressor
FKF1	FLAVIN-BINDING, KELCH REPEAT, F-BOX 1	Putative blue light photoreceptor	Floral promoter
FT	FLOWERING LOCUS T	Homology to RAF kinase inhibitor	Floral promoter
GI	GIGANTEA	Novel, large nuclear protein	Floral promoter
LHY	LATE ELONGATED HYPOCOTYL	Myb domain transcription factor	Circadian clock, floral repressor
PHYA	PHYTOCHROME A	Red/far red/blue light photoreceptor	Floral promoter
PHYB	PHYTOCHROME B	Red/far red light photoreceptor	Floral repressor
SOC1	SUPPRESSOR OF OVEREXPRESSION OF CONSTANS 1	MADS domain transcription factor	Floral promoter
TOC1	TIMING OF CHLOROPHYLL A/B BINDING PROTEIN 1	Nuclear protein with pseudo-response regulator domain and CCT domain	Circadian clock
Rice			
Ehd1	Early heading date 1	Homology to B-type response regulator	Promotion of SD flowering
Hd1	Heading date 1	CONSTANS orthologue	Promotion of SD flowering, inhibition of LD flowering
Hd3a	Heading date 3a	FT orthologue	Floral promoter
OsGI	Oryza sativa GIGANTEA	High conservation with GIGANTEA	Promotion of SD flowering, inhibition of LD flowering
Se5	Photoperiod sensitivity 5	Heme oxygenase involved in phytochrome chromophore biosynthesis	Floral repressor

A

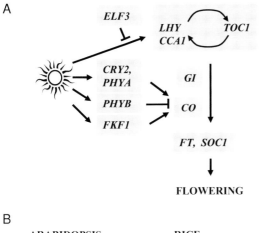

B

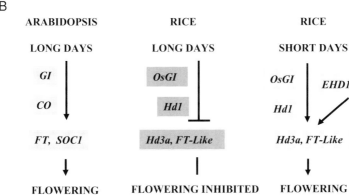

Figure 7.2 Regulatory hierarchies controlling photoperiodic flowering in Arabidopsis and rice.
(A) The photoperiod pathway in Arabidopsis that functions to promote flowering in long days.
(B) Conservation and modification of the photoperiod pathway in Arabidopsis and rice in different day lengths. Arrows indicate promotive effects, while bars indicate inhibition.

named for its conservation in other plant proteins (CO, COL, TOC1). The CCT domain contains nuclear localization signals and probably also mediates protein-protein interactions (Putterill *et al.*, 1995; Kurup *et al.*, 2000; Robson *et al.*, 2001). Both conserved domains are critical to CO function as all known *co* mutations affect one or other of them (Putterill *et al.*, 1995; Robson *et al.*, 2001). The zinc fingers are most similar to B boxes that in animals function as protein-protein interaction domains (Borden, 1998). In addition, direct binding of CO to DNA has not been shown (Hepworth *et al.*, 2002), suggesting that CO may require a DNA-binding partner for transcriptional regulation. *CO–LIKE* genes are widespread in the plant kingdom including in the moss *Physcomitrella* (Shimizu *et al.*, 2004) and *CO* is part of a family of 17 Arabidopsis *COL* genes which possess one or two B boxes and the CCT domain (Robson *et al.*, 2001). Although overexpression of *CO* causes an early-flowering phenotype, overexpression of *COL1* or *COL2* has no affect on flowering time, but causes a short-period phenotype (Ledger *et al.*, 2001). A role for other *COL* genes in control of flowering time has not been reported.

The *FT* gene encodes a 23 kD protein with amino acid sequence similarity to mammalian RAF kinase inhibitor proteins (Kardailsky *et al.*, 1999; Kobayashi *et al.*, 1999). It is part of a small gene family in Arabidopsis and other plants, some of which have been functionally characterized (Bradley *et al.*, 1996, 1997; Pnueli *et al.*, 1998; Amaya *et al.*, 1999). The molecular mode of action of these proteins in plants is still not clear, although they interact with a range of proteins associated with signaling and transcriptional regulation (Pnueli *et al.*, 2001). Nevertheless, *FT* is a powerful promoter of flowering when overexpressed (Kardailsky *et al.*, 1999; Kobayashi *et al.*, 1999) and based on genetic experiments appears to promote flowering by activating *APETALA1(AP1)* (Ruiz-Garcia *et al.*, 1997). *AP1* specifies floral-meristem identity, and is first expressed in the primordia of young flowers that initiate on the flanks of the shoot apical meristem (Mandel *et al.*, 1992).

7.4.2   Role of CO and FT in long-distance day-length signaling

The initial site of perception of day length in plants, including Arabidopsis, is the leaves (Zeevaart, 1976; Corbesier *et al.*, 1996), suggesting that a long-distance signal is transported from the leaves to the shoot apex for upregulation of *AP1* and other floral meristem identity genes and ensuing flower development. Recent work from two groups using a combination of grafting and transgenic approaches in Arabidopsis strongly support the idea that *CO* functions remotely from the shoot apex, in the phloem, and thus is likely to be responsible for generating a systemic signal that promotes flowering under LDs (An *et al.*, 2004; Ayre & Turgeon, 2004). Furthermore, *FT* is shown to be a target for CO in the phloem and required for the rapid flowering promoted by phloem-expressed *CO* (An *et al.*, 2004).

CO mRNA abundance is very low, but it is widely expressed in wild-type plants in both the shoot apical meristem and in leaves, particularly in the vascular tissue (Putterill *et al.*, 1995; Takada & Goto, 2003; An *et al.*, 2004). To define the spatial requirement for *CO* during flowering, a range of tissue-specific promoters was used to drive *CO* expression in *co* mutants (An *et al.*, 2004). *CO* expression from phloem-specific promoters, such as that of the *SUCROSE TRANSPORTER 2* (*SUC2*) gene, promoted flowering, but those that drive gene expression in the shoot apical meristem, such as from the *KNAT1* gene, did not (An *et al.*, 2004). In parallel, expression of *CO* from the promoter of the *GALACTINOL SYNTHASE* gene of melon, which is active specifically in the phloem of the minor veins, also promoted early flowering (Ayre & Turgeon, 2004). These experiments strongly suggest that *CO* promotes flowering when expressed in the phloem companion cells of either the major (*SUC2*) or minor veins (*GAS*), but not in the shoot apex and thus that *CO* may be responsible for generating a mobile signal. Grafting supported this, as stocks expressing *CO* promote flowering of *co* mutant scions (An *et al.*, 2004; Ayre & Turgeon, 2004). CO itself probably does not act as a long-distance signal, because a GFP:CO translational fusion expressed in the phloem rescued *co*, but GFP signal was observed only in the phloem and not in the shoot apex. In addition, the CO target gene *FT* is only upregulated in the phloem, suggesting that CO activates target genes cell autonomously and does not move (An *et al.*, 2004). *CO* mRNA was also not detected in *co* mutant scions that had been grafted onto *GAS::CO* stocks (Ayre & Turgeon, 2004).

In contrast to *CO*, *FT* promotes flowering when expressed from promoters active specifically in the phloem or the shoot apical meristem (An *et al.*, 2004). In wild-type plants expression of an FT::GUS fusion was detected in the vascular tissue, but not in the shoot apical meristem (Takada & Goto, 2003). This suggests that the small FT protein may be mobile and moves through the phloem from the leaf to the shoot apical meristem or that it regulates the synthesis of a mobile floral inducing substance. However, there are likely to be other targets of CO in the phloem, because the *ft* mutation only partly delayed the rapid flowering induced by *SUC2::CO*. One of these targets may be *SOC1*, a second gene that is directly upregulated by inducible CO, and which encodes a MADS box transcription factor (Borner *et al.*, 2000; Samach *et al.*, 2000; Lee *et al.*, 2000; Onouchi *et al.*, 2000). The *soc1* mutant delays flowering of wild-type plants and of *CO* overexpressors, and *SOC1* overexpression causes very early flowering. Thus, it will be interesting to test whether *SOC1* and genes upstream of *CO* in the regulatory hierarchy, such as *GI*, also promote flowering from the phloem.

What about other putative floral signals such as assimilates or hormones that were identified in physiological studies in other plants (Bernier *et al.*, 1993)? It is not yet clear whether these compounds play a role in the photoperiod pathway and long-distance day-length signaling in Arabidopsis. However, application of sucrose to the shoot apex of *co* mutants complemented the late flowering defect, but similar experiments did not rescue *ft* (Roldan *et al.*, 1999). This suggests that sucrose acts downstream of *CO* and requires functional *FT* to promote flowering. Recently, the discovery that small RNAs can enter and then move in the phloem raised the exciting new possibility that microRNAs, known regulators of plant gene expression and development, may be part of the floral stimulus (Ding *et al.*, 2003; Yoo *et al.*, 2004). In particular, microRNAs with effects on flowering time have been described, although these seem to be expressed at the shoot apex (Aukerman & Sakai, 2003).

7.5   Regulation of the Arabidopsis photoperiod pathway: molecular evidence in favor of the coincidence model

7.5.1   *CO: a major link between photoperiod and the circadian clock*

Circadian-clock based models for photoperiodic time perception predicted that some components of floral regulatory pathways should exhibit circadian regulation. This prediction was fulfilled by the observation that expression of the *CO* mRNA was rhythmic in Arabidopsis (Suarez-Lopez *et al.*, 2001). The *CO* transcript levels exhibited 24 h oscillations under diurnal light-dark cycles. These rhythmic changes in *CO* mRNA persisted when plants were transferred to constant light conditions, indicating control by a circadian clock. Under LD photoperiods, expression of *CO* mRNA began to increase approximately 8 h after lights-on and showed a broad peak between approximately 12 h and dawn, with a reproducible 'dip' at 20 h. A similar pattern was observed under SD conditions, although the waveform of *CO* expression was less obviously bimodal and peak levels were only observed between 12 and 20 h after dawn. A crucial observation from these experiments was that *CO* mRNA expression coincided with light under LDs, but not under SDs. Expression of the CO target gene, *FT*, paralleled that of *CO* under LD but transcript levels remained low at all times under SDs. These findings gave rise to the hypothesis that *CO* might link the circadian clock to floral development pathways. Light was proposed to modulate the translation, activity or degradation of the CO protein, so that transcriptional activation of *FT* can only take place when expression of the *CO* mRNA coincides with a light signal.

If this hypothesis were true, increased expression of *CO* during the light portion of a SD cycle should result in abnormally early flowering in these conditions. This prediction was verified for plants that overexpressed *CO* (*35S::CO*) (Onouchi *et al.*, 2000). Furthermore, the early flowering phenotype of the *elf3* mutant correlated with elevated levels of *CO* mRNA than wild-type plants throughout the diurnal cycle (Suarez-Lopez *et al.*, 2001). Conversely, mutations that cause reduced coincidence of *CO* expression with light should also delay flowering under normally inductive, long-day conditions. Thus, the late-flowering mutants *gi* and *lhy* exhibited low levels of *CO* mRNA (Suarez-Lopez *et al.*, 2001).

The mutations analyzed by Suarez-Lopez and colleagues caused defects in *CO* expression levels throughout the diurnal cycle, and therefore, did not truly test whether the timing of *CO* expression relative to dawn and dusk determined the floral response. This question was addressed in two subsequent studies. The first (Yanovsky & Kay, 2002) analyzed *CO* and *FT* expression in the short period mutant *toc1* under a variety of light-dark cycles. Under 24 h light-dark cycles (8 h light/16 h dark), the *toc1* mutant exhibited an earlier peak of *CO* transcription than wild-type plants, so that expression of the *CO* mRNA coincided with light for approximately 4 hours under short-day conditions. As expected, coincidence of *CO* with light was correlated with active transcription of *FT*. The abnormal phase of *CO* expression was corrected by growing *toc1* plants under SD light-dark cycles that matched the endogenous, free-running period of the *toc1* circadian clock (21 h, 7 h light/14 h dark). In these conditions, the peak of *CO* expression in the *toc1* mutant was contained within the dark interval and expression of the *FT* mRNA remained low at all times. A parallel study was carried out using wild-type plants but relied on atypical, non-24 h light-dark cycles to alter the phase of the *CO* rhythm relative to dawn and dusk (Roden *et al.*, 2002). Treatments that caused *CO* transcription to be displaced into the light portion of the cycle also accelerated flowering. Both sets of experiments supported the hypothesis that the timing of *CO* mRNA expression relative to light and dark determined the photoperiodic response. Many rhythms were altered in both studies and it remains a possibility that other rhythmically expressed genes also play a role in photoperiodic responses.

7.5.2   How does light coincidence with CO mRNA expression result in the induction of FT transcription?

Rather than modulating the activity of pre-existing CO protein, it seems that light promotes its accumulation. In *35S::CO* plants, which expressed the *CO* mRNA in a constitutive manner, the CO protein exhibited rhythmic nuclear accumulation under diurnal conditions. Gradual accumulation was observed from dawn to dusk, then CO protein levels decayed to trough levels during the dark period. Treatment of whole seedlings with proteasome inhibitors resulted in increased levels of CO protein, suggesting that it was actively degraded by this specific proteolytic pathway. Therefore, light was suggested to promote accumulation of the CO protein by preventing its degradation by the proteasome (Valverde *et al.*, 2004).

Blue and far-red light were more effective than white light at promoting accumulation of the CO protein in *35S::CO* plants and these effects were reduced by mutations in the genes encoding the CRY2 and PHYA photoreceptors, respectively (Valverde *et al.*, 2004). This was in good agreement with previous observations that these photoreceptors promote flowering in LDs and play a role in the photoperiodic induction of *FT* mRNA accumulation (Johnson *et al.*, 1994; Guo *et al.*, 1998; Yanovsky & Kay, 2002). In contrast, red light did not promote significant accumulation of CO protein and failed to activate *FT* transcription. Furthermore, the CO protein accumulated to higher levels in *phyB* mutant plants, suggesting that red light acts through PHYB to promote degradation of CO. Double mutant analyses indicated that CRY2 may promote flowering by antagonizing the inhibitory effect of PHYB (Mockler *et al.*, 1999). The balance between blue, far-red and red photoreceptors activities is therefore an important factor that determines the level and timing of CO protein accumulation. Expression of the mRNAs encoding these photoreceptors is rhythmic (Toth *et al.*, 2001) and diurnal changes in CRY2 protein levels are known to influence flowering time in Arabidopsis (El-Assal *et al.*, 2001, 2003; Mockler *et al.*, 2003). Rhythmic expression (or activity) of photoreceptors may therefore represent another layer of circadian regulation in photoperiodic time perception.

Light also regulates accumulation of the *CO* mRNA through the action of a rhythmically expressed protein known as FKF1 (Imaizumi *et al.*, 2003). FKF1 is part of a family of three Flavin-binding, Kelch repeat, F box proteins (FKF1, ZTL and LKP2, for a more detail description this family of proteins see Chapter 4) that are believed to function as photoreceptors and to regulate circadian rhythms by targeting specific proteins for degradation by the proteasome. In wild-type plants grown under long-day cycles, FKF1 protein levels reached maximum levels before dusk and the peak of FKF1 expression coincided with the first half of the bimodal peak of *CO* mRNA. This early peak in *CO* mRNA was absent in a T-DNA insertion allele, *fkf1-2*, and this correlated with low expression of *FT* and late flowering. The effect of FKF1 on *CO* transcription required exposure light, since no difference in *CO* expression patterns was observed between wild-type and *fkf1* mutant plants upon transfer to shorter photoperiods where FKF1 expression does not coincide with light.

7.5.3 Summary of the evidence in support of the external coincidence model of the photoperiodic regulation of flowering and a role for CO

In summary, results so far are consistent with the hypothesis that photoperiodic time perception in Arabidopsis is mediated according to the external coincidence model. The rhythmic accumulation of the *CO* mRNA allows differential accumulation of the CO protein in response to light signals given at different times of the day, and therefore represents a rhythm in photoresponsiveness. Photoperiodic time perception in Arabidopsis involves more than one rhythm of light sensitivity, however, since the peak of FKF1 protein must also coincide with light in order to promote *CO* transcription.

7.6 Interaction between photoperiod pathway and other flowering-time pathways in Arabidopsis

Successful sexual reproduction depends on co-coordinating flowering time with favourable external conditions and the appropriate stage of development. In Arabidopsis, a complex genetic network, involving four main pathways, regulates the transition to flowering in response to the seasonal signals of changing day length (photoperiod pathway) and winter chilling (vernalization pathway) as well as to internal signals (gibberellin and autonomous pathways) and local factors such as changes in ambient temperature or stresses such as overcrowding (Mouradov *et al.*, 2002; Simpson & Dean, 2002). A common set of genes integrates the balance of signals from these pathways to determine when flowering occurs. These genes are termed floral integrators and include *FT*, *SOC1*, and *LEAFY* (*LFY*), which encodes a transcription factor that promotes floral meristem identity. Upregulation of these genes leads to further activation of *LFY* and other floral meristem genes such as *AP1* and development of floral primordia on the flanks of the shoot apex.

7.6.1 Convergence of the photoperiod and vernalization pathways

Many Arabidopsis ecotypes are winter annuals that germinate in the first summer season, grow vegetatively over winter and then flower and set seed in the following spring and summer. These plants require 1 to 3 months exposure to cold temperatures (vernalization) before they can be induced to flower by long days, thus ensuring that they can reproduce in mild conditions. This contrasts with summer annuals, often used as model laboratory plants because of their short life cycles, which germinate in the spring and summer and flower rapidly.

Why do winter annual accessions only flower after exposure to winter conditions? The difference between winter and summer annuals largely lies in the alleles of the *FRI* and *FLC* genes that they carry. *FRI* encodes a protein of unknown biochemical function that positively regulates the level of *FLC* mRNA (Michaels & Amasino, 1999; Sheldon *et al.*, 1999; Johanson *et al.*, 2000). *FLC* encodes a MADS transcription factor whose overexpression delays flowering in summer annuals, suggesting that it functions as a strong floral repressor. *FLC* is expressed at high levels in winter annual ecotypes and this blocks flowering in the first summer (Michaels & Amasino, 2000), until vernalization leads to the modification of *FLC* chromatin and stable *FLC* gene silencing (Bastow *et al.*, 2004; Sung & Amasino, 2004). Summer annuals lack a vernalization requirement as they contain inactive *FRI* alleles, or *FLC* alleles that are only weakly expressed (Johanson *et al.*, 2000; Michaels *et al.*, 2003; Gazzani *et al.*, 2003).

How *FLC* opposes the photoperiod pathway during the first summer of growth is not yet entirely clear. However, the first insights into the molecular mechanisms involved stemmed from the discovery that both FLC and CO target the same floral integrators, *FT* and *SOC1*. CO upregulates their expression, while FLC represses it (Borner *et al.*, 2000; Lee *et al.*, 2000; Samach *et al.*, 2000; Michaels & Amasino, 2001). Subsequent analysis of the *SOC1* promoter and of gene expression in plants overexpressing *CO* and *FLC* suggested a model whereby FLC binds to the *SOC1* promoter and thus prevents recruitment or transcriptional activation by CO (Hepworth *et al.*, 2002). A second mechanism by which FLC could antagonize CO was suggested by the recent demonstration that CRY2 mRNA levels are reduced in plants expressing FLC at high levels. This suggests that FLC could have a general effect on CO function by reducing the level of CRY2 and thereby reducing the stability of CO protein. How widespread the antagonism is between FLC and CO on their transcriptional targets remains to be seen, but provides a model for understanding how two pathways, that respond to different seasonal signals, interact to regulate flowering time in winter annuals.

Many other species also have winter and summer annual varieties, but *FLC* orthologues outside of the Brassicaceae have not been identified. However, recently in wheat, a dominant repressor of flowering, *VRN2*, whose levels are reduced by vernalization was identified (Yan *et al.*, 2004). Spring varieties have deletions or mutations in *VRN2*, while RNAi suppression of *VRN2* caused winter wheat to flower earlier. *VRN2* encodes a grass-specific protein, which has a CCT domain with some similarity to CO and other CO-LIKE proteins, and a zinc finger domain at the amino terminus. Release of the floral repression exerted by VRN2 is proposed to lead to expression of a second gene, *VRN1*, whose mRNA levels are upregulated by vernalization and that promotes flowering (Yan *et al.*, 2003). VRN1 is a MADS box transcription factor similar to AP1.

7.6.2   Other pathways converging with the photoperiod pathway?

In addition to the photoperiod and vernalization pathways, *FT* expression is also regulated by other flowering-time pathways. Light quality influences flowering time and *FT* mRNA abundance. Flowering is accelerated and *FT* expression increased by overcrowding of plants, which increases the concomitant high FR to R light ratios that they are exposed to. This effect is largely regulated by phytochrome B and independently of CO (Cerdan & Chory, 2003; Halliday *et al.*, 2003). However, the light quality pathway is also likely to regulate *FT* via effects on the stability of the CO protein, as high FR to R light ratios stabilize the protein (Valverde *et al.*, 2004). The expression of some photoperiod pathway genes is also affected by factors such as salicylic acid and the growth regulator nitric oxide, suggesting that other stress pathways may also modulate the activity of the pathway (Martinez *et al.*, 2004; He *et al.*, 2004).

Are developmentally or environmentally induced changes to chromatin modification of photoperiod genes important for modulating their activity? Winter chilling leads to stable *FLC* chromatin modifications and gene silencing and some of the genes that regulate this process have been identified (Bastow *et al.*, 2004; Sung & Amasino 2004; He & Amasino, 2005). While no changes to chromatin modifications have yet been demonstrated for photoperiod pathway genes, a number of genes encoding putative chromatin modifiers do regulate their expression or activity. These include *EBS*, which represses flowering by repressing *FT* (Pineiro *et al.*, 2003) and two other chromatin regulators that repress flowering and may be particularly relevant to the photoperiod pathway as their expression overlaps with *CO* in the phloem. *TERMINAL FLOWER 2* (*TFL2*) encodes a Heterochromatin-like 1 chromatin repressor that delays flowering (Gaudin *et al.*, 2001). The early flowering of the *tfl2* mutant is *FT* dependent and correlates with increased *FT* expression in the vascular tissue of the leaves (Kotake *et al.*, 2003; Takada & Goto, 2003). Since *TFL2* reduces *35S::CO* dependent upregulation of *FT*, it is proposed to counteract CO activity and thus may minimize the response of the plant to short lived increases in CO. A second flowering time repressor, the SNF2 homolog, AtBRM, is predicted to be part of a SWI/SNF like complex, that in yeast and Drosophila are chromatin remodelling machines (Farrona *et al.*, 2004). In an *AtBRM* RNAi line, plants flower early and this correlates with strong up-regulation of both *CO* expression and its targets *FT* and *SOC1*, particularly under SD conditions. No changes to *FLC* expression were observed, raising the possibility that *AtBRM* represses the photoperiod pathway in non-inductive conditions.

7.7   Comparative analysis of photoperiod regulation in rice and Arabidopsis

Many Angiosperm species exhibit photoperiodic responses, but these differ between long and short-day plants and control diverse developmental decisions, including bud dormancy, tuberization and flowering (Thomas & Vince-Prue, 1997).

Determining whether the photoperiodic pathway identified in Arabidopsis is conserved in other species enables a comparative approach to understanding how different photoperiodic responses can be conferred by the same regulatory pathway.

Rice has emerged as a powerful model species for studies of photoperiodism in a short-day plant (Izawa *et al.*, 2003; Hayama & Coupland, 2004). Studying rice and Arabidopsis allows comparison of the regulation of photoperiodic responses between distantly related Angiosperms, from monocotyledonous and dicotyledonous species respectively, and between long and short-day response types. The availability of the rice genomic sequence (Yu *et al.*, 2002; Goff *et al.*, 2002) and the sophisticated populations generated to identify quantitative trait loci (QTL) segregating in populations made by crossing cultivars differing in their response to day length have made rice an attractive system for genetic analysis of photoperiodism (Yano *et al.*, 1997).

Three QTLs for photoperiodic flowering and segregating in the cross between indica rice cultivars Kipponbare and Kasalath were recently isolated. These QTLs, *Heading-date1* (*Hd1*), *Heading-date3a* (*Hd3a*), and *Heading-date6* (*Hd6*), were found to encode proteins similar to CO, FT, and the $\alpha\gamma$ subunit of CASEIN KINASE 2 (CK2), respectively (Yano *et al.*, 2000; Takahashi *et al.*, 2001; Kojima *et al.*, 2002). This suggested strong conservation of the mechanism underlying photoperiodic flowering, since CO and FT are also central to the control of this process in Arabidopsis (Table 7.1; Fig. 7.2), and CK2 was implicated in the control of circadian rhythms and flowering of Arabidopsis through the phosphorylation of CCA1 (Sugano *et al.*, 1998, 1999). In addition to natural genetic variation, rice mutants with altered flowering time were also recovered. The *photoperiod sensitivity5* (*se5*) mutant exhibits severe early flowering in continuous light as well as under LDs and SDs, and shows no flowering response to day length. The *Se5* gene encodes a protein similar to Arabidopsis HY1, a heme oxygenase that participates in biosynthesis of phytochrome chromophore. This result suggests that phytochrome is essential for the regulation of day-length responses of flowering in rice (Izawa *et al.*, 2000). In addition, the *GI* homolog of rice (*OsGI*) was isolated as a gene whose mRNA abundance is changed in *se5* mutants (Hayama *et al.*, 2002). In transgenic rice plants in which expression of *OsGI* mRNA was increased or reduced, flowering was delayed or accelerated, respectively (Hayama *et al.*, 2002, 2003). In addition to these components, which are closely related to genes identified in Arabidopsis, the *Early Heading Date1* (*ehd1*) gene was identified as a promoter of flowering in a cross between a japonica rice variety and Nipponbare (Doi *et al.*, 2004). The *Ehd1* gene encodes a protein containing a B-type response regulator domain, which may be involved in relaying a phosphorylation signal, and a GARP DNA binding domain. No orthologue of *Edh1* was detected in the Arabidopsis genome, and it appears to activate *Hd3a* expression independently of Hd1 (Doi *et al.*, 2004).

The proteins that confer photoperiod response on Arabidopsis and rice are closely related, so how do these proteins generate a short-day response in rice? Some of the interactions between rice and Arabidopsis flowering-time genes are conserved while others are different. The role of OsGI seems to be to promote *Hd1* transcription, as described for GI and *CO* in Arabidopsis, because in plants overexpressing *OsGI* the abundance of *Hd1* mRNA is increased (Hayama *et al.*, 2003). Furthermore, transgenic rice plants overexpressing *Hd3a* mRNA exhibit strong early flowering, similar to the effect of *FT* overexpression in Arabidopsis (Kojima *et al.*, 2002). However, although in rice and Arabidopsis FT and Hd3a act as floral promoters, *Hd3a* expression is induced specifically under SDs in rice whereas *FT* expression is induced under LDs in Arabidopsis (Kojima *et al.*, 2002).

The day-length dependent regulation of *Hd3a* and of *FT* in rice and Arabidopsis is mediated by *Hd1* and *CO*, respectively. In Arabidopsis, CO induces *FT* expression under LDs and promotes flowering. In contrast, loss of *Hd1* function causes early flowering under LDs and late flowering under SDs, suggesting that *Hd1* has two independent and opposite functions in the control of rice flowering time (Yano *et al.*, 2000). In the *Hd1* loss-of-function mutant (called *se1*), expression of *Hd3a* is consistent with the flowering-time defect under LDs and SDs, so that under LDs *Hd3a* mRNA abundance is increased in *se1* compared to wild-type plants, whereas under SDs it is decreased (Izawa *et al.*, 2002; Kojima *et al.*, 2002). Also, transcripts of *Hd1* exhibit diurnal patterns under LDs and SDs in a phase similar to those of *CO* (Izawa *et al.*, 2002; Kojima *et al.*, 2002; Hayama *et al.*, 2003). Thus, the mechanism by which Hd1 suppresses *Hd3a* and inhibits flowering under LDs may be related to the post-transcriptional regulation based mechanism by which CO activates *FT* under LDs in Arabidopsis. Under LDs, a coincidence between *Hd1* expression and exposure to light may lead to inhibition of *Hd3a* transcription and suppress flowering.

Activation of *Hd1* by light under LDs could be mediated by phytochrome, because loss of *Se5* function causes early flowering under LDs, similar to *se1*, but does not alter the diurnal pattern of *Hd1* expression or the circadian rhythms of other clock output genes, suggesting that phytochrome might affect Hd1 function at the post-transcriptional level (Izawa *et al.*, 2002). Supporting this proposal, the double mutant *se5 se1* flowered later than the *se5* mutant under LDs, indicating that in the absence of *Se5* function, *Hd1* promotes flowering under LDs (Izawa *et al.*, 2002). The role of phytochrome signaling under LDs may therefore be to convert Hd1 into a repressor of flowering, and when phytochrome signaling is impaired in the *se5* mutant, Hd1 promotes flowering as it does under SDs.

These studies allow formulation of a model for the control of day-length response in the short-day rice plant. Under LDs, Hd1 protein that is expressed at the end of the day is proposed to be activated by phytochrome so that it inhibits flowering through inactivating *Hd3a* expression. In contrast, under SDs, *Hd1* is expressed during the night, when phytochrome is proposed to be inactivated, and this allows Hd1 to induce *Hd3a* expression and promote flowering under these conditions. The *se1* mutant, in which *Hd1* is inactive, therefore exhibits early flowering under LDs because Hd1 is not present during the day to inhibit flowering and late flowering under SDs because Hd1 is not expressed in the dark, when it would promote flowering. The strong early-flowering phenotype of the *se5* mutant irrespective of the day-length conditions is due to Hd1 being present in a form that promotes flowering irrespective of the length of day or night, because of the lack of phytochrome activity required to convert Hd1 to the floral repressing form (Izawa *et al.*, 2002). Edh1 appears to act independently of Hd1, and since it is expressed specifically in SDs and leads to the activation of *Hd3a*, Edh1 probably represents a second mechanism by which *Hd3a* expression is increased in a SD-specific manner.

7.8   Perspectives

The demonstration that the GI-CO-FT module confers photoperiodic flowering in Arabidopsis, and that it is conserved in rice, has provided the first insights into the molecular mechanisms controlling photoperiodism. Analysis of the expression of these genes and of CO protein supports the external coincidence model as the general mechanism by which Arabidopsis plants distinguish LD and SD (Figs. 7.1

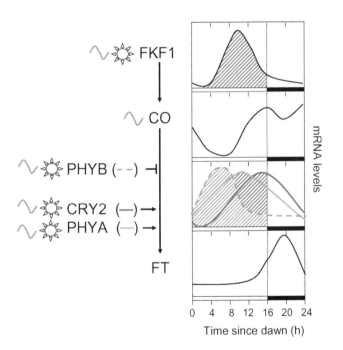

Figure 7.3 Layers of external coincidence in the photoperiodic response pathway of *Arabidopsis*. Multiple elements of the photoperiodic response pathway are expressed under the control of the circadian clock, but require light for activity. These include CO, FKF1, PHYB, CRY2 and PHYA. Coincidence of FKF1 expression with light triggers expression of CO mRNA in the early evening. CO protein accumulation in turn is prevented early in the day due to high levels of PHYB signaling (inhibitory signal shown by regions hatched in red) and promoted late in the day as a result of PHYA and CRY2 signaling (activating signal shown by regions hatched in blue). This leads to promotion of flowering in long days by upregulation of *FT*.

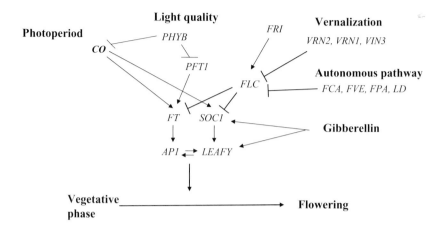

Figure 7.4 Network of interactions controlling flowering time of *Arabidopsis*. The control of flowering by photoperiod, light quality, vernalization, the autonomous pathway and gibberellins are illustrated. A more complete description of the control of photoperiod response is given in Fig. 7.2A. The autonomous pathway is illustrated as a single linear pathway for simplicity, but is actually a series of distinct processes that negatively regulate *FLC* (reviewed by He and Amasino, 2005). The logic of the network is a convergence of distinct environmental signals on the control of *FT* and *SOC1* expression, which are therefore sometimes described as floral integrator genes.

and 7.3). However, the biochemical functions of these regulatory proteins are still unclear and further components in the pathway remain to be discovered. Although both GI and CO are nuclear proteins required for expression of downstream genes, the biochemical mechanisms by which they regulate transcription are unknown. Nevertheless, an understanding of the mechanism by which CO regulates transcription is likely to be essential in explaining how it acts as a transcriptional repressor in the short-day response of rice but as a transcriptional activator during the long-day response of Arabidopsis. Similarly, none of the connections illustrated in Fig. 7.2 seems to represent direct biochemical interactions; GI and FKF1 are unlikely to be DNA binding proteins, and therefore probably regulate *CO* transcription indirectly. Similarly, CO seems to contain protein-protein interaction domains but no recognizable DNA binding domains, and consequently is likely to require protein partners to promote *FT* transcription.

A second feature of the pathway that can be deduced from the components already identified is that light signaling influences the activity of the pathway at several levels. For example, *CO* transcription is increased by light through the action of FKF1, a likely photoreceptor, and independently, CO protein is stabilized through the activity of the phyA and cry photoreceptors. Additional levels of light regulation are likely to be uncovered; for example, *GI* mRNA levels fall rapidly when the lights go off under SDs (Fowler *et al.*, 1999), suggesting that *GI* transcription is directly boosted by light. The coincidence between light signaling and circadian regulation therefore clearly acts at both the transcriptional and post-transcriptional levels on CO and may act on further components in the pathway, including GI, which has established roles in phytochrome signaling. The original co-incidence model requires only one light-sensitive circadian rhythm, for example in CO protein (Figs. 7.1 and 7.3), to coincide with light and therefore multiple layers of regulation on more than one component were not predicted. Nevertheless, the requirement for light at two levels of regulation, transcription of *CO* and stabilization of CO protein, may reduce the likelihood of CO activity occurring in the dark and inducing flowering under inappropriate conditions.

CO seems to be involved in photoperiodic responses in a wide range of Angiosperm species. In addition to flowering in Arabidopsis and rice, this seems to include tuberization of potato (Martinez-Garcia *et al.*, 2002). Both tuberization and flowering are induced by a long-distance signal produced in the leaves in response to appropriate day lengths. Therefore, the features of CO as a day-length responsive switch acting in the leaves to control a long-distance signaling pathway may have evolved early in Angiosperm evolution, before the divergence of Monocotyledonous and Dicotyledonous plants and are subsequently recruited to control different developmental responses. The appearance of the CO system early in plant evolution is supported by the existence of a *CO-like* gene that is controlled at the transcriptional level by the circadian clock in *Physcomitrella patens*, indicating that these genes were already present in the progenitor of mosses and higher plants, although there is no evidence that they control photoperiodic responses in moss. Approaching this pathway in several Angiosperm species may more efficiently identify new pathway components than relying on a single model species; for example, a role for CK2 was revealed by natural genetic variation in rice, but has not been identified by extensive genetic analysis in Arabidopsis, perhaps due to genetic redundancy. In addition, comparing the regulation of the system in several species and between accessions of the same species may provide insights into how photoperiodic responses could be engineered to alter the specific day length at which flowering is induced, or even to switch the response type of a particular species.

References

Amaya, I., Ratcliffe, O.J. & Bradley, D.J. (1999) Expression of *CENTRORADIALIS (CEN)* and *CEN*-like genes in tobacco reveals a conserved mechanism controlling phase change in diverse species. *Plant Cell*, **11**, 1405–1417.

An, H., Roussot, C., Suarez-Lopez, P., Corbesier, L., Vincent, C., Pineiro, M., Hepworth, S., Mouradov, A., Justin, S., Turnbull, C.G.N. & Coupland, G. (2004) CONSTANS acts in the phloem to regulate a systemic signal that induces photoperiodic flowering of Arabidopsis. *Development*, **131**, 3615–3626.

Aukerman, M.J. & Sakai, H. (2003) Regulation of flowering time and floral organ identity by a microRNA and its *APETALA2*-like target genes. *Plant Cell*, **15**, 2730–2741.

Ayre, B. & Turgeon, R. (2004) Graft transmission of a floral stimulant derived from CONSTANS. *Plant Physiol.*, **135**, 2271–2278.

Bastow, R., Mylne, J.S., Lister, C., Lippman, Z., Martienssen, R.A. & Dean, C. (2004) Vernalization requires epigenetic silencing of FLC by histone methylation. *Nature*, **427**, 164–167.

Bernier, G., Havelange, A., Houssa, C., Petitjean, A. & Lejeune, P. (1993) Physiological signals that induce flowering. *Plant Cell*, **5**, 1147–1155.

Borden, K.L.B. (1998) RING fingers and B-boxes: zinc-binding protein-protein interaction domains. *Biochem. Cell Biol.*, **76**, 351–358.

Borner, R., Kampmann, G., Chandler, J., Gleissner, R., Wisman, E., Apel, K. & Melzer, S. (2000) A MADS domain gene involved in the transition to flowering in Arabidopsis. *Plant J.*, **24**, 591–599.

Borthwick, H.A., Hendricks, S.B. & Parker, M.W. (1952) The reaction controlling floral initiation. *Proc. Natl. Acad. Sci. USA*, **38**, 929–934.

Bradley, D., Carpenter, R., Copsey, L., Vincent, C., Rothstein, S. & Coen, E. (1996) Control of inflorescence architecture in Antirrhinum. *Nature*, **379**, 791–797.

Bradley, D., Ratcliffe, O., Vincent, C., Carpenter, R. & Coen, E. (1997) Inflorescence commitment and architecture in Arabidopsis. *Science*, **275**, 80–83.

Bünning, E. (1936) Die endogene Tagesrhythmik als Grundlage der photoperiodischen Reaktion. *Ber. Dtsch. Bot. Ges.*, **54**, 590–607.

Cerdan, P.D. & Chory, J. (2003) Regulation of flowering time by light quality. *Nature*, **423**, 881–885.

Corbesier, L., Gadisseur, I., Silvestre, G., Jacqmard, A. & Bernier, G. (1996) Design in Arabidopsis thaliana of a synchronous system of floral induction by one long day. *Plant J.*, **9**, 947–952.

Cumming, B.G., Hendricks, S.B. & Borthwick, H.A. (1965) Rhythmic flowering responses and phytochrome changes in a selection of *Chenpodium rubrum. Can. J. Bot.*, **43**, 825–853.

Ding, B., Itaya, A. & Qi, Y.J. (2003) Symplasmic protein and RNA traffic: regulatory points and regulatory factors. *Curr. Opin. Plant Biol.*, **6**, 596–602.

Doi, K., Izawa, T., Fuse, T., Yamanouchi, U., Kubo, T., Shimatani, Z., Yano, M. & Yoshimura, A. (2004) *Ehd1*, a B-type response regulator in rice, confers short-day promotion of flowering and controls FT-like gene expression independently of Hd1l. *Genes Dev.*, **18**, 926–936.

El-Assal, S.E.D., Alonso-Blanco, C., Peeters, A.J.M., Raz, V. & Koornneef, M. (2001) A QTL for flowering time in Arabidopsis reveals a novel allele of *CRY2. Nature Genet.*, **29**, 435–440.

El-Assal, S.E.D., Alonso-Blanco, C., Peeters, A.J.M., Wagemaker, C., Weller, J.L. & Koornneef, M. (2003) The role of cryptochrome 2 in flowering in Arabidopsis. *Plant Physiol.*, **133**, 1504–1516.

Farrona, S., Hurtado, L., Bowman, J.L. & Reyes, J.C. (2004) The Arabidopsis thaliana SNF2 homolog AtBRM controls shoot development and flowering. *Development*, **131**, 4965–4975.

Fowler, S., Lee, K., Onouchi, H., Samach, A., Richardson, K., Coupland, G. & Putterill, J. (1999) *GIGANTEA*: A circadian clock-controlled gene that regulates photoperiodic flowering in Arabidopsis and encodes a protein with several possible membrane-spanning domains. *EMBO J.*, **18**, 4679–4688.

Garner, W.W. & Allard, H.A. (1920) Effect of the relative length of day and night and other factors of the environment on growth and reproduction in plants. *J. Agric. Res.*, **18**, 553–606.

Gaudin, V., Libault, M., Pouteau, S., Juul, T., Zhao, G.C., Lefebvre, D. & Grandjean, O. (2001) Mutations in LIKE HETEROCHROMATIN PROTEIN 1 affect flowering time and plant architecture in Arabidopsis. *Development*, **128**, 4847–4858.

Gazzani, S., Gendall, A.R., Lister, C. & Dean, C. (2003) Analysis of the molecular basis of flowering time variation in Arabidopsis accessions. *Plant Physiol.*, **132**, 1107–1114.

Goff, S.A., Ricke, D., Lan, T.-H., Presting, G., Wang, R., Dunn, M., Glazebrook, J., Sessions, A., Oeller, P., Varma, H., Hadley, D., Hutchison, D., Martin, C., Katagiri, F., Lange, B.M., Moughamer, T., Xia, Y., Budworth, P., Zhong, J., Miguel, T., Paszkowski, U., Zhang, S., Colbert, M., Sun, W.-l., Chen, L., Cooper, B., Park, S., Wood, T.C., Mao, L., Quail, P., Wing, R., Dean, R., Yu, Y., Zharkikh, A., Shen, R., Sahasrabudhe, S., Thomas, A., Cannings, R., Gutin, A., Pruss, D., Reid, J., Tavtigian, S., Mitchell, J., Eldredge, G., Scholl, T., Miller, R.M., Bhatnagar, S., Adey, N., Rubano, T., Tusneem, N., Robinson, R., Feldhaus, J., Macalma, T., Oliphant, A. & Briggs, S. (2002) A draft sequence of the rice genome (Oryza sativa L. ssp. japonica). *Science*, **296**, 92–100.

Guo, H.W., Yang, W.Y., Mockler, T.C. & Lin, C.T. (1998) Regulations of flowering time by Arabidopsis photoreceptors. *Science*, **279**, 1360–1363.

Halliday, K.J., Salter, M.G., Thingnaes, E. & Whitelam, G.C. (2003) Phytochrome control of flowering is temperature sensitive and correlates with expression of the floral integrator *FT. Plant J.*, **33**, 875–885.

Hayama, R. & Coupland, G. (2004) The molecular basis of diversity in the photoperiodic flowering responses of Arabidopsis and rice. *Plant Physiol.*, **135**, 677–684.

Hayama, R., Izawa, T. & Shimamoto, K. (2002) Isolation of rice genes possibly involved in the photoperiodic control of flowering by a fluorescent differential display method. *Plant Cell Physiol.*, **43**, 494–504.

Hayama, R., Yokoi, S., Tamaki, S., Yano, M. & Shimamoto, K. (2003) Adaptation of photoperiodic control pathways produces short-day flowering in rice. *Nature*, **422**, 719–22.

He, Y. & Amasino, R.M. (2005) Role of chromatin modification in flowering-time control. *Trends Plant Sci.*, **10**, 30–35.

He, Y., Tang, R.-H., Hao, Y., Stevens, R.D., Cook, C.W., Ahn, S.M., Jing, L., Yang, Z., Chen, L., Guo, F., Fiorani, F., Jackson, R.B., Crawford, N.M. & Pei, Z.-M. (2004) Nitric oxide represses the Arabidopsis floral transition. *Science*, **305**, 1968–1971.

Hepworth, S.R., Valverde, F., Ravenscroft, D., Mouradov, A. & Coupland, G. (2002) Antagonistic regulation of flowering-time gene *SOC1* by CONSTANS and FLC via separate promoter motifs. *EMBO J.*, **21**, 4327–37.

Hicks, K.A., Millar, A.J., Carre, I.A., Somers, D.E., Straume, M., MeeksWagner, D.R. & Kay, S.A. (1996) Conditional circadian dysfunction of the Arabidopsis early-flowering 3 mutant. *Science*, **274**, 790–792.

Huq, E., Tepperman, J.M. & Quail, P.H. (2000) GIGANTEA is a nuclear protein involved in phytochrome signaling in Arabidopsis. *Proc. Natl. Acad. Sci. USA*, **97**, 9789–9794.

Imaizumi, T., Tran, H.G., Swartz, T.E., Briggs, W.R. & Kay, S.A. (2003) FKF1 is essential for photoperiodic-specific light signalling in Arabidopsis. *Nature*, **426**, 302–306.

Izawa, T., Oikawa, T., Tokutomi, S., Okuno, K. & Shimamoto, K. (2000) Phytochromes confer the photoperiodic control of flowering in rice (a short-day plant). *Plant J.*, **22**, 391–399.

Izawa, T., Takahashi, Y. & Yano, M. (2003) Comparative biology comes into bloom: genomic and genetic comparison of flowering pathways in rice and Arabidopsis. *Curr. Opin. Plant Biol.*, **6**, 113–120.

Johanson, U., West, J., Lister, C., Michaels, S., Amasino, R. & Dean, C. (2000) Molecular analysis of FRIGIDA, a major determinant of natural variation in Arabidopsis flowering time. *Science*, **290**, 344–347.

Johnson, E., Bradley, M., Harberd, N.P. & Whitelam, G.C. (1994) Photoresponses of light-grown phyA mutants of Arabidopsis: Phytochrome A is required for the perception of daylength extensions. *Plant Physiol.*, **105**, 141–149.

Kardailsky, I., Shukla, V.K., Ahn, J.H., Dagenais, N., Christensen, S.K., Nguyen, J.T., Chory, J., Harrison, M.J. & Weigel, D. (1999) Activation tagging of the floral inducer *FT*. *Science*, **286**, 1962–1965.

Kobayashi, Y., Kaya, H., Goto, K., Iwabuchi, M. & Araki, T. (1999) A pair of related genes with antagonistic roles in mediating flowering signals. *Science*, **286**, 1960–1962.

Kojima, S., Takahashi, Y., Kobayashi, Y., Monna, L., Sasaki, T., Araki, T. & Yano, M. (2002) *Hd3a*, a rice ortholog of the Arabidopsis *FT* gene, promotes transition to flowering downstream of *Hd1* under short-day conditions. *Plant Cell Physiol.*, **43**, 1096–1105.

Koornneef, M., Alonso-Blanco, C., Vries, H.B.-D., Hanhart, C.J. & Peeters, A.J.M. (1998) Genetic interactions among late-flowering mutants of Arabidopsis. *Genetics*, **148**, 885–892.

Koornneef, M., Hanhart, C.J. & Van Der Veen, J.H. (1991) A genetic and physiological analysis of late flowering mutants in Arabidopsis thaliana. *Mol. Gen. Genet.*, **229**, 57–66.

Kotake, T., Takada, S., Nakahigashi, K., Ohto, M. & Goto, K. (2003) Arabidopsis *TERMINAL FLOWER 2* gene encodes a heterochromatin protein 1 homolog and represses both *FLOWERING LOCUS T* to regulate flowering time and several floral homeotic genes. *Plant Cell Physiol.*, **44**, 555–564.

Kurup, S., Jones, H.D. & Holdsworth, M.J. (2000) Interactions of the developmental regulator ABI3 with proteins identified from developing Arabidopsis seeds. *Plant J.*, **21**, 143–155.

Lee, H., Suh, S.-S., Park, E., Cho, E., Ahn, J.H., Kim, S.-G., Lee, J.S., Kwon, Y.M. & Lee, I. (2000) The AGAMOUS-LIKE 20 MADS domain protein integrates floral inductive pathways in Arabidopsis. *Genes Dev.*, **14**, 2366–2376.

Mandel, M.A., Gustafson-Brown, C., Savidge, B. & Yanofsky, M.F. (1992) Molecular characterization of the Arabidopsis floral homeotic gene *APETALA1*. *Nature*, **360**, 273–277.

Martinez, C., Pons, E., Prats, G. & Leon, J. (2004) Salicylic acid regulates flowering time and links defence responses and reproductive development. *Plant J.*, **37**, 209–217.

Martinez-Garcia, J.F., Virgos-Soler, A. & Prat, S. (2002) Control of photoperiod-regulated tuberization in potato by the Arabidopsis flowering-time gene *CONSTANS*. *Proc. Natl. Acad. Sci. USA*, **99**, 15211–6.

Michael, T.P., Salome, P.A., Yu, H.J., Spencer, T.R., Sharp, E.L., McPeek, M.A., Alonso, J.M., Ecker, J.R. & McClung, C.R. (2003) Enhanced fitness conferred by naturally occurring variation in the circadian clock. *Science*, **302**, 1049–1053.

Michaels, S.D. & Amasino, R.M. (1999) *FLOWERING LOCUS C* encodes a novel MADS domain protein that acts as a repressor of flowering. *Plant Cell*, **11**, 949–956.

Michaels, S.D. & Amasino, R.M. (2000) Memories of winter: Vernalization and the competence to flower. *Plant Cell Environ.*, **23**, 1145–1153.

Michaels, S.D. & Amasino, R.M. (2001) Loss of FLOWERING LOCUS C activity eliminates the late-flowering phenotype of FRIGIDA and autonomous pathway mutations but not responsiveness to vernalization. *Plant Cell*, **13**, 935–941.

Michaels, S.D., He, Y.H., Scortecci, K.C. and Amasino, R.M. (2003) Attenuation of FLOWERING LOCUS C activity as a mechanism for the evolution of summer-annual flowering behavior in Arabidopsis. *Proc. Natl. Acad. Sci. USA*, **100**, 10102–10107.

Mizoguchi, T., Wheatley, K., Hanzawa, Y., Wright, L., Mizoguchi, M., Song, H.R., Carre, I.A. & Coupland, G. (2002) *LHY* and *CCA1* are partially redundant genes required to maintain circadian rhythms in Arabidopsis. *Dev. Cell*, **2**, 629–41.

Mockler, T., Yang, H.Y., Yu, X.H., Parikh, D., Cheng, Y.C., Dolan, S. & Lin, C.T. (2003) Regulation of photoperiodic flowering by Arabidopsis photoreceptors. *Proc. Natl. Acad. Sci. USA*, **100**, 2140–2145.

Mockler, T.C., Guo, H.W., Yang, H.Y., Duong, H. & Lin, C.T. (1999) Antagonistic actions of Arabidopsis cryptochromes and phytochrome B in the regulation of floral induction. *Development*, **126**, 2073–2082.

Mouradov, A., Cremer, F. & Coupland, G. (2002) Control of flowering time: interacting pathways as a basis for diversity. *Plant Cell*, **14** Suppl, S111–30.

Onouchi, H., Igeno, M.I., Perilleux, C., Graves, K. & Coupland, G. (2000) Mutagenesis of plants over-expressing *CONSTANS* demonstrates novel interactions among Arabidopsis flowering-time genes. *Plant Cell*, **12**, 885–900.

Park, D.H., Somers, D.E., Kim, Y.S., Choy, Y.H., Lim, H.K., Soh, M.S., Kim, H.J., Kay, S.A. & Nam, H.G. (1999) Control of circadian rhythms and photoperiodic flowering by the *GIGANTEA* gene. *Science*, **285**, 1579–1582.

Pineiro, M., Gomez-Mena, C., Schaffer, R., Martinez-Zapater, J.M. & Coupland, G. (2003) EARLY BOLTING IN SHORT DAYS is related to chromatin remodelling factors and regulates flowering in Arabidopsis by repressing FT. *Plant Cell*, **15**, 1552–1562.

Pittendrigh, C.S. & Minis, D.H. (1964) The entrainment of circadian oscillations by light and their role as photoperiodic clocks. *The American Naturalist*, **98**, 261–322.

Pnueli, L., Carmel-Goren, L., Hareven, D., Gutfinger, T., Alvarez, J., Ganal, M., Zamir, D. & Lifschitz, E. (1998) The *SELF-PRUNING* gene of tomato regulates vegetative to reproductive switching of sympodial meristems and is the ortholog of *CEN* and *TFL1*. *Development*, **125**, 1979–1989.

Pnueli, L., Gutfinger, T., Hareven, D., Ben-Naim, O., Ron, N., Adir, N. & Lifschitz, E. (2001) Tomato SP-interacting proteins define a conserved signaling system that regulates shoot architecture and flowering. *Plant Cell*, **13**, 2687–2702.

Putterill, J., Robson, F., Lee, K., Simon, R. & Coupland, G. (1995) The *CONSTANS* gene of Arabidopsis promotes flowering and encodes a protein showing similarities to zinc finger transcription factors. *Cell*, **80**, 847–857.

Ray, P.M. & Alexander, W.E. (1966) Photoperiodic adaptation to latitude in *Xanthium strumarium*. *Amer. J. Bot.*, **53**, 806.

Redei, G.P. (1962) Supervital mutants of Arabidopsis. *Genetics*, **47**, 443–460.

Robson, F., Costa, M.M.R., Hepworth, S., Vizir, I., Pineiro, M., Reeves, P.H., Putterill, J. & Coupland, G. (2001) Functional importance of conserved domains in the flowering-time gene *CONSTANS* demonstrated by analysis of mutant alleles and transgenic plants. *Plant J.*, **28**, 619–631.

Roden, L.C., Song, H.R., Jackson, S., Morris, K. & Carre, I.A. (2002) Floral responses to photoperiod are correlated with the timing of rhythmic expression relative to dawn and dusk in Arabidopsis. *Proc. Natl. Acad. Sci. USA*, **99**, 13313–13318.

Roldan, M., Gomez-Mena, C., Ruiz-Garcia, L., Salinas, J. & Martinez-Zapater, J.M. (1999) Sucrose availability on the aerial part of the plant promotes morphogenesis and flowering of Arabidopsis in the dark. *Plant J.*, **20**, 581–590.

Ruiz-Garcia, L., Madueno, F., Wilkinson, M., Haughn, G., Salinas, J. & Martinez-Zapater, J.M. (1997) Different roles of flowering-time genes in the activation of floral initiation genes in Arabidopsis. *Plant Cell*, **9**, 1921–1934.

Samach, A., Onouchi, H., Gold, S.E., Ditta, G.S., Schwarz-Sommer, Z., Yanofsky, M.F. & Coupland, G. (2000) Distinct roles of *CONSTANS* target genes in reproductive development of Arabidopsis. *Science*, **288**, 1613–1616.

Schaffer, R., Ramsay, N., Samach, A., Corden, S., Putterill, J., Carre, I.A. & Coupland, G. (1998) The *late elongated hypocotyl* mutation of Arabidopsis disrupts circadian rhythms and the photoperiodic control of flowering. *Cell*, **93**, 1219–1229.

Searle, I. & Coupland, G. (2004) Induction of flowering by seasonal changes in photoperiod. *EMBO J.*, **23**, 1217–1222.

Sheldon, C.C., Burn, J.E., Perez, P.P., Metzger, J., Edwards, J.A., Peacock, W.J. & Dennis, E.S. (1999) The FLF MADS box gene: A repressor of flowering in Arabidopsis regulated by vernalization and methylation. *Plant Cell*, **11**, 445–458.

Shimizu, M., Ichikawa, K. & Aoki, S. (2004) Photoperiod-regulated expression of the PpCOL1 gene encoding a homolog of CO/COL proteins in the moss Physcomitrella patens. *Biochemical and Biophysical Research Communications*, **324**, 1296–1301.

Simpson, G.G. & Dean, C. (2002) Arabidopsis, the Rosetta stone of flowering time? *Science*, **296**, 285–9.

Somers, D.E., Webb, A.A.R., Pearson, M. & Kay, S.A. (1998) The short-period mutant, *toc1–1*, alters circadian clock regulation of multiple outputs throughout development in Arabidopsis thaliana. *Development*, **125**, 485–494.

Stinchcombe, J.R., Weinig, C., Ungerer, M., Olsen, K.M., Mays, C., Halldorsdottir, S.S., Purugganan, M.D. & Schmitt, J. (2004) A latitudinal cline in flowering time in Arabidopsis thaliana modulated by the flowering time gene *FRIGIDA*. *Proc. Natl. Acad. Sci. USA*, **101**, 4712–4717.

Suarez-Lopez, P., Wheatley, K., Robson, F., Onouchi, H., Valverde, F. & Coupland, G. (2001) *CONSTANS* mediates between the circadian clock and the control of flowering in Arabidopsis. *Nature*, **410**, 1116–1120.

Sugano, S., Andronis, C., Green, R.M., Wang, Z.-Y. & Tobin, E.M. (1998) Protein kinase CK2 interacts with and phosphorylates the Arabidopsis CIRCADIAN CLOCK-ASSOCIATED 1 protein. *Proc. Natl. Acad. Sci. USA*, **95**, 11020–11025.

Sugano, S., Andronis, C., Ong, M.S., Green, R.M. & Tobin, E.M. (1999) The protein kinase CK2 is involved in regulation of circadian rhythms in Arabidopsis. *Proc. Natl. Acad. Sci. USA*, **96**, 12362–12366.

Sung, S.B. & Amasino, R.M. (2004) Vernalization in Arabidopsis thaliana is mediated by the PHD finger protein VIN3. *Nature*, **427**, 159–164.

Takada, S. & Goto, K. (2003) TERMINAL FLOWER2, an Arabidopsis homolog of HETEROCHRO-MATIN PROTEIN1, counteracts the activation of *FLOWERING LOCUS T* by CONSTANS in the vascular tissues of leaves to regulate flowering time. *Plant Cell*, **15**, 2856–2865.

Takahashi, Y., Shomura, A., Sasaki, T. & Yano, M. (2001) *Hd6*, a rice quantitative trait locus involved in photoperiod sensitivity, encodes the alpha subunit of protein kinase CK2. *Proc. Natl. Acad. Sci. USA*, **98**, 7922–7927.

Thomas, B. & Vince-Prue, B. (1997) *Photoperiodism in Plants*, 2nd ed., Academic Press, San Diego, CA.

Toth, R., Kevei, E., Hall, A., Millar, A.J., Nagy, F. & Kozma-Bognar, L. (2001) Circadian clock-regulated expression of phytochrome and cryptochrome genes in Arabidopsis. *Plant Physiol.*, **127**, 1607–16.

Tseng, T.-S., Salome, P.A., McClung, C.R. & Olszewski, N.E. (2004) SPINDLY and GIGANTEA interact and act in Arabidopsis thaliana pathways involved in light responses, flowering, and rhythms in cotyledon movements. *Plant Cell*, **16**, 1550–1563.

Valverde, F., Mouradov, A., Soppe, W., Ravenscroft, D., Samach, A. & Coupland, G. (2004) Photoreceptor regulation of CONSTANS protein and the mechanism of photoperiodic flowering. *Science*, **303**, 1003–1006.

Wang, Z.-Y. & Tobin, E.M. (1998) Constitutive expression of the CIRCADIAN CLOCK ASSOCIATED 1 (CCA1) gene disrupts circadian rhythms and suppresses its own expression. *Cell*, **93**, 1207–1217.

Yan, L., Loukoianov, A., Blechl, A., Tranquilli, G., Ramakrishna, W., SanMiguel, P., Bennetzen, J.L., Echenique, V. & Dubcovsky, J. (2004) The wheat *VRN2* gene is a flowering repressor down-regulated by vernalization. *Science*, **303**, 1640–1644.

Yan, L., Loukoianov, A., Tranquilli, G., Helguera, M., Fahima, T. & Dubcovsky, J. (2003) Positional cloning of wheat vernalization gene *VRN1*. *Proc. Natl. Acad. Sci. USA*, **100**, 6263–6268.

Yano, M., Harushima, Y., Nagamura, Y., Kurata, N., Minobe, Y. & Sasaki, T. (1997) Identification of quantitative trait loci controlling heading date in rice using a high-density linkage map. *Theor. Appl. Genet.*, **95**, 1025–1032.

Yano, M., Katayose, Y., Ashikari, M., Yamanouchi, U., Monna, L., Fuse, T., Baba, T., Yamamoto, K., Umehara, Y., Nagamura, Y. & Sasaki, T. (2000) *Hd1*, a major photoperiod sensitivity quantitative trait locus in rice, is closely related to the Arabidopsis flowering time gene *CONSTANS*. *Plant Cell*, **12**, 2473–2483.

Yanovsky, M.J. & Kay, S.A. (2002) Molecular basis of seasonal time measurement in Arabidopsis. *Nature*, **419**, 308–312.

Yoo, B.C., Kragler, F., Varkonyi-Gasic, E., Haywood, V., Archer-Evans, S., Lee, Y.M., Lough, T.J. & Lucas, W.J. (2004) A systemic small RNA signaling system in plants. *Plant Cell*, **16**, 1979–2000.

Yu, J., Hu, S., Wang, J., Wong, G.K.-S., Li, S., Liu, B., Deng, Y., Dai, L., Zhou, Y., Zhang, X., Cao, M., Liu, J., Sun, J., Tang, J., Chen, Y., Huang, X., Lin, W., Ye, C., Tong, W., Cong, L., Geng, J., Han, Y., Li, L., Li, W., Hu, G., Huang, X., Li, W., Li, J., Liu, Z., Li, L., Liu, J., Qi, Q., Liu, J., Li, L., Li, T., Wang, X., Lu, H., Wu, T., Zhu, M., Ni, P., Han, H., Dong, W., Ren, X., Feng, X., Cui, P., Li, X., Wang, H., Xu, X., Zhai, W., Xu, Z., Zhang, J., He, S., Zhang, J., Xu, J., Zhang, K., Zheng, X., Dong, J., Zeng, W., Tao, L., Ye, J., Tan, J., Ren, X., Chen, X., He, J., Liu, D., Tian, W., Tian, C., Xia, H., Bao, Q., Li, G., Gao, H., Cao, T., Wang, J., Zhao, W., Li, P., Chen, W., Wang, X., Zhang, Y., Hu, J., Wang, J., Liu, S., Yang, J., Zhang, G., Xiong, Y., Li, Z., Mao, L., Zhou, C., Zhu, Z., Chen, R., Hao, B., Zheng, W., Chen, S., Guo, W., Li, G., Liu, S., Tao, M., Wang, J., Zhu, L., Yuan, L. & Yang, H. (2002) A Draft Sequence of the Rice Genome (*Oryza sativa* L. ssp. indica). *Science*, **296**, 79–92.

Zeevaart, J.A.D. (1976) Physiology of flower formation. *Ann. Rev. Plant Physiol.*, **27**, 321–348.

8 Circadian regulation of Ca^{2+} signalling

Michael J. Gardner, Antony N. Dodd, Carlos T. Hotta,
Dale Sanders and Alex A. R. Webb

8.1 Introduction

The circadian clock consists of a complex signal transduction network that assimilates information concerning the external time of day and uses that information to allow the plant to anticipate daily changes in the environment. Over the last decade, research in plants and animals has revealed that Ca^{2+} may be an essential component of this signalling network. In this chapter, we compare potential mechanisms of circadian signalling by Ca^{2+} in plants and animals. We focus upon the genetic basis for circadian rhythms in the concentration of cytosolic free calcium ($[Ca^{2+}]_{cyt}$) in plants, and exploit studies of the circadian transcriptome to explore the processes that may underlie circadian Ca^{2+} signalling.

8.2 Ca^{2+} signalling is ubiquitous and versatile

8.2.1 Transduction of extracellular signals

Ca^{2+} is an intracellular signalling molecule that transduces extracellular signals and regulates the physiology, biochemistry and development of plant and animal cells (Hetherington & Brownlee, 2004). Transient increases in $[Ca^{2+}]_{cyt}$ in response to extracellular stimuli cause increased binding of Ca^{2+} to target proteins, which include Ca^{2+}-dependent protein kinases (CDPK), ion channels, calmodulin (CaM), Ca^{2+}-calmodulin-binding proteins and transcription factors (Hirschi, 2004). The resultant reversible changes in protein conformation alter enzyme activity or binding affinity for proteins, lipids, polysaccharides and other macromolecules. Ca^{2+}-binding to these sensor proteins can directly alter physiology – for example, by activating or inhibiting ion channel activity – or more typically initiate a cascade of further post-transcriptional modifications (e.g. phosphorylation by a CDPK) that transduces and amplifies the initial signal. In plants, elevations in $[Ca^{2+}]_{cyt}$ occur in response to many signals, including the hormones abscisic acid (ABA) and indoleacetic acid, abiotic stimuli that include CO_2, heat, cold, touch, blue light and red light, and biotic stimuli such as pathogens and NOD factors (Hetherington & Brownlee, 2004).

8.2.2 Spatio-temporal dynamics of $[Ca^{2+}]_{cyt}$ increases

Stimulus-induced increases in $[Ca^{2+}]_{cyt}$ vary greatly in their spatio-temporal dynamics and can include transient spikes, longer term increases, localized or global increases and sometimes complex oscillations that have been proposed to encode information (Hetherington & Brownlee, 2004; Evans *et al.*, 2001). Additionally, intracellular gradients of Ca^{2+} regulate development of asymmetrically growing cells such as pollen tubes and root hairs (Holdaway-Clark & Hepler, 2003). Finally, there are diurnal and circadian rhythms of $[Ca^{2+}]_{cyt}$.

The stimulus-induced oscillations of $[Ca^{2+}]_{cyt}$ are a fascinating example of plant ultradian rhythms and offer a potential insight into the mechanisms underlying rhythms with a longer period, such as circadian rhythms. Specificity might be conferred to ubiquitous $[Ca^{2+}]_{cyt}$ signals by encoding stimulus-specific information within these ultradian $[Ca^{2+}]_{cyt}$ oscillations. Hence, stimulus-specific $[Ca^{2+}]_{cyt}$ oscillations that occur in single cells and exhibit defined properties are termed 'Ca^{2+} signatures' (McAinsh & Hetherington, 1998). $[Ca^{2+}]_{cyt}$ oscillations exhibit reproducible signatures following specific stimuli in several cell types, including guard cells, seedlings, roots and root hairs, reproductive organs, protoplasts and green algae (Evans *et al.*, 2001).

The guard cell is ideal for investigating signal transduction by $[Ca^{2+}]_{cyt}$ oscillations, due to the autonomy of its Ca^{2+}-based signal transduction and the clearly defined response of stomatal movement (Hetherington *et al.*, 1998). Oscillations of guard cell $[Ca^{2+}]_{cyt}$ with periods in the order of minutes are caused by abscisic acid (ABA), exogenous Ca^{2+}, H_2O_2, fungal elicitors and cold shock (Evans *et al.*, 2001; Klüsener *et al.*, 2002). The frequency and pattern of the $[Ca^{2+}]_{cyt}$ oscillations depends on the stimulus type and strength (McAinsh *et al.*, 1995; Staxén *et al.*, 1999). Also, introduction of the signalling intermediates sphingosine-1-phosphate (Ng *et al.*, 2001), cADPR (Leckie *et al.*, 1998) and Ca^{2+} (McAinsh *et al.*, 1995) into the cytosol can cause oscillations of guard cell $[Ca^{2+}]_{cyt}$.

ABA-induced stomatal closure requires an increase in $[Ca^{2+}]_{cyt}$ (Webb *et al.*, 2001) and there is a correlation between the degree of stomatal closure and the pattern of $[Ca^{2+}]_{cyt}$ oscillation (Staxén *et al.*, 1999). Once stomatal aperture has reduced, maintenance of the new 'steady-state' aperture is dependent on the correct patterning of the $[Ca^{2+}]_{cyt}$ oscillations (Allen *et al.*, 2001). For example, guard cells of the *de-etiolation3* (*det3*) mutant, which has reduced vacuolar ATPase dependent energization of the tonoplast, do not elicit $[Ca^{2+}]_{cyt}$ oscillations and do not reduce their steady-state aperture in response to external Ca^{2+} and oxidative stress (Allen *et al.*, 2000). However, both cold and ABA cause wild-type $[Ca^{2+}]_{cyt}$ oscillations and reductions in steady-state aperture in *det3* (Allen *et al.*, 2000). These data suggest a correlation between the ability of $[Ca^{2+}]_{cyt}$ to oscillate and the final steady-state aperture achieved. Similarly, in the *gca2* mutant, the period of ABA-induced $[Ca^{2+}]_{cyt}$ oscillations is less than half that in the wild-type, and stomatal closure is incomplete or short-lived. Imposition of $[Ca^{2+}]_{cyt}$ oscillations with wild-type frequency in *gca2*, by iterative transfer between hyperpolarizing and depolarizing solutions, causes normal pore closure (Allen, *et al.*, 2001). The artificial imposition of $[Ca^{2+}]_{cyt}$ oscillations demonstrated that the period and duration of the oscillation regulate the final steady-state aperture of the stomata and there are optima of both oscillation period and duration to induce the maximal reduction in steady state aperture (Allen *et al.*, 2001).

8.2.3   Circadian oscillations of cytosolic free calcium

In plants entrained to light and dark cycles and placed in to constant red or white light (LL) [Ca^{2+}]$_{cyt}$ oscillates with a period of approximately 24 h. This is consistent with a circadian rhythm (Johnson *et al.*, 1995; Wood *et al.*, 2001; Love *et al.*, 2004; Fig. 8.1A). [Ca^{2+}]$_{cyt}$ peaks during the subjective day at concentrations between

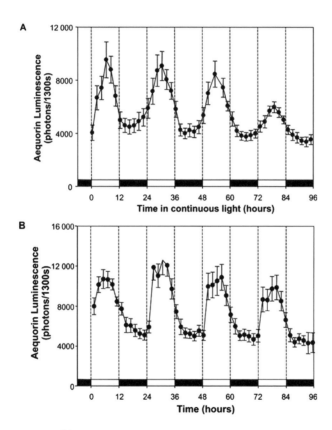

Figure 8.1 Circadian [Ca^{2+}]$_{cyt}$ oscillations in *A. thaliana*. Aequorin luminescence of 10 day old *A. thaliana*, (A) in constant light and (B) in 12 h light/12 h dark cycles, at a photon flux density of 100 μmol m^{-2} s^{-1}; mean luminescence ± standard error. White boxes represent light period, black boxes represent dark periods, and light grey boxes represent subjective night when in continuous light. Seedlings were germinated and entrained for 10 days under 12 h light/12 h dark, then dosed with 5 μM coelenterazine and transferred to constant light (A) or light/dark cycles (B), and aequorin luminescence was imaged every 2 h using a photon counting camera. High aequorin luminescence indicates relatively high [Ca^{2+}]$_{cyt}$, and low luminescence indicates relatively low [Ca^{2+}]$_{cyt}$.

350 and 700 nmol l^{-1} (Johnson *et al.*, 1995; Love *et al.*, 2004). Circadian oscillations of [Ca^{2+}]$_{cyt}$ in plants are measured using aequorin, a Ca^{2+}-sensitive photoprotein originally isolated and cloned from the jellyfish *Aequorea victoria* (Mills, 1999–2004). Aequorin consists of an apoprotein, and coelenterazine acts as a chromophoric ligand that reconstitutes the holoenzyme. Since plants do not produce coelenterazine, incubation of plants containing the *APOAEQUORIN* transgene with synthetic coelenterazine is required to reconstitute the functional bioluminescent protein. The intensity of light emitted from the plants is measured using photon counting luminometry (Johnson *et al.*, 1995) or photon counting imaging (Love *et al.*, 2004). The number of photons counted can be calibrated to [Ca^{2+}]$_{cyt}$ using empirically-derived equations (Fricker *et al.*, 1999). The intensity of light emitted is usually too low to allow imaging of single cells which can make interpretation of the physiological importance of aequorin measured Ca^{2+} signals difficult. Furthermore, calibrations might be compromised with some cells in a population being unresponsive (Dodd *et al.*, 2005).

The phase of aequorin-measured circadian [Ca^{2+}]$_{cyt}$ oscillations depends on the cell types in which the promoter used to drive aequorin expression is active. This suggests that different phases of circadian [Ca^{2+}]$_{cyt}$ oscillations occur in different plant cell types (Wood *et al.*, 2001). The different phases of circadian [Ca^{2+}]$_{cyt}$ oscillations in different cell types demonstrates that circadian rhythms are not driven by one central oscillator; instead, multiple oscillators are present in the plant. It is likely that these multiple oscillators are cell autonomous intracellular oscillators present in every cell. This may explain why rhythms of *CAB* promoter activity and [Ca^{2+}]$_{cyt}$ can desynchronize and run with different periods when measured from whole seedlings (Sai & Johnson, 1999). Cell autonomous oscillators acting in different tissues might be responsible for driving the different rhythms of *CAB* promoter activity and [Ca^{2+}]$_{cyt}$. Alternatively, cells might contain more than one circadian oscillator, one that drives rhythms of *CAB* promoter activity and another that causes [Ca^{2+}]$_{cyt}$ to oscillate. To distinguish between these possibilities, it will be necessary to determine whether circadian rhythms of [Ca^{2+}]$_{cyt}$ and *CAB* promoter activity desynchronize in the same cell (Sai & Johnson, 1999; Dodd *et al.*, 2005).

The time at which peak [Ca^{2+}]$_{cyt}$ occurs depends on the entrainment photoperiod (Love *et al.*, 2004). When *Arabidopsis* plants are entrained to short days (8 h light/16 h dark), the peak of the circadian oscillation of [Ca^{2+}]$_{cyt}$ in LL is approximately 5 h after subjective dawn. In contrast, peak circadian [Ca^{2+}]$_{cyt}$ following entrainment to long days (16L/8D) occurs later, at approximately 8 h after subjective dawn (Love *et al.*, 2004). The trough of the circadian oscillation of [Ca^{2+}]$_{cyt}$ in *Arabidopsis* is 19–20 h after subjective dawn, irrespective of photoperiod length during entrainment (Love *et al.*, 2004). Therefore, the shape of free-running oscillations of [Ca^{2+}]$_{cyt}$ becomes less symmetrical as the entrainment photoperiod increases.

Circadian oscillations in [Ca^{2+}]$_{cyt}$ occur in continuous light, but not in continuous dark (DD) (Johnson *et al.*, 1995). This contrasts with chloroplasts, where the stromal free Ca^{2+} concentration ($[Ca^{2+}]_{chl}$) does not oscillate with a circadian period in LL. However, transfer from LL to DD causes a transient elevation of $[Ca^{2+}]_{chl}$, with a peak of 5–10 μmol l^{-1}. This is followed by oscillations of $[Ca^{2+}]_{chl}$, with a period of approximately 24 h that progressively damp (Wood *et al.*, 2001; Sai & Johnson, 2002). Targeting of aequorin to the nucleus, using a nucleoplasm coding region, demonstrates that circadian oscillations of free Ca^{2+} do not occur in the nucleus in LL or DD (Wood *et al.*, 2001).

8.3   The cellular basis of circadian oscillations of plant $[Ca^{2+}]_{cyt}$

$[Ca^{2+}]_{cyt}$ is under tight control because sustained elevations of $[Ca^{2+}]_{cyt}$ can be cytotoxic. Therefore, the observation that there are 24 h oscillations of $[Ca^{2+}]_{cyt}$ indicates that there is temporal regulation of Ca^{2+} influx to the cytosol from intracellular or extracellular Ca^{2+} stores, and possibly also circadian efflux from the cytosol.

8.3.1   *Mechanisms controlling* $[Ca^{2+}]_{cyt}$ *influx*

During short-term $[Ca^{2+}]_{cyt}$ oscillations in plants, $[Ca^{2+}]_{cyt}$ is increased by influx of Ca^{2+} through ion channels in the plasma, tonoplast and endoplasmic reticulum (ER) membranes (Sanders *et al.*, 2002). Stimulus-induced influx of Ca^{2+} across the plasma membrane occurs via voltage and reactive oxygen species-regulated Ca^{2+} channels, non-specific cation channels and possibly ligand-gated Ca^{2+} channels, which include cyclic nucleotide-gated channels (Hetherington & Brownlee, 2004; Sanders *et al.*, 2002). Intracellular Ca^{2+} release pathways are thought to involve voltage-regulated non-specific cation channels in the tonoplast and also a number of ligand-gated channels in the tonoplast and ER membranes (Sanders *et al.*, 2002). Ligand-gated Ca^{2+} channels in plant endomembranes include those activated by inositol 1,4,5-trisphosphate (Ins(1,4,5)P$_3$) and cyclic adenosine diphosphate-ribose (cADPR). Ligand-gated channels that are sensitive to inositol hexakisphosphate and nicotinic acid adenine dinucleotide phosphate (NAADP) are also likely to be present (Sanders *et al.*, 2002). Sphingosine-1 phosphate is another second messenger that elevates $[Ca^{2+}]_{cyt}$, but its target(s) and the mechanism of Ca^{2+} influx are unknown (Ng *et al.*, 2001; Pandey & Assmann, 2004).

8.3.2   *How does the clock control* Ca^{2+} *influx?*

Which of these routes for Ca^{2+}-influx are clock-controlled, and drive circadian oscillations of $[Ca^{2+}]_{cyt}$ in plant cells? Presumably, all known Ca^{2+} entry routes have the potential to be regulated by the clock. Studies of the circadian regulation of $[Ca^{2+}]_{cyt}$ in neurons of the mammalian suprachiasmatic nucleus (SCN) describe mechanisms by which the circadian clock can mobilize Ca^{2+}, and suggest candidate mechanisms that might generate circadian oscillations of $[Ca^{2+}]_{cyt}$ in plants.

8.4   The cellular basis of circadian oscillations of $[Ca^{2+}]_{cyt}$ in mammals

8.4.1   The mammalian circadian clock

The master circadian clock of mammals is located in the hypothalamic suprachi-asmatic nuclei (SCN). The SCN receives input signals from the retina, to control hypothalamic and pituitary activity, which subsequently regulates circadian phys-iology. The circadian oscillator comprises several inter-locking molecular loops. Although the oscillator is probably cell autonomous, feedback and reinforcement signals may occur between pacemaker cells (Honma & Honma, 2003; Ikeda, 2004). The core molecular loop of the mammalian clock is autoregulatory, and consists of at least six genes: *CLOCK, BRAIN-MUSCLE ARNT-LIKE PROTEIN1 (BMAL1), PERIOD1 (PER1), PER2, CRYPTOCHROME1 (CRY1)* and *CRY2*. CLOCK and BMAL1 proteins are basic helix-loop-helix PAS family transcription factors that form heteromers and bind E-box enhancers in *PER* and *CRY* genes, activating their transcription. The *PER* and *CRY* gene products then form heteromers, translocate to the nucleus and inhibit the transactivation by CLOCK and BMAL1 (Honma & Honma, 2003).

8.4.2   A role for glutamate and cADPR-sensitive ryanodine receptor induced Ca^{2+} release in the entrainment of the mammalian clock

The SCN is located centrally in the brain and therefore cannot receive light signals directly. Instead, light signals received through neurons of the retinohypothalamic tract (RHT) cause glutamate release from the terminals of RHT neurons. Glutamate is sensed by N-methyl-D-asparate (NMDA) receptors, and other mechanisms, and stimulates Ca^{2+} influx across the plasma membrane of SCN neurons. This elevates $[Ca^{2+}]_{cyt}$ in the SCN (Fig. 8.2; Hannibal, 2002). Glutamate-induced elevations of $[Ca^{2+}]_{cyt}$ are amplified by Ca^{2+}-induced Ca^{2+} release through cADPR-sensitive ryanodine receptors (Fig. 8.2). Elevations of $[Ca^{2+}]_{cyt}$ are proposed to activate a Ca^{2+}/calmodulin kinase, or mitogen activate protein kinase (MAPK) cascade. This ultimately results in phosphorylated cyclic AMP-response element protein pCREB, which is a transcriptional activator of *PER1*. In subjective night, an alternative pathway is proposed to be activated by glutamate-induced increases in $[Ca^{2+}]_{cyt}$. This comprises Ca^{2+}-dependent activation of nitric oxide synthase (NOS), which elevates nitric oxide (NO) and activates cGMP-dependent protein kinase (PKG; Fig. 8.2). This could explain why the phase advance response that occurs following late subjective night stimulation is insensitive to inhibitors of ryanodine receptors but sensitive to activation of PKG (Ikeda, 2004). PKG activity is also required to allow clock progression from the 'night' phase to the 'day' phase by phosphorylation of CLOCK (Tischkau *et al.*, 2004).

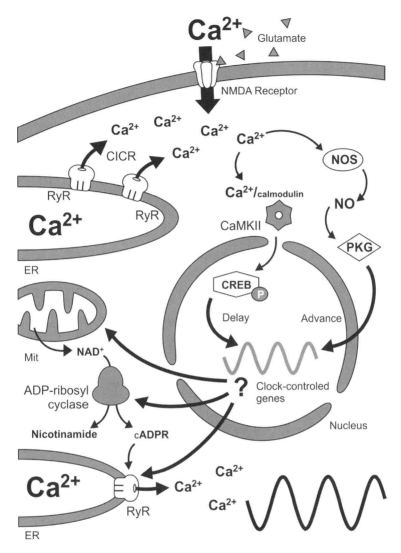

Figure 8.2 Role of [Ca^{2+}]$_{cyt}$ in the mammalian circadian clock. Ca^{2+} is believed to participate in input and output pathways of the mammalian clock (Ikeda, 2003). The SCN of the mammalian brain contains the circadian pacemaker. Glutamate released by the retinohypothalamic neurons stimulates NMDA receptors in SCN neurons. This causes a NMDA receptor-mediated Ca^{2+} influx, which is amplified by Ca^{2+}-induced Ca^{2+} release (CICR) from ryanodine receptors (RyR) in the ER. Elevation of [Ca^{2+}]$_{cyt}$ activates Ca^{2+}/calmodulin-dependent protein kinase II (CaMKII) that phosphorylates CREB. Phosphorylated CREB delays the phase of the clock. [Ca^{2+}]$_{cyt}$ increases late in the subjective night activates nitric oxide synthase (NOS), which generates NO, that in turn activates a pathway mediated by protein kinase G (PKG) and advances the phase of the clock. Circadian [Ca^{2+}]$_{cyt}$ oscillations are mediated by the cADPR pathway. Three possibilities for the mechanism by which this pathway is regulated by the circadian clock are: (i) the regulation of mitochondrial NAD$^+$ levels, (ii) the regulation of ADP-ribosyl cyclase activity and (iii) the regulation of RyR sensitivity to cADPR.

8.4.3   The source of $[Ca^{2+}]_{cyt}$ *influx*

Both glutamate-induced increases in SCN $[Ca^{2+}]_{cyt}$, and the free-running circadian oscillations of $[Ca^{2+}]_{cyt}$, are fuelled by Ca^{2+} release from cADPR- and ryanodine-sensitive intracellular stores (Ikeda *et al.*, 2003; Fig. 8.2). Other Ca^{2+} entry routes have been proposed, but the data conflict. L-type Ca^{2+} channel activity (depolarization activated) is diurnally regulated (Pennartz *et al.*, 2002), but blocking these channels, or blocking neuronal firing activity, does not alter single-cell circadian oscillations of $[Ca^{2+}]_{cyt}$. This suggests that influx across the plasma membrane is not required for the oscillations (Ikeda *et al.*, 2003). However, mean SCN cell population $[Ca^{2+}]_{cyt}$ oscillations are inhibited by blocking neuronal firing, suggesting the involvement of plasma membrane Ca^{2+} influx (Colwell, 2000). It has also been proposed that phospholipase C (PLC)-mediated increases in $Ins(1,4,5)P_3$ contribute to circadian oscillations of $[Ca^{2+}]_{cyt}$ in the SCN, based on studies in $PLC\beta4$ knock-out mice (Park *et al.*, 2003). This is unlikely, because thapsigargin, which inhibits $Ins(1,4,5)P_3$-sensitive store Ca^{2+}-ATPase activity, does not alter the $[Ca^{2+}]_{cyt}$ oscillation (Ikeda *et al.*, 2003).

The dual role of ryanodine receptor-mediated $[Ca^{2+}]_{cyt}$ increases in light/glutamate-entrainment of the clock and the circadian oscillations of $[Ca^{2+}]_{cyt}$, combined with the sensitivity of clock gene expression to elevations in $[Ca^{2+}]_{cyt}$, has led Ikeda (2003, 2004) to propose that the circadian oscillations of $[Ca^{2+}]_{cyt}$ might participate in a feedback loop regulating clock gene function. Alternatively, Ca^{2+} might act as a transducer of temporal information from the clock to regulate cell physiology (Honma & Honma, 2003). Reducing *PER2* expression had no affect on circadian oscillations of $[Ca^{2+}]_{cyt}$, which could indicate that the $[Ca^{2+}]_{cyt}$ oscillations are independent of *PER2*. However, interpretation of these data are complicated by the potential for compensatory changes in other components of the loop (Sugiyama *et al.*, 2004).

8.5   A comparison of the circadian regulation of Ca^{2+} homeostasis in both SCN and plant cells

Circadian regulation of Ca^{2+} homeostasis in both SCN and plant cells has some striking similarities. In both kingdoms, $[Ca^{2+}]$ oscillates in the cytosol but not the nucleus (Wood *et al.*, 2001; Ikeda *et al.*, 2003). The circadian oscillations of SCN $[Ca^{2+}]_{cyt}$ troughs at 120 nmol l^{-1} and peaks at 440 nmol l^{-1}. This is similar to a recent estimate of the circadian peak of $[Ca^{2+}]_{cyt}$ in leaf cells (Love *et al.*, 2004), although an earlier estimate of peak whole plant circadian oscillations of $[Ca^{2+}]_{cyt}$ was higher at approximately 700 nmol l^{-1} (Johnson *et al.*, 1995). In mammals and plants, light signals that entrain the oscillator might be transduced by elevations of $[Ca^{2+}]_{cyt}$, although the mechanisms by which $[Ca^{2+}]_{cyt}$ is elevated in response to light differ between the kingdoms. In plant cells, red (Shacklock *et al.*, 1992) and blue (Stoelzle *et al.*, 2003) light increase $[Ca^{2+}]_{cyt}$ by photoreceptor activation, and increases in $[Ca^{2+}]_{cyt}$ may entrain the clock (Gomez & Simon, 1995). Blue light-mediated $[Ca^{2+}]_{cyt}$ increases are caused by phototropin-activation of voltage-regulated plasma membrane Ca^{2+}-channels, although phototropins are not believed to entrain the clock.

8.5.1 cADPR is a key player in mammalian circadian behaviour

In mammalian SCN neurons, cADPR is key player in circadian signalling (Fig. 8.2). Could cADPR play a similar role in plants and regulate circadian oscillations of $[Ca^{2+}]_{cyt}$? There are no obvious homologues for the cADPR-generating enzyme, ADPR cyclase, or the cADPR receptor, in the *Arabidopsis* genome (Hetherington & Brownlee, 2004). However, ABA-inducible ADPR cyclase and hydrolase activities are present in *Arabidopsis*, and ABA upregulates cADPR accumulation (Sánchez *et al.*, 2004). Furthermore, cADPR elevates $[Ca^{2+}]_{cyt}$ in plants by stimulating release of Ca^{2+} from the vacuole (Leckie *et al.*, 1998) and ER (Navazio *et al.*, 2001), and transgenic elevation of cADPR levels up- and down-regulate the expression of 581 and 357 genes respectively (Sánchez *et al.*, 2004). These and other data have implicated cADPR in ABA and also defence signalling, but no role has been reported for cADPR in plant circadian signalling (Hetherington & Brownlee, 2004).

8.5.2 A role for nitric oxide in the plant clock

Intriguingly, NO, which regulates ADPR cyclase in mammals, and is involved in SCN phase regulation (Ikeda, 2004; Fig. 8.2), has been proposed to be a plant clock output (He *et al.*, 2004). NO release from leaves is higher by day than by night (Morot-Gaudry-Talarmain *et al.*, 2002). It is not known whether NO release oscillates in constant conditions, but treatments that elevate NO concentrations in the leaf reduce the amplitude of circadian outputs (He *et al.*, 2004). Sodium nitroprusside (an NO donor) decreases the amplitude of circadian expression of *CONSTANS (CO)*. Similarly, *nox1*, an NO-overproducing and hypersensitive mutant of *Arabidopsis* reduced amplitude in circadian rhythms of leaf movement and *CAB* transcript abundance. NO is likely to be an output of the clock, rather than a component of the clock mechanism, because *nox1* affected neither the period nor amplitude of the expression of the oscillator genes *TIMING OF CAB (TOC1)* and *CIRCADIAN CLOCK ASSOCIATED1 (CCA1)* (He *et al.*, 2004). It is not known whether NO regulates ADPR cyclase in plants, but NO-induced increases in guard cell $[Ca^{2+}]_{cyt}$ are reduced by inhibitors of guanylate cyclase and ryanodine, an inhibitor of cADPR action (Garcia-Mata *et al.*, 2003).

Inhibitor studies suggest that some components of the mammalian circadian signalling network are present in plants, including NO, guanylate cyclase, cADPR and Ca^{2+}. Although NO and Ca^{2+} have been implicated in plant circadian signalling, further work is essential to determine the relationships between these molecules, their contribution to circadian signalling, and the hierarchy of the signalling network.

8.6   Molecular regulation of circadian Ca^{2+} signalling in plants

Comparative physiology between kingdoms provides a tool to identify candidate signalling pathways that lead to the circadian control of $[Ca^{2+}]_{cyt}$ in plants. A complementary approach is to examine the transcriptional control of genes encoding Ca^{2+} signalling elements based on circadian microarray studies to identify potential control mechanisms.

A number of transcripts proposed to be involved in Ca^{2+} signalling, based on domain conservation or known functionality, exhibit circadian regulation of steady-state transcript abundance (Harmer *et al.*, 2000; Schaffer *et al.*, 2001). The majority exhibit maximal abundance around ZT8 (zeitgeber time 8, i.e. 8 h from the start of subjective day), which is approximately coincident with the timing of peak $[Ca^{2+}]_{cyt}$ (Fig. 8.1). We have placed these transcripts into three predicted functional groups: those which may regulate cytosolic Ca^{2+} influx, those which may regulate immediate responses to Ca^{2+}, and those predicted to regulate cytosolic Ca^{2+} efflux. To summarize our analysis of Ca^{2+} signalling elements that exhibit circadian regulation at the transcript level (Harmer *et al.*, 2000), we have arranged transcripts in diagrams that represent hypothetical cells, and highlighted proteins associated with Ca^{2+} homeostasis that might be upregulated at specific times during the circadian cycle (Fig. 8.3). We have arranged these proteins according to the pathways in which they operate, to suggest candidate mechanisms that could underlie the influx, efflux and sensing of circadian rhythms of $[Ca^{2+}]_{cyt}$ (Fig. 8.3).

8.6.1   *Circadian influx of* Ca^{2+} *to the cytosol*

In plants, inositol phosphate signalling by $Ins(1,4,5)P_3$, and phosphatidylinositol 3- and 4-phosphate ($PtdIns(3)P$ and $PtdIns(4)P$), leads to Ca^{2+}-release events that underlie stomatal movements (Gilroy *et al.*, 1990; Lee *et al.*, 1998; Staxén *et al.*, 1999; Jung *et al.*, 2002; Sanders *et al.*, 2002). At least seven proteins associated with inositol phosphate turnover and mobilization exhibit circadian regulation at the transcript level. Steady-state transcript abundance of these components peaks between ZT0 and ZT12, and the majority peak at ZT8 (Harmer *et al.*, 2000; Schaffer *et al.*, 2001). These components include two phosphatidylinositol-specific PLC transcripts, a CDP-diacylglycerol synthase (CDS; CDS-diacylglycerol being the precursor of phosphatidylinositol) homologue that peaks at ZT8, and a putative phosphatidylinositol-4-kinase (PtdIns4K) transcript that is maximally expressed at subjective dusk (ZT12, Harmer *et al.*, 2000; Schaffer *et al.*, 2001). Circadian PtdIns4K transcription may be important for circadian Ca^{2+} signalling since PtdIns4K catalyzes production of $PtdIns(4)P$. This is the precursor of phosphatidylinositol 4,5-bisphosphate ($PtdIns(4,5)P_2$), which PLC cleaves into $Ins(1,4,5)P_3$ and diacylglycerol (Fig. 8.3; Jung *et al.*, 2002). In *Drosophila melanogaster*, CDS activity is essential for $Ins(1,4,5)P_3$-mediated light-perception (Wu *et al.*, 1995). In plants, the synchronization of circadian expression of PtdIns4K, PLC and CDS, with the peak

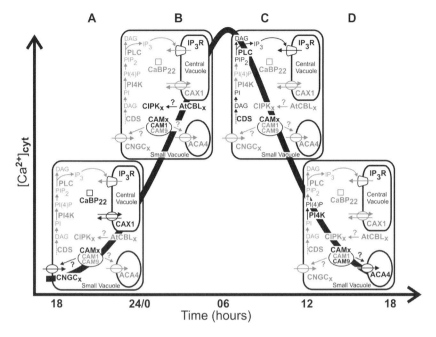

Figure 8.3 Mapping temporal expression of Ca^{2+} signalling genes onto Ca^{2+} signalling pathways. Circadian-regulated transcripts (Harmer *et al.*, 2000) that encode proteins associated with Ca^{2+} homeostasis are positioned in this model according to their cellular function and time of peak expression. Over the circadian cycle, black text indicates peak circadian abundance of transcripts, whilst grey text indicates lower expression levels of that transcript. (A) $[Ca^{2+}]_{cyt}$ rises between ZT18 and ZT24. At this time, the Ca^{2+}-binding protein CaBP22, the Ca^{2+}/H^+ antiport CAX1 and CNGCs are maximally expressed. (B) $[Ca^{2+}]_{cyt}$ approaches peak concentrations between ZT0 and ZT06. At this time, peaks occur in abundance of transcripts encoding Calmodulin-1 (CAM1), calcineurin B-like protein (AtCBL) and AtCBL interacting protein kinases (CIPK), which could represent activation of a Ca^{2+}-sensor mechanisms whilst $[Ca^{2+}]_{cyt}$ is elevated. (C) During the peak in circadian $[Ca^{2+}]_{cyt}$, between ZT6 and ZT12, transcripts encoding proteins that underlie inositol phosphate signalling reach maximum abundance. Also, a CDP-diacylglycerol synthase (CDS) homologue and two phospholipase C (PLC) homologues are maximally expressed. (D) $[Ca^{2+}]_{cyt}$ decreases between ZT12 and ZT18. This might be due to induction of ACA4 transcripts. Calmodulin-9 (CAM9) is also induced during this period. Induction of phosphatidilinositol-4-kinase (PI4K) transcripts might be important for phosphatidyl inositol (1,5)-bisphosphate (PIP_2) regeneration during this period. A transcript for the Ins(1,4,5) P_3 receptor (IP_3R) has not been identified. In this model we assume the receptor is always available.

in $[Ca^{2+}]_{cyt}$ might implicate the Ins(1,4,5)P_3-based signalling in circadian rhythms of $[Ca^{2+}]_{cyt}$. This is supported by experiments in which inhibition of PtdIns(4)P synthesis reduces stomatal opening during the early part of the photoperiod, suggesting a link between phosphoinositide metabolism, Ca^{2+} signals, and circadian control of stomatal movements (Jung *et al.*, 2002).

Circadian expression of genes encoding Ca^{2+} channels represents another mechanism that could regulate circadian $[Ca^{2+}]_{cyt}$ oscillations. Few Ca^{2+}-permeable channels appear to be transcriptionally regulated by the clock. However, at least one cyclic nucleotide-gated ion channel (CNGC) has circadian rhythms of transcript abundance, with transcript levels peaking prior to subjective dawn (Schaffer *et al.*, 2001). In animals, cyclic nucleotides, and the CNGCs that they target, have well-defined roles in photoperception and clock entrainment (Ivanova & Iuvone, 2003; Ko *et al.*, 2004; Golombek *et al.*, 2004). CNGCs contain a C-terminal calmodulin binding domain, and can cause plasma membrane Ca^{2+} fluxes following activation by binding of cAMP and/or cGMP (Volotovski *et al.*, 1998; Leng *et al.*, 1999, 2002; Chan *et al.*, 2003; Lemtiri-Chlieh & Berkowitz, 2004). A possible explanation for the small number of clock-controlled Ca^{2+} channels in *Arabidopsis* is that channel activity, rather than channel protein abundance, underlies circadian Ca^{2+} fluxes. It is likely that ligand synthesis, or expression of interacting proteins, have circadian rhythmicity, and subsequently drive changes in channel activity.

8.6.2   Circadian efflux of Ca^{2+} from the cytosol

Re-establishment of resting $[Ca^{2+}]_{cyt}$ after stimulus-induced elevations requires rapid removal of Ca^{2+} from the cytosol (Hirschi, 2004). Similarly, the decline in $[Ca^{2+}]_{cyt}$ during the subjective night must be achieved through the activity of Ca^{2+} transporters. Several such transporters exhibit circadian rhythms of transcript abundance. *ARABIDOPSIS AUTO-INHIBITED Ca^{2+}-ATPase 4 (ACA4)* transcripts, which encode a P-type Ca^{2+}-ATPase, are expressed with a circadian rhythm with peak transcript levels at ZT12 (Geisler *et al.*, 2000; Harmer *et al.*, 2000). This is consistent with the possibility that it might function to restore low $[Ca^{2+}]_{cyt}$, since $[Ca^{2+}]_{cyt}$ is decreasing by ZT12 (Fig. 8.1; Love *et al.*, 2004). Similarly, CALCIUM EXCHANGER 1 (CAX1), which is a high-affinity tonoplast H^+/Ca^{2+} antiporter that maintains low $[Ca^{2+}]_{cyt}$ by driving vacuolar calcium accumulation (Hirschi, 1999, 2001), is maximally expressed toward the end of the subjective night (ZT20-24, Schaffer *et al.*, 2001) and could contribute to the nocturnal $[Ca^{2+}]_{cyt}$ trough (Figs. 8.1 and 8.3). Therefore, circadian induction of two transcripts encoding Ca^{2+}-efflux proteins, one of which is located at the plasma membrane and one at the tonoplast, is synchronized with the circadian decline and trough in $[Ca^{2+}]_{cyt}$ (Fig. 8.3).

8.6.3 Ca^{2+} sensor proteins

Decoding of Ca^{2+} signals is believed to begin by binding of Ca^{2+} to sensor protein(s), which subsequently interact with target proteins to propagate the signal. Known *Arabidopsis* Ca^{2+}-sensors include calcineurin B-like (AtCBL) proteins (Kudla *et al.*, 1999), calmodulin and CaM-related proteins (Zielinski, 1998), and CDPKs (Roberts & Harmon, 1992). CaMs interact with a large range of targets, whereas AtCBLs interact with a subset of 10 protein kinases, known as calcineurin B-like interacting protein kinases (CIPKs) (Kim *et al.*, 2000). At least two putative AtCBL and two CIPK (CIPK4, CIPK15) transcripts exhibit circadian rhythms of abundance, all of which peak at ZT0 (Harmer *et al.*, 2000; Schaffer *et al.*, 2001).

Circadian control of calcineurin-related proteins could contribute to circadian Ca^{2+} signalling, since in *Arabidopsis* guard cells, calcineurin regulates Ca^{2+}-permeable slow vacuolar ion channel activity.

Several CaM and CaM-binding proteins also have circadian expression patterns. CaM1 transcript levels peak at ZT0, CaM9 and a CaM-like Ca^{2+}-binding protein transcripts peak at ZT12, and CA^{2+}-BINDING PROTEIN 22 expression peaks at ZT20 (Harmer *et al.*, 2001). One explanation for circadian expression of these Ca^{2+} sensors is that they might link the core oscillator with clock-controlled physiology. Alternatively, they could contribute to regulation of the clock. For example, in mammals a Ca^{2+}/CaM-dependent protein kinase II (CaMKII) contributes to resetting of the clock by light (Golombek & Ralph, 1994, 1995; Agostino *et al.*, 2004). In *Neurospora crassa*, Ca^{2+}/CaM-dependent protein kinase (CAMK-1) phosphorylates the clock protein FREQUENCY, and the phase and period of conidiation rhythms are altered in CAMK-1 null mutants (Yang *et al.*, 2001). Importantly, circadian-regulated CNGCs, and ACA4, contain CaM-binding domains, indicating the potential for Ca^{2+}-based regulation of circadian Ca^{2+} transport.

Physiological studies have demonstrated that [Ca^{2+}]$_{cyt}$ oscillates with a circadian rhythm. Transcript analysis suggests that a number of pathways in the cytosol and various membranes might function in concert to bring about these changes in Ca^{2+} homeostasis. In the next section, we consider the possible roles for circadian oscillations of [Ca^{2+}]$_{cyt}$.

8.7   Physiological targets for circadian oscillations of [Ca^{2+}]$_{cyt}$

8.7.1   Difficulties in assigning of physiological function to circadian oscillations in [Ca^{2+}]$_{cyt}$

At the time of writing, a physiological target for circadian oscillations of [Ca^{2+}]$_{cyt}$ in plants and mammals has not been identified (Ikeda, 2004; Dodd *et al.*, 2005). This is a major impediment to understanding the role and significance of the circadian regulation of Ca^{2+} signalling. Providing a link between circadian [Ca^{2+}]$_{cyt}$ oscillations and the regulation of cell physiology is technically demanding because the long timescales involved are not readily compatible with cell physiological techniques. Non-invasive imaging technologies (Ikeda, 2003; Love *et al.*, 2004) and specially designed physiology instrumentation (Dodd *et al.*, 2004) offer great promise in dissecting the role of circadian [Ca^{2+}]$_{cyt}$ oscillations.

A second limitation to predicting and testing the role of circadian [Ca^{2+}]$_{cyt}$ oscillations in plant physiology is that the circadian regulation of [Ca^{2+}]$_{cyt}$ has been measured in populations of cells, rather than individual cells (Johnson *et al.*, 1995; Wood *et al.*, 2001; Love *et al.*, 2004). The mean population Ca^{2+} signal may not always reflect Ca^{2+} signals in single cells (Dodd *et al.*, in preparation). Recently, we proposed four non-exclusive models of circadian [Ca^{2+}]$_{cyt}$ dynamics in individual plant cells (Dodd *et al.*, 2005). [Ca^{2+}]$_{cyt}$ may oscillate with a period of 24 h in all cells, and those cells might all be in phase, or some cells may oscillate out of phase (this is the situation in the mammalian SCN (Ikeda *et al.*, 2003)). Alternatively, the mean population $[Ca^{2+}]_{cyt}$ changes may not reflect the activity of single plant cells and instead single cells might exhibit rapid transients ('spiking') of $[Ca^{2+}]_{cyt}$ whose amplitude, frequency or both incorporate circadian modification. Testing these models will require long-term imaging of single plant cells, such as has been achieved using recombinant Cameleon YC2.1 fluorescent reporter of Ca^{2+} in neurons of the SCN (Ikeda *et al.*, 2003).

Almost any cellular event that is regulated by $[Ca^{2+}]_{cyt}$ might be controlled by circadian oscillations of $[Ca^{2+}]_{cyt}$. For example, cellular machinery regulated by the Ca^{2+}-based signalling cascades that are activated by red light, blue light, ABA, cold, touch and NOD factors are potential candidates for regulation by circadian $[Ca^{2+}]_{cyt}$ oscillations. Previously, we have suggested that stomatal movements, leaf movements in legumes and Crassulacean acid metabolism (CAM) are strong candidates for regulation by circadian $[Ca^{2+}]_{cyt}$ oscillations (Webb, 2003; Dodd *et al.*, 2005). The cellular events that underlie the circadian physiology of these cell types are dependent on regulation by $[Ca^{2+}]_{cyt}$ (see Webb (2003) for detail).

8.7.2   A role for Ca^{2+} in the photoperiodic control of flowering

Ca^{2+} might participate in the network by which the circadian clock contributes to the photoperiodic control of flowering. Information concerning day length is encoded in both circadian oscillations of $[Ca^{2+}]_{cyt}$ and the oscillations of $[Ca^{2+}]_{cyt}$ that occur during day–night cycles (Love *et al.*, 2004; Fig. 8.1B). In long days, the $[Ca^{2+}]_{cyt}$ in *Arabidopsis* leaf cells reaches a maximum several hours later than in short days (Love *et al.*, 2004). Furthermore, $[Ca^{2+}]_{cyt}$ is high at dusk in short days, but in long days, $[Ca^{2+}]_{cyt}$ is low at dusk. Love *et al.*, (2004) therefore proposed that information encoded in oscillations of $[Ca^{2+}]_{cyt}$ could contribute to photoperiodic control of flowering. Interestingly, pharmacological manipulations that interfere with Ca^{2+} signalling cause phase-specific reductions in the number of flowers produced in *Pharbitis nil* (Friedman *et al.*, 1992), but since *P. nil* is a short-day plant the role of Ca^{2+} signals in control of flowering may differ markedly from the long-day plant *Arabidopsis*.

Two Ca^{2+} sensing proteins might have a role in regulating the transition from vegetative growth to flowering: a CDPK (Jaworski *et al.*, 2003), and a cell-surface sensor of extracellular Ca^{2+} (Han *et al.*, 2003). The relationship between the extracellular Ca^{2+} sensor and circadian $[Ca^{2+}]_{cyt}$ oscillations is unclear, but this cell-surface sensor also has a role in generating short period oscillations of $[Ca^{2+}]_{cyt}$ in guard cells in response to external Ca^{2+}. NO is also involved in the network by which the clock contributes to the control of flowering. NO elevates $[Ca^{2+}]_{cyt}$ in plants (Garcia-Mata *et al.*, 2003; Lamotte *et al.*, 2004) and reduces the amplitude of circadian outputs (He *et al.*, 2004). Additionally, NO delays flowering in *Arabidopsis* in all photoperiods (He *et al.*, 2004). The effects of NO on flowering were attributed to reductions in *LEAFY (LFY)* expression, which is a regulator of flowering that increases in abundance before flowering. NO also affects the expression of other genetic determinants of flowering, including increasing expression of *FLOWERING LOCUS (FLC)* and decreasing the expression of *CONSTANS (CO)*, which is thought to provide a link between the circadian clock and photoperiodic control of flowering. Expression of *GIGANTEA (GI)*, which acts genetically upstream from CO, is also reduced by NO (He *et al.*, 2004).

8.8   Conclusions and future prospects

Our analysis has demonstrated that relatively few Ca^{2+} signalling components show circadian patterns of steady-state transcript abundance. This might appear puzzling, given that Ca^{2+} is a powerful regulator of physiological processes, and many aspects of physiology are clock-controlled. Although circadian regulation of signalling might occur by modulation of transcript abundance, post-transcriptional mechanisms such as rhythmic translation, RNA interference, protein phosphorylation, ligand gating (including Ca^{2+}-regulation) of channels, and targeted protein degradation, have the potential to drive and decode circadian [Ca^{2+}]$_{cyt}$ rhythms. Exhaustive analyses at many levels of molecular and biochemical regulation are required for full identification of components that contribute to circadian control of physiology. Tissue- and cell-specific circadian rhythms (Sai & Johnson, 1999; Wood *et al.*, 2001) might complicate the search for components of the circadian Ca^{2+} signalling network, meaning that it will be essential to establish the nature of this system in single cells (Dodd *et al.*, 2005). Since the biological clocks of at least two kingdoms incorporate circadian rhythms of Ca^{2+}, we hope that the next decade might reveal why circadian Ca^{2+} rhythms are so fundamental as to be recruited more than once into clock architecture.

Acknowledgements

AARW, DS and AD thank the BBSRC for funding. MJG thanks the Gates Trust for award of a scholarship, and CTH thanks Coordenadoria de Aperfeiçoamento de Pessoal de Nível Superior of Brazil for a studentship award. AARW is grateful to the Royal Society of London for the award of a University Research Fellowship. The authors thank Dr. Mei-See Man (Department of Anatomy, Cambridge) for discussions relating to mammalian neurobiology.

Abbreviations

ABA;   abscisic acid
[Ca^{2+}]$_{cyt}$;   the concentration of cytosolic free Ca^{2+}
CAB;   *CHLOROPHYLL A BINDING*
cADPR;   cyclic adenosine diphosphate-ribose

CAM;	Crassulacean acid metabolism
CCA1;	*CIRCADIAN CLOCK ASSOCIATED*
CDPK;	Ca^{2+}-dependent protein kinase
CO;	*CONSTANS*
D;	dark
DD;	continuous darkness
$Ins(1,4,5)P_3$;	inositol (1,4,5) trisphosphate
L;	light
LD;	light dark cycle
LL;	continuous light
NAADP;	nicotinic acid adenine dinucleotide phosphate
PLC;	phospholipase C
SCN;	suprachiasmatic nucleus
SNP;	sodium nitoprusside
TOC1;	*TIMINIG OF CAB1*

References

Agostino, P.V., Ferreyra, G.A., Murad, A.D., Watanabe, Y. & Golombek, D.A. (2004) Diurnal, circadian and photic regulation of calcium/calmodulin-dependent kinase II and neuronal nitric oxide synthase in the hamster suprachiasmatic nuclei. *Neurochem. Int.*, **44**, 61–69.

Allen, G.J., Chu, S.P., Harrington, C.L., Schumacher, K., Hoffman, T., Tang, Y.Y., Grill, E. & Schroeder, J.I. (2001) A defined range of guard cell calcium oscillation parameters encodes stomatal movements. *Nature*, **411**, 1053–1057.

Allen, G.J., Chu, S.P., Schumacher, K., Shimazaki, C.T., Vafeados, D., Kemper, A., Hawke, S.D., Tallman, G., Tsien, R.Y., Harper, J.F., Chory, J. & Schroeder, J.I. (2000) Alteration of stimulus-specific guard cell calcium oscillations and stomatal closing in Arabidopsis *det3* mutant. *Science*, **289**, 2338–2342.

Chan, C.W., Schorrak, L.M., Smith Jr, R.K., Bent, A.F. & Sussman, M.R. (2003) A cyclic nucleotide-gated ion channel, CNGC2, is crucial for plant development and adaptation to calcium stress. *Plant Physiol.*, **132**, 728–731.

Colwell, C.S. (2000) Circadian modulation of calcium levels in cells in the suprachiasmatic nucleus. *Eur. J. Neuroscience*, **12**, 571–576.

Dodd, A.N., Parkinson, K. & Webb, A.A.R. (2004) Independent circadian regulation of assimilation and stomatal conductance in the *ztl-1* mutant of Arabidopsis. *New Phyt.*, **162**, 63–70.

Dodd, A.N., Love, J. & Webb, A.A.R. (2005) The plant clock shows its metal: circadian regulation of cytosolic free Ca^{2+}. *Trends Plant. Sci.* **10**, 15–21.

Evans, N.H., McAinsh, M.R. & Hetherington A.M. *et al.* (2001) Calcium oscillations in higher plants. *Current Curr. Op. Plant Biol.*, **4**, 415–420.

Fricker, M.D., Plieth, C., Knight, H., Blancaflor, E., Knight, M.R., White, N.S. & Gilroy, S. (1999) Fluorescence and luminescence techniques to probe ion activities in living plant cells. In: *Fluorescent and Luminescent Probes for Biological Activity* second edition (ed W.T. Mason). Academic Press, San Diego.

Friedman, H., Goldschmidt E.E., Spiegelstein, H. & Halevy, A.H. *et al.* (1992) A rhythm in the flowering response of photoperiodically-induced *Pharbitis nil* to agents affecting cytosolic calcium and pH. *Physiol. Plant*, **85**, 57–60.

Garcia-Mata, C., Gay, R., Sokolovski, S., Hills, A., Lamattina, L. & Blatt, M.R. (2003) Nitric oxide regulates K^+ and Cl^- channels in guard cells through a subset of abscisic acid-evoked signalling pathways. *PNAS Proc. Natl. Acad. Sci. USA*, **100**, 11116–11121.

Geisler, M., Frangne, N., Gomes, E., Martinoia, E. & Palmgren, M.G. (2000) The ACA4 gene of Arabidopsis encodes a vacuolar membrane calcium pump that improves salt tolerance in yeast. *Plant Physiol.*, **124**, 1814–1827.

Gilroy, S., Read, N.D. & Trewavas, A.J. (1990) Elevation of cytoplasmic calcium by caged calcium or caged inositol trisphosphate initiates stomatal closure. *Nature*, **346**, 769–771.

Golombek, D.A. & Ralph, M.R. (1994) KN-62, an inhibitor of Ca^{2+}/calmodulin kinase II, attenuates circadian responses to light. *Neuroreport*, **5**, 1638–1640.

Golombek, D.A. & Ralph, M.R. (1995) Circadian responses to light: the calmodulin connection. *Neurosci Lett.*, **192**, 101–104.

Golombek, D.A., Agostino, P.V., Plano, S.A. & Ferreyra, G.A. (2004) Signalling in the mammalian circadian clock: the NO/cGMP pathway. *Neurochem Int.*, **45**, 929–936.

Gómez, L.A. & Simón, E. (1995) Circadian rhythm of *Robinia pseudoacacia* leaflet movement: Role of calcium and phytochrome. *Photochem. Photobiol.*, **61**, 210–215.

Han, S., Tang, R., Anderson, L.K., Woerner, T.E. & Pei Z-M. (2003) A cell surface receptor mediates extracellular Ca^{2+} sensing in guard cells. *Nature*, **425**, 196–200.

Hannibal, J. (2002) Neurotransmitters of the retino-hypothalamic tract. *Cell and Tissue Res.*, **309**, 73–88.

Harmer, S.L., Hogenesch, J.B., Straume, M., Chang, H.S., Han, B., Zhu, T., Wang, X., Kreps, J.A. & Kay, S.A. (2000) Orchestrated transcription of key pathways in Arabidopsis by the circadian clock. *Science*, **290**, 2110–2113.

Harmer, S.L., Panda, S. & Kay, S.A. (2001) Molecular basis of circadian rhythms. *Ann. Rev. Cell Dev. Biol.*, **17**, 215–253.

He, Y., Tang, R-H., Hao, Y., Stevens, R.D., Cook, C.W., Ahn, S.M., Jing, L., Yang, Z., Chen, L., Guo, F., Fiorani, F., Jackson, R.B., Crawford, N.M. & Pei, Z-M. (2004) Nitric oxide represses the Arabidopsis floral transition. *Science*, **305**, 1968–1971.

Hetherington, A.M., Gray, J.E., Leckie, C.P., McAinsh, M.R., Ng, C., Pical, C., Priestley, A.J., Staxén, I. & Webb, A.A.R. (1998) The control of specificity in guard cell signal transduction. *Philosophical Transactions of the Royal Society of London Series B-Biological Sciences*, **353**, 1489–1494.

Hetherington, A.M. & Brownlee, C. (2004) The generation of Ca^{2+} signals in plants. *Ann. Rev. Plant Biol.*, **55**, 401–427.

Hirschi, K.D. (1999) Expression of Arabidopsis CAX1 in tobacco: Altered calcium homeostasis and increased stress sensitivity. *Plant Cell*, **11**, 2113–2122.

Hirschi, K. (2001) Vacuolar H^{+}/Ca^{2+} transport: Who's directing the traffic? *Trends Plant Sci.*, **6**, 100–104.

Hirschi, K.D. (2004) The calcium conundrum. Both versatile nutrient and specific signal. *Plant Physiol.*, **136**, 2438–2442.

Holdaway-Clark, T.L. & Hepler, P.K. (2003) Control of pollen tube growth: the role of ion gradients and fluxes. *New Phytol.*, **159**, 539–563.

Honma, S. & Honma, K. (2003) The biological clock: Ca^{2+} links the pendulum to the hands. *Trends Neurosci.*, **26**, 650–653.

Ikeda, M. (2003) Response to Honma and Honma: Do circadian rhythms in cytosolic Ca^{2+} modulate autonomous gene transcription cycles in the SCN? *Trends Neurosci.*, **26**, 654.

Ikeda, M. (2004) Calcium dynamics and circadian rhythms in suprachiasmatic nucleus neurons. *The Neuroscintist*, **10**, 315–324.

Ikeda, M., Sugiyama, T., Wallace, C.S., Gompf, H.S., Yoshioka, T., Miyawaki, A. & Allen, C.N. (2003) Circadian dynamics of circadian and nuclear Ca^{2+} in single suprachiasmatic nucleus neurones. *Neuron*, **38**, 253–263.

Ivanova, T.N. & Iuvone, P.M. (2003) Circadian rhythm and photic control of cAMP level in chick retinal cell cultures: a mechanism for coupling the circadian oscillator to the melatonin-synthesizing enzyme, arylalkylamine N-acetyltransferase, in photoreceptor cells. *Brain Res.*, **991**, 96–103.

Jaworski, K., Szmidt-Jaworska, A., Tretyn, A. & Kopcewicz, J. (2003) Biochemical evidence for a calcium-dependent protein kinase from *Pharbitis nil* and its involvement in photoperiodic flower induction. *Phytochemistry*, **62**, 1047–1055.

Johnson, C.H., Knight, M.R., Kondo, T., Masson, P., Sedbook, J., Haley, A. & Trewavas, A. (1995) Circadian oscillations of cytosolic and chloroplastic free calcium in plants. *Science*, **269**, 1863–1865.

Jung, J.Y., Kim, Y.W., Kwak, J.M., Hwang, J.U., Young, J., Schroeder, J.I., Hwang, I. & Lee, Y. (2002) Phosphatidylinositol 3- and 4-phosphate are required for normal stomatal movements. *Plant Cell*, **14**, 2399–2412.

Kim, K.N., Cheong, Y.H., Gupta, R. & Luan, S. (2000) Interaction specificity of Arabidopsis calcineurin B-like calcium sensors and their target kinases. *Plant Physiol.*, **124**, 1844–1853.

Klüsener, B., Young, J.J., Murata, Y., Allen, G.J., Mori, I.C., Hugouvieux, V. & Schroeder, J.I. (2002) Convergence of calcium signalling pathways of pathogenic elicitors and abscisic acid in Arabidopsis guard cells. *Plant Physiol.*, **130**, 2152–2163.

Ko, G.Y., Ko, M.L. & Dryer, S.E. (2004) Circadian regulation of cGMP-gated channels of vertebrate cone photoreceptors: Role of cAMP and Ras. *J. Neurosci.*, **24**, 1296–1304.

Kudla, J., Xu, Q., Harter, K., Gruissem, W. & Luan, S. (1999) Genes for calcineurin B-like proteins in Arabidopsis are differentially regulated by stress signals. *Proc. Natl. Acad. Sci. USA*, **96**, 4718–4723.

Lamotte, O., Gould, K., Lecourieux, D., Sequeira-Legrand, A., Lebrun-Garcia, A., Durner, J., Pugin, A. & Wendehenne, D. (2004) Analysis of nitric oxide signalling functions in tobacco cells challenged by the elicitor cryptogein. *Plant Physiology Physiol.*, **135**, 516–529.

Leckie, C.P., McAinsh, M.R., Allen G.J., Sanders, D. & Hetherington, A.M. (1998) Abscisic acid-mediated stomatal closure mediated by cyclic ADP-ribose. *Proc. Natl. Acad. Sci. USA*, **95**, 15837–15842.

Leckie, C.P., McAinsh, M.R., Allen, G.J., Sanders, D. & Hetherington, A M. (1998) Abscisic acid-induced stomatal closure mediated by cyclic ADP-ribose. *Proc. Natl. Acad. Sci. USA*, **95**, 15837–15842.

Lee, J.Y., Yoo, B .C. & Harmon, A.C. (1998) Kinetic and calcium-binding properties of three calcium-dependent protein kinase isoenzymes from soybean. *Biochemistry*, **37**, 6801–6809.

Lemtiri-Chlieh, F. & Berkowitz, G.A. (2004) Cyclic adenosine monophosphate regulates calcium channels in the plasma membrane of Arabidopsis leaf guard and mesophyll cells. *J. Biol. Chem.*, **279**, 35306–35312.

Leng, Q., Mercier, R.W., Yao, W.Z. & Berkowitz, G.A. (1999) Cloning and first functional characterization of a plant cyclic nucleotide-gated cation channel. *Plant Physiol.*, **121**, 753–761.

Leng, Q., Mercier, R.W., Hua, B-G., Fromm, H. & Berkowitz, G.A. (2002) Electrophysiological analysis of cloned cyclic nucleotide-gated ion channels. *Plant Physiol.*, **128**, 400–410.

Love, J., Dodd, A.N. & Webb, A.A.R. (2004) Circadian and diurnal calcium oscillations encode photoperiodic information in Arabidopsis. *Plant Cell*, **16**, 956–966.

McAinsh, M.R., Webb, A.A.R., Taylor, J.E. & Hetherington, A.M. (1995) Stimulus-induced oscillations in guard cell cytoplasmic free calcium. *Plant Cell*, **7**, 1207–1219.

McAinsh, M.R. & Hetherington, A.M. (1998). Encoding specificity in Ca^{2+} signalling systems. *Trends Plant Sci.*, **3**, 32–36.

Mills, C.E. 1999–2004. Bioluminescence of *Aequorea*, a hydromedusa. Electronic internet document available at http://faculty.washington.edu/cemills/Aequorea.html. Published by the author, web page established June 1999, last updated 2004.

Morot-Gaudry-Talarmain, Y., Rockel, P., Moureaux, T., Quillere, I., Leydecker, M.T., Kaiser, W.M. & Morot-Gaudry, J.F. (2002) Nitrite accumulation and nitric oxide emission in relation to cellular signalling in nitrite reductase antisense tobacco. *Planta*, **215**, 708–715.

Navazio, L., Mariani, P. & Sanders, D. (2001) Mobilization of Ca^{2+} by cyclic ADP-ribose from endoplasmic reticulum of cauliflower florets. *Plant Physiol.*, **125**, 2129–21238.

Ng, C.K.Y., Carr, K., McAinsh, M.R., Powell, B. & Hetherington, A.M. (2001) Drought-induced guard cell signal transduction involves sphingosine-1-phosphate. *Nature*, **410**, 596–599.

Pandey, S. & Assmann, S.M. (2004) The Arabidopsis putative G protein-coupled receptor GCR1 interacts with the G protein alpha subunit GPA1 and regulates abscisic acid signalling. *Plant Cell*, **16**, 1616–1632.

Park, D., Lee, S., Jun, K., Hong, Y.M., Kim, D.Y., Kim, Y.I. & Shin, H.S. (2003) Translation of clock rhythmicity into neural firing in the suprachiasmatic nucleus requires mGluR-PLCβ4 signalling. *Nature Neurosci.*, **6**, 571–576.

Pennartz, C.M.A., de Jeu, M.T.G., Bos, N.P.A., Schaap, J. & Geurtsen, A.M.S. (2002) Diurnal modulation of pacemaker potentials and calcium current in the mammalian circadian clock. *Nature*, **416**, 286–290.

Roberts, D.M. & Harmon, A.C. (1992) Calcium modulated protein targets of intracellular calcium signals in higher plants. *Ann. Rev. Plant. Physiol. Plant Mol. Biol.*, **43**, 375–414.

Sai, J. & Johnson, C.H. (1999) Different circadian oscillators control Ca^{2+} fluxes and *Lhcb* gene expression. *Proc. Natl. Acad. Sci. USA*, **96**, 11659–11663.

Sai, J.Q. & Johnson, C.H. (2002) Dark-stimulated calcium ion fluxes in the chloroplast stroma and cytosol. *Plant Cell*, **14**, 1279–1291.

Sánchez, J-P, Duque, P. & Chua, N-H. (2004) ABA activates ADPR cyclase and cADPR induces a subset of ABA-responsive genes in Arabidopsis. *Plant J.*, **38**, 381–395.

Sanders, D., Pelloux, J., Brownlee, C & Harper, J.F. (2002) Calcium at the crossroads of signalling. *Plant Cell*, **14**, S401–S417.

Schaffer, R., Landgraf, J., Accerbi, M., Simon, V., Larson, M. & Wisman E. (2001) Microarray analysis of diurnal and circadian-regulated genes in Arabidopsis. *Plant Cell*, **13**, 113–123.

Shacklock, P.S., Read, N.D. & Trewavas, A.J. (1992) Cytosolic free calcium mediates red light-induced photomorphogenesis. *Nature*, **358**, 753–755.

Staxèn, I., Pical, C., Montogomery, L.T., Gray, J.E., Hetherington, A.M. & McAinsh, M.R. (1999) Abscisic acid induces oscillations in guard cell cytosolic free calcium that involve phosphoinositide-specific phospholipase C. *Proc. Natl. Acad. Sci. USA*, **96**, 1779–1784.

Stoelzle, S., Kagawa, T., Wada, M., Hedrich, R. & Dietrich, P. (2003) Blue light activates calcium-permeable channels in Arabidopsis mesophyll cells via the phototropin signalling pathway. *Proc. Natl. Acad. Sci. USA*, **100**, 1456–1461.

Sugiyama, T., Yoshioka, T. & Ikeda M. (2004) *mPer2* antisense oligonucleotides inhibit *mPer2* expression but not circadian rhythms of physiological activity in cultured suprachiasmatic nucleus neurones. *Biochem. Biophys. Res. Comm.*, **323**, 479–483.

Tischkau, S.A., Mitchell J.W., Pace, L.A., Barnes, J.W., Barnes, J.A. & Gillette, M.U. (2004) Protein kinase G Type II is required for night-to-day progression of the mammalian circadian clock. *Neuron*, **43**, 539–549.

Volotovski, I.D., Sokolovsky, S.G., Molchan, O.V. & Knight, M.R. (1998) Second messengers mediate increases in cytosolic calcium in tobacco protoplasts. *Plant Physiol.*, **117**, 1023–1030.

Webb, A.A.R. (2003) The physiology of circadian rhythms in plants. *New Phytologist*, **160**, 281–303.

Webb, A.A.R., Larman, M.G., Montgomery, L.T., Taylor, J.E. & Hetherington, A.M. (2001) The role of calcium in ABA-induced gene expression and stomatal movements. *Plant Journal*, **26**, 351–362.

Wood, N.T., Haley, A., Viry-Moussaïd, M., Johnson, C.H., van der Luit, A.H. & Trewavas, A.J. (2001) The calcium rhythms of different cell types oscillate with different circadian phases. *Plant Physiol.*, **125**, 787–796.

Wu, L., Niemeyer, B., Colley, N., Socolich, M. & Zucker C.S. (1995) Regulation of PLC-mediated signalling in vivo by CDP-diacylglycerol synthase. *Nature*, **373**, 216–222.

Yang, Y., Cheng, P., Zhi, G. & Liu, Y. (2001) Identification of a calcium/calmodulin-dependent protein kinase that phosphorylates the *Neurospora* circadian clock protein FREQUENCY. *J. Biol. Chem.*, **276**, 41064–41072.

Zielinski, R.E. (1998) Calmodulin and calmodulin-binding proteins in plants. *Ann. Rev. Plant Physiol. Plant Mol. Biol.*, **49**, 697–725.

9 The circadian clock in CAM plants

James Hartwell

9.1 Introduction

The majority of higher plants fix CO_2 using the C_3 pathway of photosynthesis, whereby the Calvin cycle enzyme ribulose bisphosphate carboxylase oxygenase (Rubisco) catalyzes CO_2 fixation during the day. The oxygenase activity of Rubisco makes C_3 plants particularly ill-equipped for survival in high temperatures, because increasing leaf temperatures favour the oxygenase activity over the carboxylase. Approximately 10% of the angiosperms have evolved enhancements of photosynthetic metabolism that permit them to fix CO_2 efficiently via Rubisco at higher temperatures (Sage, 1999; Winter & Smith, 1996). These plants use either the C_4 pathway or the Crassulacean acid metabolism (CAM) pathway. C_4 and CAM plants initially fix CO_2 into the four-carbon product malic acid via the enzyme phosphoenolpyruvate carboxylase (PEPc) and concentrate CO_2 around Rubisco during secondary CO_2 fixation. C_4 plants perform primary and secondary CO_2 fixation in the light, with a spatial separation of the two steps into two cell types. CAM plants perform both CO_2 fixation steps in the same cell, but separate them temporally, with primary CO_2 fixation by PEPc occurring at night. These CO_2 concentrating mechanisms favour the carboxylase activity of Rubisco, even at higher temperatures. The operation of the CAM pathway leads to a large increase in water use efficiency, because CAM plants only need to open their stomatal pores at night when temperatures are lower and thus transpirational water loss is kept to a minimum. This chapter reviews the large amount of research carried out on the circadian clock in CAM species; the role of the circadian clock in C_4 species has been very little studied to date.

9.2 The daily cycle of metabolism in CAM plants

CAM leaves open their stomatal pores at night and PEPc catalyzes primary CO_2 fixation (Fig. 9.1). PEPc uses phosphoenolpyruvate and bicarbonate as its substrates and generates the four-carbon product oxaloacetate (OAA). OAA is rapidly reduced to malate by NADP-malate dehydrogenase and this malate accumulates in the vacuole throughout the night as malic acid. The malic acid gives CAM plants a characteristic acidic taste at dawn; a facet of diurnal CAM physiology that has been recognized since Roman times. Following sunrise, the accumulated malic acid floods out of the vacuole into the cytosol where it is decarboxylated by either NAD- and/or NADP-malic enzyme or PEP carboxykinase, depending on the species. CO_2

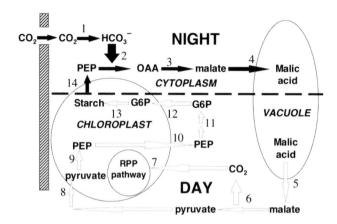

Figure 9.1 The Crassulacean acid metabolism pathway in an NADP-ME species. Black filled arrows represent dark reactions whilst unfilled arrows represent light reactions. The dashed line running across the centre separates night at the top from day at the bottom. The hashed boxes on the left of the cartoon represent the leaf epidermis with the gap representing a stomatal pore. The enzymes that catalyze the reactions are as follows: 1. carbonic anhydrase; 2. phosphoenolpyruvate carboxylase; 3. malate dehydrogenase; 4. voltage-gated malate anion channel; 5. unknown protein proposed to mediate malate efflux; 6. NADP-malic enzyme; 7. ribulose bisphosphate carboxylase oxygenase; 8. unknown pyruvate transporter on the inner envelope membrane of the chloroplast; 9. pyruvate orthophosphate dikinase; 10. phosphoenolpyruvate phosphate translocator; 11. gluconeogenesis; 12. glucose 6-phosphate phosphate translocator (GPT); 13. starch synthesis beginning with ADP-glucose pyrophosphorylase; 14. nocturnal starch breakdown possibly via chloroplastic starch phosphorylase and the export of G6P via the GPT followed by glycolytic conversion of G6P to PEP which is then used by PEPc as a substrate for CO_2 fixation.

builds up to high internal partial pressures that mediate stomatal closure and is refixed by ribulose bisphosphate carboxylase oxygenase (Rubisco) using energy from the light reactions of photosynthesis. This intricate temporal separation of primary and secondary CO_2 fixation within single photosynthetic cells requires tight control to avoid a futile cycle of simultaneous malate synthesis and degradation. In the handful of CAM species that have been studied in sufficient detail, it is clear that a circadian clock maintains strict temporal control of the CAM pathway.

9.2.1 Discovery of a circadian rhythm of CO_2 fixation in CAM species

Kalanchoë fedtschenkoi R. Hamet et Perrier (family: Crassulaceae, order: Saxifragales) is an obligate CAM plant that is endemic to Madagascar. CAM is a physiological advantage to the plant in this arid environment with low night temperatures and high day temperatures. Since the late 1950s a great deal of research has been carried out on the persistent circadian rhythm of carbon dioxide exchange observed in detached leaves of *K. fedtschenkoi* (Wilkins, 1992).

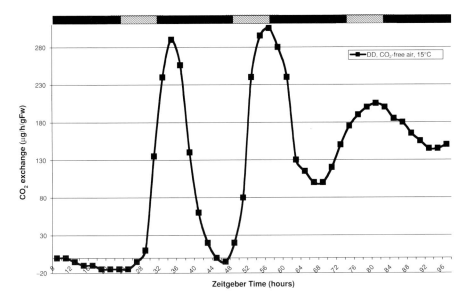

Figure 9.2 The rhythm of CO_2-output generated by detached leaves of *Kalanchoë fedtschenkoi* in DD, CO_2-free air at 15°C. ZT 8 is dusk.

9.2.2   *CAM CO_2 output rhythms in constant darkness, CO_2-free air, 15°C*

When *K. fedtschenkoi* leaves were detached from plants entrained in LD 8:16 at lights off, and placed at 15°C in continuous darkness (DD) and an initially CO_2-free air stream, a circadian rhythm of CO_2 output was observed (Fig. 9.2) (Wilkins, 1959, 1962). This rhythm persisted for 3–4 days, was robust between temperatures of 10–28°C and showed substantial temperature compensation (Wilkins, 1959, 1962). The observed rhythm of CO_2 output was generated by an oscillation of flux through PEPc. This was inferred from the observation that incorporation of trace $^{14}CO_2$, in an otherwise CO_2-free air stream, resulted in the periodic labeling of malate (Warren and Wilkins, 1961). When the enzyme was active during the subjective dark period, respired CO_2 was refixed and this generated corresponding troughs in the measured CO_2 output (Fig. 9.2). By contrast, when PEPc was inactive during the subjective light period, respired CO_2 escaped from the leaf generating corresponding peaks in CO_2 output (Fig. 9.2).

Malate synthesized during periods of PEPc activity accumulates in the cytosol, inhibiting PEPc and curtailing the troughs in CO_2 output. However, the malate is pumped into the vacuole and this then allows PEPc to be active again leading to a second trough in the CO_2 output. In conditions of constant darkness and CO_2-free air, the rhythm of CO_2 output is thought to disappear after 3 to 4 days due to the vacuole reaching its capacity for malate. Under these circumstances, the malate synthesized during periods of PEPc activity will remain in the cytoplasm and therefore prevent the refixation of respired CO_2 by PEPc. The rhythm of cytosol CO_2 output from the leaves therefore damps after 3 to 4 days, as indicated by the constant escape of respired CO_2 from the leaves. Exposure of such leaves to a 4 h period of illumination allows all of the malate to come out of the vacuole into the cytoplasm, where it is decarboxylated, causing a large peak in CO_2 output. The rhythm will then reinitiate until the vacuole has once more become saturated with malate (Wilkins, 1992).

It should be noted that although this theory has a number of attractive features, such as the fact that reinitiation of the rhythm by light can be explained by decarboxylation of the accumulated malate, other factors must also be important. For example, the total malate content of leaves that have been in CO_2-free air at 15°C in the dark for 3–4 days is much lower than the malate content of leaves after a single normal night. Stored starch provides the PEP that is the substrate for PEPc activity and the possibility that the rhythm damps out after 3–4 days due to starch depletion has not been tested. Another aspect of the DD, CO_2-free air rhythm that has been somewhat overlooked is that the rhythm persisted for much longer than 3–4 days if 15 minutes of white or red light were provided once every 24 h (Harris & Wilkins, 1978b). It seems unlikely that the leaf replenished a significant proportion of its starch pool during this short period of low intensity red light or that significant malate decarboxylation occurred in only 15 minutes. Thus, rhythms in DD and CO_2-free air may damp out due to a lack of phytochrome activation in constant darkness and not due to a buildup of malate or a dearth of starch.

In constant darkness and normal air at temperatures between 0°C and 20°C, *K. fedtschenkoi* leaves performed a single period of CO_2-fixation followed by a prolonged period of relatively constant CO_2 output (Carter *et al.*, 1995b). However, at 25°C and 30°C a distinct rhythm of CO_2-output was evident with the strongest rhythmicity at 30°C. The rhythm was less distinct at 35°C. The rhythm at 25°C was due to periods of CO_2-fixation that caused a trough in the CO_2-output from the leaf and these periods coincided with the subjective dark period, suggesting they may be mediated by activation of PEPc. Interestingly, the rhythm of CO_2 exchange in DD, normal air at 30°C showed a trough at the beginning of the subjective dark period and peak CO_2 output from the middle of the subjective dark period until subjective dawn. This revealed that increasing the temperature from 25°C to 30°C induced a significant phase advance in the rhythm of CO_2 fixation in constant darkness and normal air.

9.2.3   *CAM CO_2-assimilation rhythms in constant light and normal air*

In continuous light and a stream of normal air, detached *K. fedtschenkoi* leaves showed a rhythm of CO_2 assimilation, rather than output, which persisted for at least 10 days (Fig. 9.3) (Wilkins, 1984; Anderson & Wilikins, 1989). Rhythms of CO_2 assimilation in constant light (LL) and normal air have also been observed in other CAM species in the genus *Kalanchoë* (Nuernbergk, 1961; Lüttge & Ball, 1978). More recently, LL rhythms have been reported in the inducible-CAM species *Mesembryanthemum crystallinum,* in which CAM has evolved independently (Dodd *et al.*, 2003). At any particular temperature, with the exception of 31°C, the period of the *K. fedtschenkoi* rhythm in LL was shorter than the rhythm in darkness and a CO_2-free air stream (Wilkins, 1992). When PEPc was least active, then CO_2 assimilation

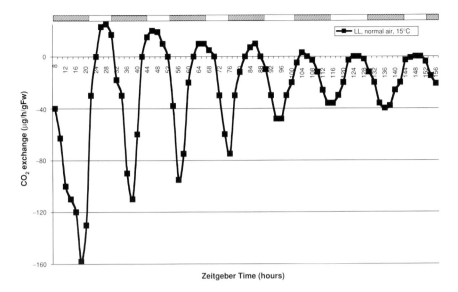

Figure 9.3 The rhythm of CO_2 assimilation generated by detached leaves of *Kalanchoë fedtschenkoi* in LL, normal air at 15°C. ZT8 is dusk.

became negligible or CO_2 output was detected and CO_2 assimilation was occurring only via the Calvin–Benson cycle (Wilkins, 1992).

There is no evidence that the rate of photosynthesis is regulated by a circadian rhythm in *K. fedtschenkoi* leaves (Wilkins, 1992). Wilkins (1984) proposed that the rhythm of CO_2 assimilation by leaves in continuous light and a stream of normal air continued for up to 10 days because malate did not accumulate in the vacuole. Continuous illumination caused malate to equilibrate between the vacuole and cytoplasm, so the vacuole was not available as a malate storage compartment. Wilkins (1984) suggested that malate accumulated during the first period of PEPc activity inhibited the enzyme and then the malate was removed via decarboxylation. This theory requires the circadian oscillator to drive a circadian alteration of the properties of the malate removal system (i.e. malic enzyme) and there is some evidence for this (Cook *et al.*, 1995). The CO_2 released during decarboxylation may be refixed by Rubisco or escape from the leaf. Hence, the malate synthesized during each period of PEPc activity is subsequently decarboxylated and so each cycle of the rhythm begins with the leaf having a low malate content.

9.3   Phase responses of CAM CO_2 rhythms

9.3.1   Light induced phase shifts

The CAM rhythm of CO_2 fixation in DD, CO_2-free air at 15°C can be perturbed by phase-resetting inputs in the form of light and temperature treatments. The response to light required a saturating light treatment of 1 h or 3 h; above this threshold, the phase response was determined not by the duration of the treatment but by the time at which the treatment ended (Wilkins, 1992). It was the length of the light treatment that was important; low fluence rates were sufficient to cause phase shifting when provided for these durations.

The effectiveness of a 3 h light treatment in inducing phase shifts depended on the time in the cycle when it was provided. The phase response curve for 3 h red light treatments revealed that the rhythm was phase delayed when red light treatments were applied in the first half of the subjective dark period, whilst it was phase advanced when red light was provided in the second half of the subjective dark period (Wilkins, 1989). The largest phase shifts occurred in the middle of the night when CO_2 fixation via PEPc was at a maximum, whilst there was no phase shift when the light treatment ended at subjective lights off.

Experiments to determine which wavelengths of the visible spectrum were active in phase shifting CO_2 rhythms revealed that only red wavelengths of the spectrum were effective with a peak at 640–660 nm (Wilkins, 1992). This suggested that phytochrome was the photoreceptor responsible for light input to this rhythm. The role of phytochrome was confirmed by experiments that demonstrated far-red reversibility of the effect of red light in entrainment studies (Harris & Wilkins, 1976, 1978a,b). Blue light had no detectable effect on the CO_2 fixation rhythm in *K. fedtschenkoi* indicating that cryptochrome did not play a role in light input to the oscillator driving CAM rhythms in this species (Wilkins, 1992).

9.3.2 Temperature induced phase shifts

The rhythm in darkness and CO_2-free air was also phase-shifted by temperature treatments. Phase response curves were generated for high temperature treatments (35°C) that lasted for 1 h or 3 h (Wilkins, 1983). The response to 35°C treatments was very similar to the response to red light: greatest phase shifts were recorded for treatments applied in the middle of the subjective night when CO_2-fixation by PEPc was at its maximum. As with the response curve for red light, it was the time at which a high temperature treatment ended that determined the magnitude of the phase shift. The one point in the cycle that did not generate a phase response was subjective lights off. Low temperature treatments (2°C) were also effective at inducing phase shifts but the times for inducing maximum and minimum phase shifts were the opposite of those for light or high temperature (Wilkins, 1962). Low temperature mediated greatest phase shifts during the subjective light period and had least effect during the subjective dark period.

The rhythm of CO_2 assimilation in constant light and normal air was also subject to phase shifts in response to temperature and dark treatments. Again, high temperatures caused the greatest phase shift during the subjective dark period; low temperatures and dark mediated maximum phase shifting during the subjective light period (Anderson & Wilikins, 1989, 1989a,b).

The rhythm of CO_2 assimilation in constant light and normal air occurred between 10°C and 30°C, but over this temperature range the period changed from

15.7 h at 10°C to 23.3 h at 30°C. The temperature coefficients (period at t-10°C/period at t°C) calculated for three 10°C intervals ranged from 0.8 for the 10/20°C interval to 0.89 for the 15/25°C interval (Anderson & Wilkins, 1989a). The closeness of these coefficients to 1 indicated that the rhythm showed a considerable degree of temperature compensation (Anderson & Wilkins, 1989a). The rhythm of CO_2 exchange in LL in *Kalanchoë daigremontiana* grown at 25/15°C L/D 12:12 became arrhythmic above 30°C (Grams *et al.*, 1995). However, when the plants were grown at 34/25°C L/D 12:12 for 4 weeks prior to the release into LL, they did perform a rhythm of CO_2 exchange at 30°C (Grams *et al.*, 1995). However, these leaves were not rhythmic at 34°C. Nevertheless, these data demonstrated that temperature compensation within the CAM clock can adapt to daily temperature rhythms in growth conditions and increase the upper limit of the temperature compensated range of the clock. It will be very important to follow these findings up with direct assessment of the adaptation of other clock-controlled processes, such as leaf movements and oscillations in clock-controlled genes, to high growth temperatures in CAM species and in C_3 and C_4 plants.

9.4 The biochemical basis for the circadian rhythm of CO_2 fixation

9.4.1 PEPc phosphorylation rhythms

When the circadian changes in CO_2 metabolism in CAM leaves were studied at the biochemical level, it was discovered that the sensitivity of PEPc to the feedback inhibitor L-malate also changed between day and night (Winter, 1982; Nimmo *et al.*, 1984). Rapidly desalted extracts prepared from *K. fedtschenkoi* during the day contained PEPc with a low apparent K_i for malate of 0.3 mM. The PEPc in desalted extracts made from leaves at night showed an apparent K_i tenfold greater at 3.0 mM (Nimmo *et al.*, 1984). The specific activity of PEPc within the extracts did not vary over a 24 h period indicating that the amount of PEPc protein in the leaves remained constant throughout. This was confirmed immunologically (Nimmo *et al.*, 1986), implying that PEPc was regulated post-translationally.

In order to determine the nature of the post-translational regulation of PEPc, detached leaves were allowed to take up [32]P_i for 72 hours followed by rapid extraction and immunoprecipitation of PEPc (Nimmo *et al.*, 1984). The immunoprecipitated PEPc was separated by SDS-polyacrylamide gel electrophoresis (SDS-PAGE). Autoradiography of the resulting gel demonstrated that the night form of PEPc contained [32]P_i but the day form did not (Nimmo *et al.*, 1984). The phosphate group was shown to be bound to serine using two-dimensional thin-layer electrophoresis of hydrolysed ^{32}P-labeled PEPc (Nimmo *et al.*, 1986). This suggested that protein phosphorylation was the post-translational modification responsible for controlling the circadian rhythm of PEPc activity.

Further investigations led to the purification of the phosphorylated night form and dephosphorylated day form of PEPc from *K. fedtschenkoi* leaves and the subsequent characterization of the enzyme in its two phosphorylation states (Nimmo *et al.*, 1986). Both the phosphorylated and dephosphorylated forms of the enzyme contained two proteins, a major one (subunit Mr 112 000) and a minor one (subunit M_r 123 000), as determined by SDS-PAGE. The two PEPc subunits were related as judged by proteolysis patterns and this was not an artefact of *in vitro* proteolysis (i.e. both subunits exist in planta). Both subunits became phosphorylated at night (Nimmo *et al.*, 1986). Alkaline phosphatase was found to remove the phosphate group from the night form PEPc *in vitro*. This gave a concomitant increase in the apparent K_i of the PEPc for malate to that of the day form (Nimmo *et al.*, 1984).

Analysis of PEPc from detached *K. fedtschenkoi* leaves maintained in continuous darkness and a stream of CO_2-free air, showed a good correlation between the time when the leaves were fixing respiratory CO_2 and when the PEPc was in the phosphorylated night form (Nimmo *et al.*, 1987). With all the evidence pointing to phosphorylation as the cause of the circadian rhythm of PEPc activity in *K. fedtschenkoi*, the search for the enzyme(s) responsible for regulating the phosphorylation state of PEPc began. There were two possible candidates for the role: the kinase and the phosphatase.

9.4.2   Regulation of PEPc phosphorylation

The first evidence as to the nature of the phosphatase that dephosphorylates PEPc at the end of the dark period came from experiments in which the purified catalytic subunits of rabbit skeletal muscle type 1 and 2A protein phosphatases (PP1 and PP2A) were tested for *in vitro* activity against phosphorylated PEPc. This showed that phosphorylated PEPc from *K. fedtschenkoi* could be dephosphorylated by PP2A but not by PP1 (Carter *et al.*, 1990). Furthermore, by using specific inhibitors, (Carter *et al.*, 1990) were able to demonstrate that *K. fedtschenkoi* leaves contained a PP2A that dephosphorylated PEPc. Examination of the specific activity of this PP2A over a 24 h period showed that there was no significant diurnal variation in its activity (Carter *et al.*, 1991).

Subsequent work in the same laboratory led to the seminal discovery of a highly specific protein kinase in *K. fedtschenkoi* leaves capable of phosphorylating PEPc (Carter *et al.*, 1991). In plants under short day conditions (LD 8:16), the kinase activity appeared between ZT 12–14 and persisted at a high level throughout the middle of the dark period, but was virtually undetectable by ZT 22 (Carter *et al.*, 1991). Presence of kinase activity correlated with a high K_i of PEPc for malate. This pattern of activity indicated that it was the kinase and not the phosphatase that was responsible for the observed circadian rhythm of PEPc phosphorylation.

9.4.3   Regulation of PEPc kinase activity

This raised the question as to what was regulating the activity of the kinase such that it was active at night but not in the day. Detached *K. fedtschenkoi* leaves were allowed to take up protein synthesis inhibitors (puromycin or cycloheximide) or RNA synthesis inhibitors (cordycepin and actinomycin D) during the day in an attempt to ascertain whether the nocturnal appearance of PEPc kinase activity was controlled at the level of protein synthesis and/or RNA synthesis (Carter *et al.*, 1991, 1996; Nimmo, 1993). Rapidly desalted extracts prepared from these leaves in the middle of the following night were found to lack PEPc kinase activity. This was true regardless of whether the inhibitor was applied during the previous day or just 1–2 h prior to kinase assays in the middle of the night (Nimmo, 1993; Carter *et al.*, 1996).

The PEPc in these extracts was dephosphorylated, as judged by its malate sensitivity. This work suggested that both *de novo* protein synthesis and an increase in the steady state level of an mRNA, possibly PEPc kinase mRNA, were necessary for the appearance of PEPc kinase activity. Furthermore, the fact that application of the inhibitors only 1–2 h prior to assay led to loss of activity suggested that kinase activity was being rapidly turned over. This indicated that the activity of PEPc kinase played a key role in the control of the circadian rhythm of PEPc activity and CO_2 metabolism in *K. fedtschenkoi*.

Work on the effect of temperature on PEPc kinase activity and the phosphorylation state of PEPc in *K. fedtschenkoi* has brought to light some interesting effects that are of great significance to the physiology of the plant in the wild (Carter *et al.*, 1995a,b). At 30°C a distinct circadian rhythm of CO_2 output was observed in the absence of a detectable rhythm in PEPc kinase activity. Therefore, it would appear that kinase activity is not essential for the generation of a circadian rhythm in PEPc activity and CO_2 output (Carter *et al.*, 1995b). Rather, circadian changes in kinase activity may increase the amplitude of the rhythm.

An alternative explanation for the existence of a rhythm of CO_2 exchange at 30°C in DD and normal air, not investigated by Carter *et al.* (1995b), is that the rhythm is caused by rhythms of CO_2 fixation by Rubisco, not by rhythms in PEPc kinase expression and activity driving rhythmic PEPc activity. In DD and normal air at 30°C, the rhythm reaches its trough (corresponding to maximum CO_2 fixation by the leaf) much earlier than in DD and CO_2-free air at 15°C (Section 9.2.2). This phase is more consistent with the activity of a carboxylating enzyme that becomes active earlier in the day, such as Rubisco, rather than with CO_2 fixation by PEPc. At 30°C PEPc kinase rhythms were lost, so PEPc could only be activated by malate removal to the vacuole (Carter *et al.*, 1995a). It has been proposed recently that the CAM rhythm in LL normal air switches rapidly from a PEPc rhythm to a Rubisco rhythm in *M. crystallinum* and *K. daigremontiana* (Dodd *et al.*, 2003; Wyka & Lüttge, 2003; Wyka *et al.*, 2004).

9.4.4   The contribution of PEPc and Rubisco to CAM CO_2 fixation rhythms

Earlier work supported a significant role for PEPc in the rhythm of CO_2 assimilation in constant light and normal air in both *K. daigremontiana* and *K. tubiflora* (Grams *et al.*, 1997; Ritz & Kluge, 1987). Furthermore, both titratable acidity (indicative of whole leaf malic acid levels) and malate levels, determined by enzymatic assay, oscillated with low amplitude in LL in CAM-induced *M. crystallinum* leaves (Dodd *et al.*, 2003). We have determined that these malate oscillations are robust over at least 4 days with peaks occurring at the end of the subjective night (J.M. Foster, S.F. Boxall & J. Hartwell, unpublished). This is consistent with fixation of CO_2 by PEPc during the subjective dark period in LL conditions. It seems most likely that a combination of Rubisco and PEPc activity is required to account for CO_2 assimilation rhythms in CAM leaves in LL. In LL, Rubisco remained active during the subjective night and probably accounted for the majority of the amplitude of the CO_2 rhythm (Wyka & Lüttge, 2003). Oscillations in the total leaf malate level and online $^{13}CO_2$ discrimination in LL revealed that PEPc also became active during the subjective night and contributed to nocturnal CO_2 fixation (Dodd *et al.*, 2003; Grams *et al.*, 1997). Clearly further experimentation is required to establish the relative contributions of PEPc and Rubisco to the CO_2 exchange rhythms performed by CAM leaves under a range of free-running conditions.

9.4.5   *The influence of temperature on rhythms of PEPc activity*

Decreasing temperature increases the apparent K_i of PEPc for malate irrespective of whether or not it is phosphorylated. At $3°C$ the apparent K_i of dephosphorylated PEPc for malate is 9 mM whereas the K_i of the phosphorylated form is 20 mM. These values are markedly higher than the values of 0.3 mM and 3.0 mM respectively obtained at $25°C$ (Carter *et al.*, 1995a). The catalytic activity of PEPc is greatly reduced at $3°C$. However, the increase in the K_i of PEPc for malate at lower temperatures would render PEPc physiologically active and may therefore compensate for this *in vivo*.

This finding is supported by measurements of the malate concentration in leaves of *K. fedtschenkoi* across a range of temperatures (Carter *et al.*, 1995b). Between 10 and $25°C$ leaf malate concentrations were around 50 mM after 72 h in the dark, with a supply of normal air (Carter *et al.*, 1995b). At $3°C$ and $0°C$ the malate concentration was much higher, reaching up to 90 mM (Carter *et al.*, 1995b). This supported the finding that the PEPc in leaves held at $0°C$ or $3°C$ was stabilized in the phosphorylated form and achieved a K_i of around 20 mM (Carter *et al.*, 1995a,b). *K. fedtschenkoi* is endemic to Madagascar and in this natural environment it is subjected to low nocturnal temperatures that promote maximal fixation of CO_2 into malate throughout the night. This allows the plant to make optimum use of the high concentration of CO_2 that occurs due to nocturnal stomatal opening. Increasing temperature has the reverse effect and thus helps prevent any futile recycling of CO_2 by PEPc in the day (Carter *et al.*, 1995b).

In C_4 plants a number of possible links in the light-induced signal transduction cascade that regulates PEPc kinase activity have been suggested (Vidal & Chollet, 1997). However, in *Kalanchoë* the signal transduction process is less well understood. Nimmo and colleagues speculated that, given the effects of RNA and protein synthesis inhibitors, the kinase itself could be synthesized or degraded in response to a circadian oscillator (Nimmo, 1993). However, the inhibitors used were not specific to PEPc kinase regulation (i.e. they inhibit all protein and RNA synthesis).

Therefore, it was equally feasible that some other protein or proteins were required to activate and/or inactivate the kinase.

The determination of which of these various possibilities explained the role of transcription and translation in the circadian control of the level of PEPc kinase activity in a leaf required several breakthroughs. First, Hartwell *et al.* (1996) developed a novel assay for measuring the level of PEPc kinase translatable mRNA in isolated samples of leaf RNA. This entailed *in vitro* translation of isolated RNA samples using rabbit reticulocyte lysate followed by direct assay of PEPc kinase activity in the translation products. This provided a means to study the level of PEPc kinase translatable RNA in CAM leaves under a variety of conditions in a manner analogous to using Northern blotting to monitor the steady-state level of a transcript. Using this assay, it was determined that PEPc kinase translatable RNA level was under diurnal and circadian control in *K. fedtschenkoi* leaves with its peak in the middle of the subjective night in DD, CO_2-free air, $15°C$ (Hartwell *et al.*, 1996).

The circadian increase in PEPc kinase translatable RNA required protein and RNA synthesis, indicating that two protein synthesis steps were required for the appearance of PEPc kinase. The first was the translation of PEPc kinase itself and the other was part of the chain of events leading to the appearance of kinase translatable mRNA, for example, the operation of the circadian clock.

9.4.6 Cloning the PEPc kinase gene from CAM plants

PEPc kinase is an extremely low abundance protein and this thwarted efforts to purify it to homogeneity. The protein still has never been purified from a CAM species. After more than 10 years of effort, PEPc kinase was eventually cloned by two groups in quick succession. Both groups used ingenious approaches to clone the gene from a CAM species.

Nimmo and co-workers, who had first identified PEPc kinase activity in desalted extracts of CAM leaves (Carter *et al.*, 1991), were also the first group to clone a PEPc kinase gene (Hartwell *et al.*, 1999). Hartwell *et al.* (1999) used an *in vitro* translation assay, developed for measuring the PEPc kinase translatable mRNA level in leaf RNA samples, to identify *in vitro* transcribed RNAs, generated from pools of cDNA clones, that directed the synthesis of active PEPc kinase. By screening pools and sub-pools of a *K. fedtschenkoi* dark leaf cDNA library, they were able to identify a full-length cDNA encoding a PEPc kinase, *PPCK*. Subsequent characterization of the regulation of this gene confirmed that its transcript was under circadian control in *K. fedtschenkoi*, and changes in the transcript level accounted for the changes in the activity of PEPc kinase and the K_i of PEPc for malate (Hartwell *et al.*, 1999).

Meanwhile, a second group used a protein kinase targeted differential display RT-PCR approach to identify protein kinase transcripts that were more abundant during the night in CAM-induced *Mesembryanthemum crystallinum* plants (Taybi *et al.*, 2000). *M. crystallinum* is an inducible CAM plant and CAM is routinely induced using 0.5 M NaCl. This allowed Taybi *et al.* (2000) to find protein kinase RT-PCR fragments whose transcripts were only detectable in salt-induced CAM leaves in the dark.

Both groups demonstrated that *PPCK* transcript levels oscillated in constant conditions. Hartwell *et al.* (1999) reported *K. fedtschenkoi PPCK* transcripts cycling in constant dark in CO_2-free air at 15°C whilst Taybi *et al.* (2000) demonstrated that, in normal air, *M. crystallinum PPCK* cycles in LL but not in DD.

9.4.7   Regulation of PPCK

The circadian control of CAM *PPCK* transcript levels can be overridden by experimental perturbations that influence the subcellular localization of malate (Borland *et al.*, 1999). Borland *et al.* (1999) enclosed *Kalanchoë daigremontiana* leaves in nitrogen at night to prevent nocturnal malate accumulation. When these leaves were released into normal air at dawn, *PPCK* transcript levels remained high for a prolonged period relative to the control. They proposed that, since the leaves had not been able to fix CO_2 and accumulate malate, *PPCK* transcript levels did not decline after dawn because cytosolic malate levels did not increase. In the control leaves, a high concentration of malate accumulated in the vacuole throughout the night and moved out into the cytosol at dawn. It was hypothesized that this malate, or a related metabolite, signaled the destruction of *PPCK* transcripts and activity (Borland *et al.*, 1999).

The malate-overrides-the-clock theory was supported by experiments in which the temperature was increased in the middle of the night. A nocturnal temperature increase in CAM leaves is believed to cause malate to equilibrate between the vacuole and cytosol, although this has proved difficult to confirm experimentally (Hartwell *et al.*, 2002). The increased cytosolic concentration of malate would inhibit PEPc and thus decrease CO_2 fixation. A high cytosolic concentration of malate is also believed to play a signaling role and override the circadian control of *PPCK*, causing its premature destruction. Increasing the temperature of *K. fedtschenkoi* leaves from 15°C to 30°C for 2 h in the middle of the night caused the disappearance of PPCK activity and transcripts and premature dephosphorylation of PEPc (Hartwell *et al.*, 1996). Decreasing the temperature to 4°C for 6 h stabilized PPCK activity and transcripts and maintained PEPc in the phosphorylated state with a high K_i for malate at a time when circadian control normally signals the disappearance of PPCK (Hartwell *et al.*, 1996). The mechanism of malate signaling to the nucleus to override circadian control and switch off *PPCK* transcription is currently unknown.

It should also be noted that this paradigm may not hold true in all CAM species. In particular, we have recently found that in CAM leaves of *M. crystallinum* a nocturnal temperature increase from 23°C to 30°C for 4 h applied in the second half of the dark period or in the second half of a subjective dark period in LL did not cause the premature disappearance of *PPCK* transcripts or override their circadian control. This was true even though the malate level was high in these leaves (S.F. Boxall, J.M. Foster & J. Hartwell, unpublished). Clearly, we cannot generalize on the ways in which CAM is coupled to the circadian clock via *PPCK* regulation based on studies of only two relatively distantly related CAM species.

Cytological studies with inhibitors of common signal transduction steps have shed some light on the signal transduction pathway that carries output signals from a central oscillator and leads to the nocturnal increase in PPCK activity in CAM leaves. When leaves of *K. fedtschenkoi* were detached at dusk after an LD 8:16 cycle and placed with their petiole in a solution of the calmodulin inhibitor W7 or the protein phosphatase inhibitor cantharidin, *PPCK* transcripts and activity did not increase in the middle of the subsequent night (Hartwell *et al.*, 2002; Nimmo *et al.*, 2001). However, the broad spectrum protein kinase inhibitor staurosporin and the calcium channel agonist Bay K8644 had no detectable effect at the concentrations tested. These two inhibitors slightly increased the nocturnal level of PPCK and the K_i of PEPc for malate relative to the control leaves (J. Hartwell & H.G. Nimmo, unpublished). Thus, in *K. fedtschenkoi* leaves, the circadian control of the nocturnal increase in *PPCK* transcripts and activity requires a calmodulin step, possibly in the form of a calcium-dependent protein kinase (CDPK), and a protein dephosphorylation step.

In CAM-induced leaves of *M. crystallinum*, inhibitor feeding experiments with detached leaves also implicated a calcium-calmodulin interaction in the signal transduction cascade leading to the circadian-controlled, nocturnal increase in PPCK activity (Bakrim *et al.*, 2001). Bakrim *et al.* (2001) also found evidence for the involvement of a phosphoinositide-dependent phospholipase C and inositol 1,4,5 phosphate-gated tonoplast calcium channels. They proposed that the increase in cytosolic calcium could activate a Ca^{2+}-dependent protein, possibly a CDPK. This signal transduction pathway is similar to the C_4 *PPCK* signaling pathway that is activated by light rather than circadian control. In C_4 leaves, photosynthesis may mediate an increase in cytosolic pH and in C_4 mesophyll protoplasts this was mimicked by supplying NH_4Cl (Giglioli-Guivarc'h *et al.*, 1996). No such alkalinization of the cytosol has been demonstrated as an early step in the nocturnal PPCK signaling cascade in CAM leaves and protoplasts (Bakrim *et al.*, 2001; Paterson & Nimmo, 2000).

9.5   Multiple points of clock control within the CAM pathway?

Affymetrix gene chip microarray transcriptomic studies on *Arabidopsis* reveal that some 6% of genes are under circadian control at the level of oscillations in the corresponding transcript abundance (Harmer *et al.*, 2000). These genes include those encoding enzymes responsible for multiple steps in important metabolic pathways such as phenylpropanoid biosynthesis, sugar and starch metabolism, nitrogen assimilation and sulphur assimilation (Harmer *et al.*, 2000). In the CAM circadian system, only a small number of clock-controlled genes have been reported in the literature. Oscillations of *PPCK* transcript and activity and the resulting oscillation in the phosphorylation state of PEPc were proposed to drive the rhythm of CO_2 fixation (Hartwell *et al.*, 1999). Importantly, circadian control of CAM CO_2 fixation is believed to be mediated by circadian control of *PPCK*.

More recently, transcript cycling of a cytosolic starch phosphorylase gene and a β-amylase gene that was not predicted to target to the chloroplast was reported in C_3 and CAM leaves of *M. crystallinum* (Dodd *et al.*, 2003). However, it is difficult to place these two genes accurately within the CAM circadian system until we understand the functions of cytosolic starch phosphorylase and of the numerous β-amylase genes that are not predicted to encode chloroplast-targeted proteins.

My group has recently carried out a much more extensive transcriptomics style analysis of clock-controlled genes in *M. crystallinum* leaves utilizing C_3 or CAM pathways using a high-throughput semi-quantitative RT-PCR screening approach. We have surveyed the transcript abundance profiles of 180 genes in both LD cycles and LL ($550\ \mu M/m^2/s$) at $23^{\circ}C$. The genes we have examined include all known steps of the CAM pathway, genes encoding cytosolic and chloroplast-targeted glycolytic enzymes, starch metabolism genes and photorespiratory pathway genes as well as over 70 uncharacterised transcription factors, protein kinases and transporter proteins.

This analysis has revealed that, in this species, the majority of CAM-associated genes are clock-controlled at the level of their transcript abundance including major metabolic enzymes such as PEPc, malate dehydrogenase, malic enzyme and enolase (S.F. Boxall & J. Hartwell, unpublished). Perhaps more intriguingly, many of these CAM-associated, clock-controlled genes are only coupled to the circadian clock in CAM induced leaves. This indicates that the connection by which CAM is regulated by the clock is made at or following CAM induction and is absent in C_3 leaves. This is one of the clearest examples of a plastic coupling mechanism linking the plant circadian clock to one of its outputs.

Another striking outcome of this detailed transcriptomic analysis of CAM rhythms is that several clock-controlled genes show a marked phase shift in the timing of their expression peak when *M. crystallinum* leaves switch from C_3 to CAM. Both malic enzyme (malate oxidoreductase, *MOD1*) and the gene that encodes the major chloroplastic β-amylase show a 6 h phase advance in their transcript peak when CAM-induced leaves are compared to C_3 leaves (S.F. Boxall & J. Hartwell, unpublished). This indicates that the switch to CAM requires earlier activation of both daytime malate decarboxylation and nocturnal starch degradation. Clearly an important goal for the future is to understand how the expression of these genes is phase advanced coincident with the induction of CAM whilst other genes maintain their C_3 phase.

9.6   The molecular identity of the central oscillator in CAM species?

The detailed understanding of the biochemical basis for the circadian control of CO_2 rhythms in CAM plants raises a fundamental question: what is the identity of the underlying central oscillator that coordinates the daily (24 h) CAM cycle and drives rhythms in PEPc activity? Many of the experimental perturbations of the CAM circadian rhythm, including light and temperature treatments, can be interpreted in terms of their influence on the subcellular localization of malate. For example, light and high temperatures cause malate to equilibrate between the vacuole and cytosol whilst low temperatures maintain high vacuolar malate concentrations and significantly increase the K_i of PEPc for malate regardless of its phosphorylation state (Carter et al., 1995a).

9.6.1   The tonoplast-as-oscillator?

This recurrent theme has led a number of authors to propose that there is an underlying oscillator based on the movement of malate into and out of the vacuole and the properties of the tonoplast membrane (Wilkins, 1983; Lüttge, 2000). Cytosolic malate feeds back to inhibit PEPc and thus leads to a decrease in CO_2 fixation. Light and high temperature cause phase delays when applied in the first half of the subjective dark period because they both cause malate efflux from the vacuole into the cytosol resulting in PEPc inhibition. After the perturbation, the accumulation of malate in the vacuole must start afresh and thus a delay occurs. Conversely, light and high temperature applied in the second half of the subjective dark period cause phase advances because they advance the oscillator to the point where malate has left the vacuole and equilibrated with the cytosol. This state normally occurs at dawn.

Fundamental to this 'tonoplast-as-oscillator' hypothesis is a hysteresis switch controlling whether malate is actively pumped into the vacuole or passively equilibrates between the cytosol and chloroplast. This metabolic or biophysical oscillator hypothesis has at its core the fact that the flux of malate into the vacuole is an active, energy-dependent process that occurs via a voltage-gated anion channel (Hafke et al., 2003). This channel is energized by the vacuolar H^+ pumps (V-ATPase and V-PP$_i$ase) that transport protons into the vacuole and establish a proton electrochemical potential difference across the tonoplast. Malate efflux is proposed to occur via passive diffusion that is favoured by changes in the fluidity of the tonoplast membrane as temperature increases at the beginning of the light period. A thermodynamic model was developed for the dependency of the lipid-matrix order on vacuolar volume and temperature (Neff et al., 1998). This model predicted that at intermediate temperatures the tonoplast can flip between malate influx and efflux and that this switch could be under rhythmic control. Integration of this switch into a computer-based autonomous model of CAM generated a model that exhibited robust oscillations at intermediate temperatures but stopped with a vacuole filled with or emptied of malate at low or high temperatures, respectively (Blasius et al., 1999).

The Lüttge group that generated the computer model of the tonoplast-as-oscillator theory recently tested their model against experimental manipulations that prevented nocturnal CO_2 fixation and malate accumulation. They achieved this by providing the leaves with CO_2-free air or encapsulating them in a nitrogen atmosphere for three nights prior to the onset of continuous illumination (Wyka et al., 2004). Whilst the computer model predicted a phase delay in rhythms of CO_2 uptake, stomatal conductance and leaf-internal CO_2 partial pressure in response to supplying leaves with CO_2-free air, the gas-exchange experiments on *Kalanchoë daigremontiana* leaves did not reveal significant phase shifts. The experiments enclosing leaves in an atmosphere of nitrogen at $33°C$ on three consecutive nights revealed that preventing CO_2-fixation and respiration for three nights did not greatly perturb the phase of rhythms of CO_2-fixation and transpiration rate upon release into LL, $24°C$ (Wyka *et al.*, 2004). Rather than a phase delay as predicted by the computer model, the experimental data revealed a 4 h phase advance. The authors concluded that their experimental data are consistent with the fact that nocturnal malate accumulation during the dark period is not providing phase information to drive self-sustaining oscillations in LL (Wyka *et al.*, 2004). Wyka *et al.* (2004) suggested that the present model needs to be extended to incorporate other factors and discussed the possibility that the entire CAM carboxylation system may be responding to a molecular-genetic clock or clocks similar to the consensus oscillator in *Arabidopsis,* rather than a biophysical oscillator centred on the tonoplast. Thus, the tonoplast-as-oscillator theory for the circadian control of CAM cannot account for all of the experimental data suggesting that a clock similar to the well-known molecular clock that operates in the nucleus of every plant cell may provide the temporal signals that coordinate the CAM cycle.

9.6.2   Is the CCA1/LHY-TOC1 oscillator controlling CAM rhythmicity?

This raises the question of whether the plant molecular oscillator as defined in *Arabidopsis thaliana,* driven by a transcription-translation negative feedback loop consisting of *CCA1/LHY* and *TOC1*, is present in CAM species. Furthermore, if CAM species possess a *CCA1/LHY-TOC1* type oscillator, does it operate in a similar way to the *Arabidopsis* oscillator and does it provide the temporal signals that coordinate the daily CAM cycle and thereby prevent a futile cycle of simultaneous malate synthesis and degradation? These questions have recently been addressed directly by Boxall *et al.* (2005) using the inducible CAM halophyte *Mesembryanthemum crystallinum.*

   M. crystallinum leaves perform C_3 photosynthesis when the plant is well watered but undergo a transition to CAM metabolism following salt, drought or high light stress. CAM is induced by watering plants with 0.5 M NaCl. Boxall *et al.* (2005) succeeded in cloning orthologues of seven *Arabidopsis* clock-associated genes from *M. crystallinum* namely, *McCCA1/LHY, McTOC1, McELF4, McELF3, McZTL, McFKF1* and *McGI*. A key difference between the regulation of these genes in *Arabidopsis* and their regulation in *M. crystallinum* is that the transcript abundance of *McZTL* is under circadian control in *M. crystallinum* whilst in *Arabidopsis* ZTL is only under circadian control at the level of the turnover of the encoded protein (Kim *et al.*, 2003b); *ZTL* transcript levels do not oscillate in LD or LL (Somers *et al.*, 2000). Furthermore, *McZTL* oscillations are more robust in CAM-induced leaves of *M. crystallinum* (Boxall *et al.*, 2005).

   The remainder of the clock-associated genes identified in *M. crystallinum* display robust oscillations of transcript levels that are largely unaffected by salt-stress or

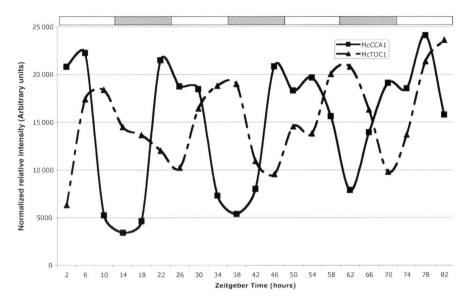

Figure 9.4 Reciprocal regulation of central clock genes in the CAM plant *Mesembryanthemum crystallinum*. *McCCA1/LHY* and *McTOC1* transcript levels were determined by semi-quantitative RT-PCR and normalized to the abundance of transcripts encoding a polyubiquitin gene that was used as a loading control. The RNA samples were isolated from CAM-induced *M. crystallinum* leaves that had been entrained in 12:12 LD and then released into LL 23°C at the start of the experiment. An almost identical trend of both genes was seen in a full biological replica experiment. ZT 12 is dusk.

development (Boxall *et al.*, 2005). These transcript oscillations peak at the same times as those of the orthologous genes in *Arabidopsis* and, importantly, *McCCA1/LHY* and *McTOC1* display peaks in antiphase consistent with the reciprocal regulation that occurs between these two genes in *Arabidopsis* (Fig. 9.4) (Alabadi *et al.*, 2001). Hence, at least one CAM species possesses a molecular clock very similar to the clock in *Arabidopsis* that could provide the temporal information that synchronizes CAM CO_2-fixation rhythms without requiring a specialized or novel oscillator.

Although the oscillations of *McCCA1/LHY* and *McTOC1* transcript levels are reciprocated on a gross scale (i.e. *McTOC1* peaks occur during *McCCA1/LHY* troughs and vice versa), it is clear that *McTOC1* transcript levels increase to a significant percentage of their peak level at a time when *McCCA1/LHY* levels are still very high (compare *McCCA1/LHY* and *McTOC1* levels at ZT 6 in Fig. 9.4). If McCCA1/LHY protein levels track *McCCA1/LHY* transcript levels with a lag of approximately 2 h (Kim *et al.*, 2003a) then it is difficult to reconcile the known direct repression of the *TOC1* promoter by CCA1 and LHY in *Arabidopsis* with these *M. crystallinum* expression patterns. We have determined that the *McTOC1* gene has two circadian regulated evening elements (Harmer *et al.*, 2000) in its promoter starting at −311 and −238 (J. Hartwell, unpublished). It is has been shown that in *Arabidopsis CCA1* and *LHY* can bind to the single evening element (EE) of the *TOC1* promoter and this is proposed to repress *TOC1* expression (Alabadi *et al.*, 2001). Other factors must mediate a lag between *McCCA1/LHY* transcript levels and *McTOC1* transcript levels. For example, perhaps *McCCA1/LHY* transcript levels are still high at ZT 6 but the protein level has already declined or the protein has been dephosphorylated and inactivated (Daniel *et al.*, 2004). Another notable anomaly in the timing of the reciprocating patterns of *McCCA1/LHY* and *McTOC1* is the lag between increasing levels of *McTOC1* and the beginning of the increase in *McCCA1/LHY* levels. If *McTOC1* acts as a positive element in the expression of *McCCA1/LHY* then it difficult to understand why *McCCA1/LHY* levels do not begin to rise much earlier. There must either be a significant lag in the translation of McTOC1 of at least 8 h following the increase in *McTOC1* transcripts or other proteins must be involved in mediating the lag between rising *McTOC1* levels and the induction of *McCCA1/LHY*. One possible protein that might mediate the lag is ZTL which targets TOC1 for degradation in *Arabidopsis* (Más *et al.*, 2003b). However, *McZTL* transcript levels track those of *McTOC1* with a transcript peak at ZT10 (S.F. Boxall & J. Hartwell, unpublished). Assuming it takes a similar amount of time for both *McZTL* and *McTOC1* transcripts to be translated, McZTL will be present in the cell at the same time as McTOC1 and could therefore target it to the proteasome from the moment it is synthesized, as occurs in *Arabidopsis* (Mas *et al.*, 2003b). This suggests that McZTL may be subject to post-translational activation, perhaps via its LOV domains. However, it is difficult to reconcile the proposed activation of the LOV domains by blue light with the expression of the gene and translation of the protein during the dark period. In *Arabidopsis,* ELF4, GI and ELF3 have been suggested as proteins that may act in concert with TOC1 as positive elements in the induction of CCA1 and LHY (Eriksson & Millar, 2003).

The discovery of all of the components of the molecular clock in a CAM species then leads us on to whether this clock underlies CAM rhythms. For example, are the robust oscillations in a plethora of CAM-associated genes, PEPc activity and CO_2-fixation that are occurring in CAM species driven by the *CCA1/LHY-TOC1* oscillator?

There are several possible experimental approaches to address this question. First, knockout mutants in either *TOC1* or *CCA1/LHY* in a CAM species would allow this question to be addressed. Mutant populations of *M. crystallinum* exist but they have not been screened to identify clock mutants and in the absence of circadian promoter::luciferase transgenic reporter lines, a forward-genetic screen would be a major challenge. Second, both overexpression and RNAi approaches could be used in transgenic *M. crystallinum* plants to manipulate the operation of the central clock.

Despite extensive recent efforts in our laboratory and in a number of others over many years, there are no reports of stable transformation and regeneration of *M. crystallinum* (J.M. Foster & J. Hartwell, unpublished). However, several groups have successfully transformed *Kalanchoë* species (Jia *et al.*, 1989; Aida & Shibata, 1996; Truesdale *et al.*, 1999). My laboratory has transformed *K. fedtschenkoi* with *McCCA1/LHY* and *McTOC1* under the control of the constitutive 35S *CaMV* promoter to achieve high-level constitutive expression (J.M. Foster & J. Hartwell, unpublished). Overexpression of *CCA1* causes arrhythmia in the central clock and a wide range of clock-controlled outputs in *Arabidopsis* (Wang & Tobin, 1998). Likewise, overexpression of *TOC1* causes arrhythmia of *CAB2::LUC* transcription in constant white, red and blue light (Más *et al.*, 2003a). The *K. fedtschenkoi* transgenic lines will allow us to test the role of the central clock in the circadian control of CAM. We are also in the process of cloning fragments of the *K. fedtschenkoi CCA1/LHY* and *TOC1* genes in order to generate RNAi transgenic lines with reduced expression of the endogenous clock genes (S.F. Boxall & J. Hartwell, unpublished). *K. fedtschenkoi* RNAi lines with efficient gene silencing should be equivalent to the *cca1*, *lhy* and *toc1* knockout mutants that have been characterized in the C_3 plant *Arabidopsis*.

The third approach for testing the role of the central oscillator in the control of CAM involves phase shifting experiments. Phase shifting can provide a significant insight into whether two clock outputs are responding to the same central oscillator. We have performed high temperature phase shifting experiments on the LL rhythm in CAM-induced leaves of *M. crystallinum* grown at 23°C. Plants were transferred to LL after LD 12:12 entrainment and phase shifts were induced using a 4 h, 30°C temperature treatment in the second half of the first subjective night. We predicted that the plants will perceive this temperature pulse as an early dawn and the central clock will phase advance in response.

We examined circadian oscillations of a number of central clock genes, as markers of phase within the clock, and clock-controlled, CAM-associated genes, as markers of circadian control within the CAM pathway. Our results revealed a consistent phase advance in the rhythms of all genes following the temperature treatment (S.F. Boxall, J.M. Foster & J. Hartwell, unpublished).

We also examined rhythms in the total leaf malate concentration in the same leaf samples used for the transcript abundance analysis and we saw the same phase advance in malate oscillations. Malate oscillations track oscillations in CO_2-fixation by PEPc and malate decarboxylation by malic enzyme and are therefore a good marker for circadian-regulated CO_2-fixation via the CAM pathway. The malate oscillations in LL demonstrated that the rhythms of gene expression of the CAM-associated genes are driving circadian rhythms of CAM and these rhythms are coordinated with the operation of the central clock. This represents the first compelling evidence linking a *CCA1/LHY/TOC1* oscillator directly to the control of oscillations in the CAM pathway.

9.7   Future perspectives and unanswered questions

My laboratory's high temperature phase shift experiments suggest that the CCA1/LHY/TOC1 oscillator is providing the temporal information for the coordination of CO_2 fixation by the CAM pathway. However, we must await the detailed analysis of our transgenic lines constitutively overexpressing *CCA1/LHY* and *TOC1* to determine categorically whether or not CAM rhythms are arrhythmic in the absence of the *CCA1/LHY/TOC1* clock.

If CAM does continue to oscillate in the absence of a functional clock then a novel oscillator must be driving CAM. The best candidate for this would be the biophysical oscillator that has long been favoured by the CAM research community and which still has many attractive properties in terms of explaining the experimental data relative to the subcellular localization of malate. However, to account for all of the experimental data, this oscillator requires additional elements (Wyka *et al.*, 2004).

If CAM ceases to oscillate in transgenic lines with an arrhythmic *CCA1/LHY/TOC1* oscillator, then CAM rhythms will become the latest addition to the long list of plant clock outputs that have been shown to be responding to the central molecular clock in the nucleus. The next step will be to determine the components of the output pathway that links the central clock to the circadian control of CAM. Key challenges will include understanding the mechanisms that allow such remarkable plasticity in the coupling of CAM to the clock. We will need to reinterpret existing data in terms of influence on the *CCA1/LHY/TOC1* oscillator and the CAM output pathway rather than the subcellular localisation of malate. It will be important to discover how circadian control is overridden by manipulations that influence the subcellular localisation of malate. We need to understand the mechanism by which cytosolic malate, or a related metabolite, is sensed and the elements of the signal transduction pathway that transmits the metabolic signal to the nucleus to switch off circadian control of *PPCK*.

Another vital question concerns the mechanism by which the clock controls whether or not malate accumulates in the vacuole or equilibrates between the vacuole and cytosol. There is evidence that the transcript level of a V-ATPase subunit c gene oscillates in CAM-induced leaves of *M. crystallinum*, but these oscillations are not as robust as those of other CAM-associated genes (J. Hartwell, unpublished). We still do not know the identity of the gene that encodes the tonoplast voltage-gated anion channel responsible for the accumulation of malate in the vacuole during the night (Hafke *et al.*, 2003). Once this has been identified, it will be important to understand how the gene and its protein product are regulated, particularly whether they are subject to circadian control. It may be that this channel is constitutively present in the tonoplast membranes of CAM mesophyll cells and is simply activated at night by the development of a proton gradient across the tonoplast membrane due to V-ATPase activity. One valuable approach would be to manipulate the activity of the V-ATPase in transgenic CAM plants in order to gain a detailed understanding of the role of the vacuolar proton gradient in the control of malate accumulation in the vacuole, and thus the CAM pathway.

9.7.1   Starch synthesis rhythms in CAM plants

We also need to understand how the circadian clock controls starch synthesis and breakdown in CAM leaves. Starch breakdown provides the PEP required for PEPc

to fix CO_2 at night. We have determined that a number of genes encoding proteins involved in starch synthesis and breakdown are under circadian control in
CAM-induced leaves of *M. crystallinum* (S.F. Boxall & J. Hartwell, unpublished).
Furthermore, we have identified CAM-specific genes encoding cytosolic isoforms
of every glycolytic enzyme, and all of them come under circadian control to some
extent in CAM-induced leaves of *M. crystallinum* (S.F. Boxall & J. Hartwell, unpublished). It is now important that we determine whether these transcript oscillations
drive changes in the activities of the encoded enzymes and ultimately drive oscillations in starch synthesis and breakdown that persist in DD or LL. CAM places an
increased demand on the starch reserves of a photosynthetic cell because the cell
must use its starch reserves to supply PEP as the substrate for PEPc to fix CO_2,
in addition to fuelling normal nocturnal metabolism. In order to be energetically
capable of performing CAM, a leaf must achieve a pool of starch at the end of
each day that is sufficient to drive both processes. *M. crystallinum* plants are most
competent to induce CAM in response to stress once they have developed 4 to 5
pairs of large primary leaves (Adams *et al.*, 1998). In *K. fedtschenkoi*, CAM is under
developmental control and is fully expressed in the 4th to 5th leaf pair down the
stem from the shoot apex and in older leaves (Hartwell *et al.*, 1999). The youngest
pair of leaves developing at the shoot apex do not perform significant CAM and
are believed to perform C_3 photosynthesis. Thus, in these species that have evolved
CAM independently, there is circumstantial evidence that a leaf must develop into
a mature source leaf before it can perform efficient CAM. If mature CAM-induced
leaves of *M. crystallinum* are prevented from fixing exogenous CO_2 for the whole
of a dark and light period by supplying them with CO_2-free air, starch levels at the
end of the day are depleted by almost 50%, and this has a stoichiometric effect on
the amount of CO_2 that is fixed by PEPc when the leaves are released into normal air
(Dodd *et al.*, 2003). These results indicate that substrate supply limits the capacity
of a CAM leaf to fix CO_2 at night and that no amount of increased phosphorylation
of PEPc by PPCK can compensate for this.

    We also need a much more detailed understanding of the circadian control of
Rubisco in CAM leaves. It is clear that Rubisco makes a significant contribution
to the rhythm of CO_2 exchange in continuous light and normal air (Dodd *et al.*,
2003; Wyka & Lüttge, 2003; Wyka *et al.*, 2004). The contribution of Rubisco to the
rhythm in constant darkness and normal air at 25°C–30°C has not been examined,
presumably because Rubisco is assumed to be inactive in the dark. Perhaps Rubisco
does become active in DD if the temperature is permissive (e.g. 25°C to 30°C). We
know that Rubisco activase is under circadian control in CAM-induced leaves of *M.
crystallinum*, but this has not been tested in DD (J. Hartwell, unpublished).

9.7.2 Stomatal rhythms in CAM plants

A further important factor is the rhythm of stomatal conductance. In DD CO_2-
free air, *K. fedtschenkoi* leaves with their epidermis removed perform a rhythm of
CO_2 output at 25°C but not at 15°C (Wilkins, 1991). This indicates that isolated

mesophyll cells perform a rhythm of CO_2 output in DD CO_2-free air at 25°C that is not dependent on the stomata. This confirms that the CAM rhythm of CO_2-output driven by circadian control of PEPc activity in mesophyll cells can be regarded as a self-sufficient rhythm that does not depend on stomatal rhythms. However, without their epidermis, leaves do not perform a CO_2 rhythm in constant light and normal air at either 15°C or 25°C (Wilkins, 1992). This suggests that stomata are essential for the rhythm in LL and raises the possibility that the observed rhythm is a result of the combined action of the stomatal rhythm, the Rubisco rhythm and the PEPc rhythm. Future work in this area should aim to dissect the relative contribution of these three rhythms to the observed rhythm of CO_2-exchange. We already have the molecular tools to inhibit Rubisco rhythms by using RNAi constructs to silence endogenous Rubisco activase or to inhibit PEPc rhythms by using RNAi to target *PPCK*. Our limited understanding of how stomatal rhythms are generated means that we are not in a position to knock out the stomatal rhythm, although targeted RNAi of *PPCK* using a guard cell specific promoter may block stomatal rhythms by preventing a rhythm of PEPc activity in guard cells. PEPc generates malate as a counter anion during stomatal opening. Furthermore, by knocking out both the Rubisco and PEPc rhythm we could determine the relative importance of the stomatal rhythm. It may be that Rubisco and PEPc drive stomatal rhythms by altering the internal partial pressure of CO_2. Stomata are believed to sense CO_2 and open in response to low internal partial pressures of CO_2 and close in response to high internal partial pressures. This rationale has long been used to explain stomatal closure during the day in CAM species because the decarboxylation of malate generates very high concentrations of CO_2 inside the leaf and this is believed to signal stomatal closure. We must develop a much better understanding of stomatal control in CAM species before we can fully understand the basis of the rhythms in LL.

9.7.3   Multiple origins of CAM

One of the most important and exciting areas for future research on the CAM circadian system is the evolutionary importance of circadian control in CAM species. CAM is found in at least 328 genera in 33 families suggesting that CAM may have evolved up to 33 times (Smith & Winter, 1996). However, this is most probably a gross underestimate because CAM is known to have evolved at least three times in the Bromeliaceae (Crayn *et al.*, 2004). It is widely accepted that C_3 plants possess all of the enzymatic machinery required for CAM and that one of the key requirements for CAM evolution is a change in the temporal control of a handful of important enzymes. It has been proposed that this occurs via gene duplication events, which generate new genes that evolve into CAM-specific isoforms. The classic example of this phenomenon is the CAM-specific PEPc isoform from *M. crystallinum* (Cushman & Bohnert, 1999). This suggests that plants can evolve CAM with relative ease in response to the correct set of driving forces which include low atmospheric CO_2 and arid environments.

Subsequent introduction of circadian controlled motifs into the promoters of the CAM-specific isoforms of these genes would permit circadian control of their transcript abundance. Adaptation of pre-existing mechanisms of post-translational regulation of certain enzymes (e.g. PEPc) could provide an additional means of fine-tuning circadian control. There are two *PPCK* genes in *Arabidopsis* and neither of them is under circadian control in leaves (Fontaine *et al.*, 2002). However, a clock controlled C_3 *PPCK* gene has recently been identified in leaves of soybean (*Glycine max*) (Fontaine *et al.*, 2002; Sullivan *et al.*, 2004). Interestingly, the transcript is not under clock control in roots even though its relative abundance is comparable to that in leaves (Sullivan *et al.*, 2004). The discovery of a circadian regulated *PPCK* gene in soybean reveals that C_3 progenitor species of CAM species may have already possessed circadian control of the protein kinase responsible for the circadian control of PEPc; it must be noted that CAM does not occur within the plant family Fabaceae. It is important to determine whether circadian control of *PPCK* occurs in C_3 species that are closely related to existing CAM species. A study of four species in the genus *Clusia* revealed that two C_3 species have relatively high and constitutive expression of *PPCK* throughout the light/dark cycle, whilst the constitutive CAM species *Clusia rosea* only expressed *PPCK* during the night (Taybi *et al.*, 2004). The drought-inducible CAM species *Clusia minor* accumulated *PPCK* transcripts at night both as a well-watered plant and following 5 days of drought stress. However, drought stress prolonged *PPCK* expression at night in *C. minor*. This correlated well with a sustained, high level of CO_2 fixation throughout the night and a higher concentration of malate at dawn (Taybi *et al.*, 2004). Taybi *et al.* (2004) proposed that, within the genus *Clusia*, C_3 species already possess high levels of *PPCK* transcripts throughout the 24 h cycle and therefore the CAM-specific expression of *PPCK* only requires the suppression of *PPCK* transcript levels during the day rather than the induction of the gene at night. These findings in C_3 and CAM *Clusia* species strengthen the argument that we need to study the circadian control of CAM in species that have evolved the pathway independently to understand whether or not circadian control is always required for CAM.

9.7.4 Insights from CAM plants

The study of the molecular and biochemical basis of the circadian control of CO_2 fixation in CAM species has led to a number of significant breakthroughs in our understanding of the regulation of plant metabolism. The regulatory phosphorylation of PEPc was first discovered in a CAM species (Nimmo *et al.*, 1984). The first *PPCK* genes were identified in CAM species by virtue of their high expression at night (Hartwell *et al.*, 1999; Taybi *et al.*, 2000).

The study of CAM has also led to the identification of a voltage-gated anion channel responsible for malate uptake into the vacuole (Hafke *et al.*, 2003). This gene has not yet been cloned from any species; the *PPCK* paradigm suggests it will be most straightforward to clone it from a CAM species because of the accentuated role it plays in CAM. Undoubtedly, this malate channel will also be important in

C_3 and C_4 vacuolar malate accumulation, particularly in stomatal guard cells. In fact, it has been proposed that CAM evolved from the pre-existing biochemistry in guard cells where PEPc produces malate to act as an counter-anion during opening (Teeri, 1982; Cockburn, 1983). For CAM to evolve from the metabolism in stomatal guard cells would simply require the inversion of the cycles of malate synthesis and decarboxylation so that malate is produced at night and decarboxylated during the day.

It is not an exaggeration to say that plant biology research is currently in an 'Arabicentric' era. Whilst many groups are beginning to capitalize on the remarkable breakthroughs made possible by the use of molecular-genetic approaches in *Arabidopsis* and move their discoveries into model crops such as rice, we must not lose sight of the great diversity of the plant kingdom and the novel biology waiting to be discovered. CAM plants have significantly improved water use efficiency relative to C_3 and C_4 plants and we overlook such valuable adaptations at our own peril, particularly in the face of current predictions concerning global warming. Whilst it may currently seem a fanciful proposal, scientific advances will undoubtedly permit inducible-CAM to be engineered into C_3 crop species so that our future crops can switch into CAM and survive short periods of drought or salt stress. Engineering inducible-CAM into crops would allow the plants to subsequently switch back into C_3 when rain returns. This ability to adapt could make the difference between failing crops and salvaged yield in the face of temporary drought conditions. If we are to place ourselves in a strong position to take advantage of this remarkable plant adaptation to drought, then we need to drastically improve our knowledge of the regulation of CAM and this will require a comprehensive understanding of the circadian control of the CAM pathway.

Acknowledgements

I gratefully acknowledge the many colleagues who have contributed to the ideas expressed above, particularly Hans Bohnert, Anne Borland, John Cushman and Hugh Nimmo. My research is supported by a David Phillips fellowship (grant no. JF14818) awarded to me by the Biotechnology and Biological Sciences Research Council, UK.

References

Adams, P., Nelson, D.E., Yamada, S., Chmara, W., Jensen, R.G., Bohnert, H.J. & Griffiths, H. (1998) *New Phytologist*, **138**, 171–190.
Aida, R. & Shibata, M. (1996) *Plant Science*, **121**, 175–185.
Alabadi, D., Oyama, T., Yanovsky, M.J., Harmon, F.G., Más, P. & Kay, S.A. (2001) *Science*, **293**, 880–883.
Anderson, C.M. & Wilikins, M.B. (1989) *Planta*, **177**, 401–408.
Anderson, C.M. & Wilkins, M.B. (1989a) *Planta*, **177**, 456–469.

Anderson, C.M. & Wilkins, M.B. (1989b) *Planta*, **180**, 61–73.

Bakrim, N., Brulfert, J., Vidal, J. & Chollet, R. (2001) *Biochem. Biophys. Res. Commun.*, **286**, 1158–1162.

Blasius, B., Neff, R., Beck, F. & Lüttge, U. (1999) *Proc. Roy. Soc. Lond. Ser. B*, **266**, 93–101.

Borland, A.M., Hartwell, J., Jenkins, G.I., Wilkins, M.B. & Nimmo, H.G. (1999) *Plant Physiol.*, **121**, 889–896.

Boxall, S.F., Foster, J.M., Bohnert, H.J., Cushman, J.C., Nimmo, H.G. & Hartwell, J. (2005) *Plant Physiol.*, **137**, 969–982.

Carter, P.J., Fewson, C.A., Nimmo, G.A., Nimmo, H.G. & Wilkins, M.B. (1996) In *Crassulacean Acid Metabolism. Biochemistry, Ecophysiology and Evolution.* Vol. 114 (eds K. Winter and J.A. Smith). Springer-Verlag, Berlin, pp. 46–52.

Carter, P.J., Nimmo, H.G., Fewson, C.A. & Wilkins, M.B. (1990) *FEBS Lett.*, **263**, 233–236.

Carter, P.J., Nimmo, H.G., Fewson, C.A. & Wilkins, M.B. (1991) *Eur. Mole. Biol. Org. J.*, **10**, 2063–2068.

Carter, P.J., Wilkins, M.B., Nimmo, H.G. & Fewson, C.A. (1995a) *Planta*, **196**, 375–380.

Carter, P.J., Wilkins, M.B., Nimmo, H.G. & Fewson, C.A. (1995b) *Planta*, **196**, 381–386.

Cockburn, W. (1983) *Plant Cell Environ.*, **6**, 275–279.

Cook, R.M., Lindsay, J.G., Wilkins, M.B. & Nimmo, H.G. (1995) *Plant Physiol.*, **109**, 1301–1307.

Crayn, D.M., Winter, K. & Smith, J.A.C. (2004) *Proc. Nat. Acad. Sci. USA*, **101**, 3703–3708.

Cushman, J.C. & Bohnert, H.J. (1999) *Ann. Rev. Plant Physiol. Plant Mol. Biol.*, **50**, 305–332.

Daniel, X., Sugano, S. & Tobin, E.M. (2004) *Proc. Nat. Acad. Sci. USA*, **101**, 3292–3297.

Dodd, A.N., Griffiths, H., Taybi, T., Cushman, J.C. & Borland, A.M. (2003) *Planta*, **216**, 789–797.

Eriksson, M.E. & Millar, A. (2003) *Plant Physiol.*, **132**, 732–738.

Fontaine, V., Hartwell, J., Jenkins, G.I. & Nimmo, H.G. (2002) *Plant Cell Environ.*, **25**, 115–122.

Giglioli-Guivarc'h, N., Pierre, J.-N., Brown, S., Chollet, R., Vidal, J. & Gadal, P. (1996) *Plant Cell*, **8**, 573–586.

Grams, T.E.E., Borland, A.M., Roberts, A., Griffiths, H., Beck, F. & Lüttge, U. (1997) *Plant Physiol.*, **113**, 1309–1317.

Grams, T.E.E., Kluge, M. & Lüttge, U. (1995) *J. Exp. Bot.*, **46**, 1927–1929.

Hafke, J.B., Hafke, Y., Smith, J.A.C., Lüttge, U.E. & Thiel, G. (2003) *Plant J.*, **35**, 116–128.

Harmer, S.L., Hogenesch, J.B., Straume, M., Chang, H.-S., Han, B., Zhu, T., Wang, X., Kreps, J.A. & Kay, S.A. (2000) *Science*, **290**, 2110–2113.

Harris, P.J.C. & Wilkins, M.B. (1976) *Planta*, **129**, 253–258.

Harris, P.J.C. & Wilkins, M.B. (1978a) *Planta*, **143**, 323–328.

Harris, P.J.C. & Wilkins, M.B. (1978b) *Planta*, **138**, 271–272.

Hartwell, J., Gill, A., Nimmo, G.A., Wilkins, M.B., Jenkins, G.I. & Nimmo, H.G. (1999) *Plant J.*, **20**, 333–342.

Hartwell, J., Nimmo, G.A., Jenkins, G.I., Wilkins, M.B. & Nimmo, H.G. (2002) *Funct. Plant Biol.*, **29**, 663–668.

Hartwell, J., Smith, L.H., Wilkins, M.B., Jenkins, G.I. & Nimmo, H.G. (1996) *Plant J.*, **10**, 1071–1078.

Jia, S., Yang, M., Ott, R. & Chua, N.H. (1989) *Plant Cell Rep.*, **8**, 336–340.

Kim, J.Y., Song, H.R., Taylor, B.L. & Carré, I.A. (2003a) *EMBO J.*, **22**, 935–944.

Kim, W.Y., Geng, R. and Somers, D.E. (2003b) *Proc. Nat. Acad. Sci. USA*, **100**, 4933–4938.

Lüttge, U. (2000) *Planta*, **211**, 761–769.

Lüttge, U. & Ball, E. (1978) *Z. Pflanzenphysiol. Bildung*, **90**, 69–77.

Más, P., Alabadi, D., Yanovsky, M.J., Oyama, T. & Kay, S.A. (2003a) *Plant Cell*, **15**, 223–236.

Más, P., Kim, W.Y., Somers, D.E. & Kay, S.A. (2003b) *Nature*, **426**, 567–570.

Neff, R., Blasius, B., Beck, F. & Lüttge, U. (1998) *J. Membrane Biol.*, **165**, 37–43.

Nimmo, G.A., Nimmo, H.G., Fewson, C.A. & Wilkins, M.B. (1984) *Federation of European Biochemical Societies Letters*, **178**, 199–203.

Nimmo, G.A., Nimmo, H.G., Hamilton, I.D., Fewson, C.A. & Wilkins, M.B. (1986) *Biochem. J.*, **239**, 213–220.

Nimmo, G.A., Wilkins, M.B., Fewson, C.A. & Nimmo, H.G. (1987) *Planta*, **170**, 408–415.

Nimmo, H.G. (1993) In *Post-translational modifications in plants*, Vol. Society for Experimental Biology Seminar Series: 53 (eds, Battey, N.H., Dickinson, H.G. & Hetherington, A.M.) Cambridge University Press, Cambridge, pp. 161–170.

Nimmo, H.G., Fontaine, V., Hartwell, J., Jenkins, G.I., Nimmo, G.A. & Wilkins, M.B. (2001) *New Phytologist*, **151**, 91–97.

Nuernbergk, E.L. (1961) *Planta*, **56**, 28–70.

Paterson, K.M. & Nimmo, H.G. (2000) *Plant Sci.*, **154**, 135–141.

Ritz, D. & Kluge, M. (1987) *J. Plant Physiol.*, **131**, 285–296.

Sage, R.F. (1999) In C_4 *plant biology* (Eds, Sage, R.F. & Monson, R.K.) Academic Press, San Diego, CA, USA, pp. 551–584.

Smith, J.A.C. & Winter, K. (1996) In *Crassulacean Acid Metabolism – Biochemistry, Ecophysiology and Evolution.*, Vol. 114 (Eds, Winter, K. & Smith, J.A.C.) Springer Verlag, Berlin Heidelberg, pp. 427–436.

Somers, D.E., Schultz, T.F., Milnamow, M. & Kay, S.A. (2000) *Cell*, **101**, 319–329.

Sullivan, S., Jenkins, G.I. & Nimmo, H.G. (2004) *Plant Physiol.*, **135**, 2078–2087.

Taybi, T., Nimmo, H.G. & Borland, A.M. (2004) *Plant Physiol.*, **135**, 587–598.

Taybi, T., Patil, S., Chollet, R. & Cushman, J.C. (2000) *Plant Physiol.*, **123**, 1471–1481.

Teeri, J.A. (1982) In *Crassulacean acid metabolism* (Eds, Ting, I.P. and Gibbs, M.) American Society of Plant Physiologists, Rockville, pp. 244–259.

Truesdale, M.R., Toldi, O. & Scott, P. (1999) *Plant Physiol.*, **121**, 957–964.

Vidal, J. & Chollet, R. (1997) *Trends Plant Sci.*, **2**, 230–237.

Wang, Z.Y. & Tobin, E.M. (1998) *Cell*, **93**, 1207–1217.

Warren, D.M. & Wilkins, M.B. (1961) *Nature*, **191**, 686–688.

Wilkins, M.B. (1959) *J. Exp. Bot.*, **10**, 377–390.

Wilkins, M.B. (1962) *Proc. Roy. Soc. Lond.*, **156**, 220–241.

Wilkins, M.B. (1983) *Planta*, **157**, 471–480.

Wilkins, M.B. (1984) *Planta*, **161**, 381–384.

Wilkins, M.B. (1989) *J. Exp. Bot.*, **40**, 1315–1320.

Wilkins, M.B. (1991) *Planta*, **185**, 425–431.

Wilkins, M.B. (1992) *New Phytologist*, **121**, 347–375.

Winter, K. (1982) *Planta*, **154,** 298–308.

Winter, K. & Smith, J.A.C. (1996) In *Crassulacean Acid Metabolism – Biochemistry, Ecophysiology and Evolution.*, Vol. 114 (eds, Winter, K. & Smith, J.A.C.) Springer Verlag, Berlin Heidelberg, pp. 1–13.

Wyka, T.P., Bohn, A., Duarte, H.M., Kaiser, F. & Lüttge, U.E. (2004) *Planta*, **219**, 705–713.

Wyka, T.P. & Lüttge, U.E. (2003) *J. Exp. Bot.*, **54**, 1471–1479.

10 Clock evolution and adaptation: whence and whither?

Carl Hirschie Johnson and Charalambos P. Kyriacou

10.1 Introductory quotation

Mechanism is not the biologist's only business. A biologist is, or should be, concerned also with questions of both function and history.
—Colin Pittendrigh, 1966

10.2 Setting the stage: the appearance of circadian clocks

Although the current volume is about endogenous rhythms in plants, this chapter on the evolution of the circadian clock will range beyond plants, because some of the key concepts are best illustrated from studies on other organisms. Nevertheless, we will use examples from photosynthetic organisms whenever possible. In fact, the earliest evolution of a circadian clock probably involved the photosynthetic cyanobacteria, the 'simplest' organisms that have been demonstrated to have ~24 h cycles. We know that cyanobacteria date back to the dawn of life, about 3.8 billion years ago. Even then the Earth was spinning on its axis, providing relentless cycles of light and dark. Because light is the energy that fuels life for photosynthetic organisms, these daily cycles would have imposed strong selection. Perhaps the first clocks were thus 'invented' during that golden age of cyanobacteria around three billion years ago? However, circadian clocks may have evolved much later, and so be a relatively recent acquisition, a possibility we shall discuss later in this chapter.

Indeed, conditions on earth have changed dramatically over the past 3.5 billion years, partly for geological/astrophysical reasons, and partly due to the atmospheric changes wrought by cyanobacteria. The daily cycle has not always been 24 h (or even close!). Fossils of corals that display annual cycles of daily rings in their exoskeletons demonstrate that as recently as the middle Devonian (about 375 million years ago) there were 400 days in the year (Wells, 1963). Since the time required for the Earth to make a complete annual cycle around the sun appears to have remained constant over the past 4 billion years (Laskar, 1999), the solar day must have been about 21.9 h in the middle Devonian in order to achieve a year of 400 days (Laskar, 1999). Therefore, the earth's rotation is slowing down. If we reach further back, the change in Earth's rotational velocity becomes astonishing (Tauber *et al.*, 2004). Due to tidal drag and changes in the distance between the Earth and its moon, Earth may have revolved with a period of only 4 h in the era when cyanobacteria evolved ! (Krasinsky, 2002; Lathe, 2004). Therefore, if cyanobacteria did develop an endogenous daily clock at the beginning of their evolution, it may have had a 4 h free-running period (FRP), which is much easier to imagine from a biochemical perspective than the current $\sim$24 h FRP. In fact, it has always been a bit of a puzzle to understand how the current model for circadian clocks – namely an autoregulatory transcriptional/translational feedback loop – could have evolved to generate a precise 24 h periodicity (see Chapter 2, this volume). Such a biochemical feedback loop could just as easily generate a much shorter FRP. This puzzle is perhaps partially explained if the early evolution of circadian clocks initially generated a $\sim$4 h FRP, the subsequent slowing of the Earth's spin could then have provided a selective pressure that gradually stretched the 4 h biochemical cycle to a 24 h FRP (Tauber *et al.* 2004). Although this may be an attractive hypothesis, the circadian clock in cyanobacteria was probably not the basis for circadian clocks in eukaryotes (see Section 10.5). If eukaryotes developed circadian clocks independently at a later date, then the daily cycles of sunrises and sunsets might have already been 16–20 h (Brosche *et al.*, 1989; Krasinsky, 2002; Lathe, 2004). In this case, the evolution of an autoregulatory transcriptional/translational feedback loop with such a long time constant still remains a mystery.

In addition to the speed of Earth's rotation, another set of relevant conditions has changed dramatically on Earth over the past 3.5 billion years due to a biotic factor, namely cyanobacteria and other photosynthetic bacteria. These bacteria are single-handedly responsible for the transfiguration of Earth's atmosphere from re-ducing (nitrogen, hydrogen, methane and carbon dioxide) to oxidizing by virtue of their emission of photosynthetically derived oxygen. This atmospheric change had at least three profound consequences. First, it allowed the evolution of aerobic metabolism, which can maintain higher metabolic rates and allowed larger cells and more complex organisms to evolve. Second, the oxygen formed ozone in the upper atmosphere. The ozone gradually formed a dense layer that acts as a shield against ultraviolet light (UV). Because UV damages DNA, the formation of the ozone shield probably reduced UV irradiation to levels that allowed organisms to leave the pro-tection of the ocean and other bodies of water to exploit terrestrial niches. Third, the oxygen killed off much of the anaerobic competition for the cyanobacteria.

10.3   Putative selective pressures

What were the selective forces that encouraged the original evolution of circadian timers? The first appearance of these timekeepers may have been intrinsically linked with photosynthesis and/or conditions on Earth before the introduction of oxygen. In many plants and algae, the circadian clock regulates photosynthetic capacity (but see below for the case of the cyanobacterium *Synechococcus elongatus*). An endogenous timer that turns on metabolic processes in anticipation of environmental changes (such as dawn for photosynthesis) would be expected to be adaptive. Although often mentioned in the context of clock evolution, the 'anticipation' hypothesis has never been experimentally tested, and – like many other evolutionary ideas that are eminently reasonable – remains just a good idea.

10.3.1 Temporal separation of metabolic events

Another possible selective pressure is that an endogenous timekeeper could have evolved to temporally separate mutually incompatible metabolic events to occur at different times. This temporal programming has been specifically proposed in the context of photosynthesis versus nitrogen fixation in cyanobacteria. Low levels of oxygen inhibit nitrogen fixation, which poses a dilemma for nitrogen-fixing photosynthetic bacteria because photosynthesis generates oxygen. In unicellular nitrogen-fixing bacteria, nitrogen fixation is often phased to occur at night. Mitsui and co-workers postulated that the nocturnal phasing of nitrogen fixation was an adaptation to permit N_2 fixation to occur when photosynthesis was not evolving oxygen (Mitsui *et al.*, 1986). This hypothesis predicts that cyanobacterial growth in constant light should be slower than in a light/dark cycle because nitrogen fixation would be inhibited under these conditions and therefore the growing cells might rapidly become starved for metabolically available nitrogen.

The problem with this hypothesis is that cyanobacteria grow perfectly well in constant light – in fact, they grow faster in constant light than in light/dark cycles, presumably because of the extra energy they derive from the added time they are able to photosynthesize. This result is inconsistent with the 'temporal separation' hypothesis, although it remains possible that under appropriate (but as yet untested) conditions, the growth characteristics of cyanobacteria will support the theory. Other cases of temporal separation regulated by the circadian clock in photosynthetic organisms are phototaxis versus chemotaxis in the eukaryotic alga *Chlamydomonas* (Byrne *et al.*, 1992) and various forms of catalase in the plant *Arabidopsis* (Zhong & McClung, 1996) (see also Chapter 10, this volume).

10.3.2 'Escape from light'

Another hypothetical selective pressure that does not exclude the two ideas above dates back 40 years ago to Colin Pittendrigh, who suggested the 'escape from light' hypothesis (Pittendrigh, 1965, 1993). This hypothesis postulates that a strong initial driving force for the early evolution of circadian clocks could have been the advantage inherent in phasing cellular events that are sensitive to the daily bombardment of sunlight to occur at night. Cells exhibit many light-sensitive processes. For example, there are many pigments in cells that probably do not act as sensory 'photoreceptors' but which nevertheless absorb light because of obligatorily associated cofactors (e.g. cytochromes), and whose activity is thereby modulated by light and dark. Moreover, DNA can be mutated by exposure to UV, and as described above, UV irradiation was more potent before the ozone shield formed and during the period of time when cyanobacteria were evolving. The genome may be more sensitive to UV when irradiation occurs at specific phases of the cell division

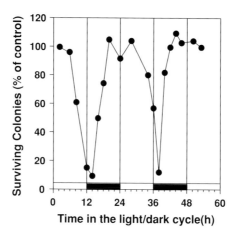

Figure 10.1 Survival of the eukaryotic alga *Chlamydomonas* after irradiation by ultraviolet light as a function of the time in a light/dark cycle. *Chlamydomonas* cultures were plated onto agar medium and treated with equal amounts of ultraviolet light at different phases of a 12 h light:12 h dark cycle. Survival was measured as the colony-forming ability of cells following treatment as compared to that of cells that were not irradiated with ultraviolet light (modified from Nikaido & Johnson, 2000).

cycle (e.g. during S phase when DNA is partially unwrapped from histones to allow replication). In fact, there are numerous examples of microorganisms with 24 h cell division cycles in which DNA replication and cell division occur during the night (Edmunds, 1988), which is consistent with predictions of the 'escape from light' hypothesis. Therefore, cellular metabolism will be sensitive to sunlight, especially to UV and high-energy blue light.

Predictions generated by the 'escape from light' hypothesis can be tested by examining whether vestiges of nocturnal programming of light-sensitive processes have been retained in present-day organisms. Such a test was performed in *Chlamydomonas* (Nikaido & Johnson, 2000). As shown in Fig. 10.1, these eukaryotic algae are more sensitive to UV light near sunset and into the early night. The rhythmic sensitivity persists in constant conditions, albeit with a reduced amplitude. Even though there is some sensitivity in the late daytime, UV light is strongly scattered at twilight and is therefore nearly absent from sunlight around the time of sunset (and sunrise), so the UV sensitivity of *Chlamydomonas* near dusk would not be expected to pose a significant problem in the natural environment. In *Chlamydomonas*, the circadian clock regulates the timing of the cell division cycle (Goto & Johnson, 1995), and the UV-sensitive phases correspond with the times in which S/G2 would be expected to occur. These data are consistent with the 'escape from light' hypothesis that the daily cycle of UV radiation may have created a selective pressure favouring the evolution of circadian clocks (Pittendrigh, 1993).

The hypothesis dovetails with the discovery of a role for cryptochromes in circadian systems. Cryptochromes are pigmented photoreceptors that are involved in blue-light mediated entrainment and photoperiodism. Cryptochromes share sequence homology to, and probably evolved from, DNA photolyase, which also uses blue light to repair UV-induced damage of DNA. Thus, an ancestral photolyase that repaired DNA damage inflicted by the daily cycle of UV light may have been enlisted for duties in biological timing mechanisms, and later evolved into cryptochrome (Nikaido & Johnson, 2000; Gehring & Rosbash, 2003). Indeed, based on the 'escape from light' hypothesis, a clock-related role for a DNA photolyase-type enzyme was anticipated and predicted by Pittendrigh (Pittendrigh, 1965).

10.4 Cyanobacteria and the first clock genes

Is the clockwork in cyanobacteria the ancestral clock for all subsequent circadian timekeepers? We can address this question by comparing the genes encoding circadian cogs in the various taxa. In *Synechococcus elongatus,* there is a single cluster of three *kai* genes (*kaiABC*, Ishiura *et al.*, 1998; Johnson, 2004), whereas in other species such as *Nostoc linckia* and *Synechocystis*, there appear to have been several duplications of this cluster (Dvornyk *et al.*, 2003). Bioinformatic analysis of 70 complete prokaryotic genomes revealed that most of the major Archaeal groupings contain *kaiC* sequences, but the gene is less well represented in eubacterial taxa (Dvornyk *et al.*, 2003). For example, *kaiC* homologs are found in halophilic, thermophilic, and methanogenic archaebacteria (Johnson & Golden 1999; Das-Sarma *et al.*, 2001; Dvornyk *et al.*, 2003). *kaiC* sequences are ubiquitous only in the cyanobacterial species and appear rather sporadically in the other kingdoms. Among the three kingdoms, cyanobacteria, proteobacteria and archaea, lateral gene transfers of *kaiC* were also evident, some of which may be quite recent.

The *kaiB* gene has a different pattern of distribution, and is found in a few Proteobacteria. For example, the purple photosynthetic proteobacteria *Rhodopseudomonas palustris* and *Rhodobacter sphaeroides* have both the *kaiB* and *kaiC* genes (Dvornyk *et al.*, 2003; Larimer *et al.*, 2004). There is intriguing evidence for circadian rhythms in *R. sphaeroides* under aerobic conditions (Dr. Hongtao Min, personal communication). Among the archaea, *kaiB* is found only in *Methanobacterium*. In contrast to the *kaiB* and *kaiC* genes, *kaiA* is exclusively cyanobacterial. Thus, we would imagine that *kaiC* is the oldest gene within the *kaiABC* cluster as it is the only one also found in all three prokaryotic kingdoms, and *kaiA* must be the youngest.

The *kaiC* gene in *Synechococcus* encodes a 'double-domain' structure, with each domain containing a Walker A ATP-binding site (Pattanayek *et al.*, 2004), whereas most of the Archaeal homologues have a single domain (Leipe *et al.*, 2000; Ditty *et al.*, 2003; Golden and Canales, 2003). This *kaiC* duplication event presumably occurred in the last common ancestors of the Bacteria and Archaea, which would be between 3.8 and 3.5 Ga (billion years ago) according to the fossil record. The *kaiB* gene was next, evolving about 2.3 Ga, with *kaiA* clocking in at 1 Ga (Dvornyk *et al.*, 2003). Could a single or double domain encoding version of *kaiC* have played a circadian role 3.5 billion years ago in cyanobacteria? *KaiC* is a member of the

ATP-dependent RecA recombinase/DnaB replication fork helicase family (Leipe *et al.*, 2000), so its pre-clock function would probably have been in chromosomal mechanics, perhaps in the condensation of chromosomes (Mori & Johnson, 2001). From this original function, *kaiC* could then have evolved a clock role. Interestingly, a similar ancestral function has been suggested for the TIMELESS clock protein from studies in nematodes (Chan *et al.*, 2003).

When might the latter clock function have developed? During the time of the atmospheric change from reducing to oxidizing (2.5 to 2 Ga), a circadian clock using KaiC (and possibly KaiB) would, it has been argued, have conferred a particularly adaptive phenotype to the photosynthetic cyanobacteria, allowing exploitation and colonization of the Earth during their golden age of around 2.5 Ga (Tauber *et al.*, 2004). As mentioned above (Section 10.2), the initial free-running period of the clock in cyanobacteria may have been as short as 4 h, reflecting the Earth's rotation two billion years ago (Krasinsky, 2002). Subsequently, cyanobacteria presumably incorporated progressive delays into its clockwork over the millennia as the Earth slowed its whirl, maybe by recruiting the by now extant KaiB, followed by KaiA over the last billion years, until we reach the current 24 h geophysical period (Tauber *et al.*, 2004). In support of this scenario, the KaiA/B/C proteins interact both homo and heterotypically (Iwasaki *et al.*, 1999); mutations in *kaiA* and *kaiC* that modulate the strength of interaction between KaiA and KaiC alter the circadian period (Taniguchi *et al.*, 2001; Nishimura *et al.*, 2002). Therefore, it is conceivable that the addition of KaiA as a new cog to the circadian clockwork played a role in lengthening the period of the clock in response to the selective pressure of a decelerating day (Tauber *et al.*, 2004). The functional interactions between these proteins would mean that the genes have coevolved together over the last billion years. Unfortunately, this lengthy coevolution precludes any simple test of this 'period lengthening' hypothesis, whereby removing each gene in turn from the youngest to the oldest should give shorter and shorter periods.

10.5   The appearance of clocks in photosynthetic eukaryotes

Circadian programming of cellular events is common among the representatives of photosynthetic eukaryotes, from algae to plants. Plants will be discussed in the next section (10.6), but it is useful to remember that there is a broad range of eukaryotic algae – some of which share ecological niches with cyanobacteria – that exhibit well-documented circadian outputs. These include the gigantic alga *Acetabularia*, the marine dinoflagellate *Lingulodinium* (formerly named *Gonyaulax*), the green alga *Chlamydomonas*, and marine macroalgae (Edmunds, 1988; Suzuki & Johnson, 2001; Jacobsen *et al.*, 2003). The very first unicellular organism to have its circadian clock unveiled was the photosynthetic protist *Euglena* (Edmunds, 1988). Might these photosynthetic and eukaryotic unicells have provided an evolutionary 'stepping stone' for a clock mechanism between the prokaryotic cyanobacteria and eukaryotic plants?

The discovery of circadian clock genes in cyanobacteria (Ishiura *et al.*, 1998) led to the fervent hope that this clockwork might have been a protoclock for the evolution of all other clocks. For example, the endosymbiotic hypothesis for the origin of the organelles of eukaryotic cells proposes a cyanobacterium-like prokaryote was the progenitor of the chloroplast of higher plants. If so, the cyanobacterial clock genes might have been transferred to the nucleus, just as many plastid genes translocated to the nuclear genome. If that had occurred, we might have found homologies between cyanobacterial clock genes and the clock genes of eukaryotes, especially of higher plants, and the cyanobacterial clockwork could have provided a 'Rosetta Stone' to decipher the clockwork of plants and possibly of other eukaryotes (Johnson *et al.*, 1996).

10.5.1 Absence of clock gene homologues across taxa

Unfortunately, no such luck! Putative homologues of the *kaiABC* clock genes do not appear in the chloroplast or nuclear genomes of the plants that have been completely sequenced (nuclear genomes: *Arabidopsis* and rice [complete nuclear genomes], tomato [partial nuclear genome]; chloroplast genomes: *Arabidopsis*, rice, tobacco, and *Medicago*; mitochondrial genome: *Arabidopsis*). Nor were the *kaiABC* genes found in the chloroplast or nuclear genomes of the eukaryotic alga *Chlamydomonas* (Mittag *et al.*, 2005). Was the clock in the endosymbiosed cyanobacterium/chloroplast cast away? The possibility remains that the clock genes of cyanobacteria did provide the building blocks for eukaryotic clocks, but that the sequences have diverged to such an extent that makes their relationship unrecognizable. While this may be the case, there is at present no persuasive evidence for the clock genes of cyanobacteria and plants appearing in each other's genomes.

Even more surprisingly, there does not appear to be much overlap between the clock genes of plants and their 'simpler' relatives, the eukaryotic green algae. Now that the genome of the eukaryotic alga *Chlamydomonas* has been completely sequenced, we can make comparisons among the cyanobacterial, *Chlamydomonas*, and plant genomes for the presence of circadian clock genes (Mittag *et al.*, 2005). Of the genes that are thought to encode central clock proteins in plants, there are no putative homologues in the *Chlamydomonas* genome to the following genes: *elf3, gi, pif3, tej*, and only barely similar putative homologues to these genes: *cca1, lhy, co1, ztl, ado1, ado2* (*lkp2*), *ado3* (*fkf1*), *prr1, prr3, prr5, prr7, and prr9* (Mittag *et al.*, 2005). The *Chlamydomonas* genome also does not appear to have phytochrome homologues. On the other hand, it does have *cryptochrome (cry)* homologues, but there is an unexpected twist. *Chlamydomonas* has two *cry* genes; one is most similar to *Arabidopsis cry1/cry2*, but the other putative homologue encodes a protein that is most similar to animal cryptochrome, especially *Drosophila's cry* gene! (Mittag *et al.*, 2005). The role of cry in plants is thought to be essentially as a clock photoreceptor, but in animals it may act as a photoreceptor and as an essential clock component. What cry does in the *Chlamydomonas* clock system (or whether it does anything at all) is presently unknown.

10.5.2   Regulatory homologues do exist across taxa

On the other hand, there are homologues in the *Chlamydomonas* genome for kinases and phosphatases that have been recruited for clock duty in other organisms. For example, casein kinases (CK1 and CK2) are involved in the circadian systems of *Neurospora/Drosophila/*mammals and *Arabidopsis* (Daniel *et al.*, 2004), and they are found in the *Chlamydomonas* genome (Mittag *et al.*, 2005). Furthermore, the kinase SHAGGY, which has been shown to phosphorylate clock proteins in *D. melanogaster*, is also conserved in the *Chlamydomonas* genome. Protein phosphatases (PP1 and PP2A) appear to be crucial for circadian timekeeping in *Drosophila* (Sathyanarayanan *et al.*, 2004), *Neurospora* (Yang *et al.*, 2004), and appear in the *Arabidopsis* and *Chlamydomonas* genomes (Mittag *et al.*, 2005). However, all these regulatory molecules have a broad range of functions in the organisms where they have been studied. Mutations in these genes are often lethal, speaking to their 'housekeeping' function within cells, so it may be of no great surprise that these molecules are also found in *Chlamydomonas*, and that they do not necessarily bespeak a clock role.

In summary, there are few central clock components that spring from the comparison of the prokaryotic cyanobacterial (*Synechococcus*), eukaryotic algal (*Chlamydomonas*) and plant (*Arabidopsis*) genomes. There may be similarities at the level of the kinases and phosphatases that act on the clock-specific components, but these kinases and phosphatases act in the fungal and animal systems as well as on many targets unrelated to clocks. Are the clock systems of cyanobacteria, algae, and plants no more related to each other than they are to the clockworks of fungi and animals? At the present time, there is persuasive genetic evidence for common clock origins among animals, but bacteria, fungi, plants, and animals all appear to have evolved clock systems independent from each other (Rosato & Kyriacou, 2002). The new evidence from the *Chlamydomonas* genome suggests that algae – not to be outdone – selected an entirely different set of cogs when evolution assembled the algal clock. Therefore, cyanobacteria have not been a 'Rosetta Stone' for deciphering the clock components of higher organisms nor are eukaryotic algae likely to have been a 'stepping stone' in the evolution of clocks from prokaryotes to eukaryotes. Nonetheless, it is still to be hoped that unveiling the basic principles of circadian biochemistry in cyanobacteria and algae will put us within a 'stone's throw' of understanding the clockwork of plants and animals.

10.6   Clocks are widespread in plants: mosses, gymnosperms and angiosperms

Most of the circadian research on plants has been conducted on angiosperms, from the simple duckweed *Lemna* to trees (Kondo, 1982; Sweeney, 1987; Lumsden & Millar, 1998). It was in angiosperms that the first study of endogenous timing mechanisms was performed. In 1729, the French astronomer de Mairan reported that 'sensitive plants' (probably the angiosperm *Mimosa*) maintained their 24 h rhythms of leaf movement when placed in constant darkness where there were no light/dark cues as to the time of day (de Mairan, 1729). Subsequent studies by Charles Darwin on leaf movments extended de Mairan's observations (Darwin, 1880). Starting in 1928, the circadian pioneer, Erwin Bünning, made careful measurements of the leaf movement rhythms of the common bean *Phaseolus*. Bünning found that the leaf movements of beans oscillated in constant darkness with a period of 25.4 hours, thereby establishing the salient property of circadian clocks; they 'free-run' in constant conditions with a period close to, but never exactly 24 hours, hence their name: 'circa' for about, and 'dies' for a day (reviewed in Sweeney, 1987). Those beginnings have blossomed into our current understanding of circadian clocks in plants, where the molecular analysis is most advanced in the angiosperm *Arabidopsis thaliana*.

10.6.1   Clocks in 'primitive' plants

Are angiosperms the only plants with clocks? The rhythmic properties of gymnosperms have been much less studied than those of angiosperms, but not totally neglected in that there is a report of rhythms of *Lhcb* mRNA abundance expressed by the ancient gymnosperm *Ginkgo* (Christensen & Silverthorne, 2001). The most primitive plant to be studied to date is the moss *Physcomitrella*. Messenger RNA abundances of the *Lhcb2* gene in *Physcomitrella* exhibited daily rhythms in constant darkness at a permissive concentration of sugar in the medium (Shimizu *et al.*, 2004). These rhythms also can entrain to light/dark cycles. Is this clockwork more plant-like, or more alga-like? It is too soon to definitively answer this question, but there are EST sequences in the *Physcomitrella* genome that have a high similarity to *cca1*, *lhy* (Shimizu *et al.*, 2004), and *co* (Masashi *et al.*, 2004) from *Arabidopsis*. Ongoing research will hopefully answer whether the proteins encoded by these genes are clock components in *Physcomitrella*, or whether they have a different function. However, their presence in the moss genome (and essential absence from the genome of *Chlamydomonas*) implies that the moss clock is probably more similar to that of *Arabidopsis* than to the clock of *Chlamydomonas*.

10.7   Rhythms controlled by the 'Clockwork Green'

The energy source of photosynthetic organisms depends upon sunlight, which is available on a 24 h cycle. Therefore, it is not difficult to imagine that this environmental cycle provided a strong selective pressure for the evolution of a system that rhythmically mobilizes many processes so as to optimize use of the sun's energy. Indeed, there is a long list of circadian-controlled processes in photosynthetic organisms. Examples among physiological functions include circadian rhythms of leaf movements, photosynthetic capacity, chloroplast movements, nitrogen fixation, hypocotyl elongation, stomatal conductance, phototaxis, chemotaxis, cell division, flower movements, fragrance emissions, ion fluxes, bioluminescence, and many others (Sweeney, 1987; Lumsden & Millar, 1998; Johnson, 2001). There are also many genes whose expression is regulated by the circadian clock; in fact, it was in photosynthetic organisms that the first rhythms of abundance of specific proteins and mRNAs were described (Johnson *et al.*, 1984; Kloppstech, 1985). Control of transcription in photosynthetic organisms appears to be pervasive: in the cyanobacterium *S. elongatus*, the activity of essentially all promoters is controlled by the clock (Liu *et al.*, 1995), and in the plant *Arabidopsis*, approximately 6–10% of genes express rhythms of mRNA abundance (Harmer *et al.*, 2000) (see Chapter 6, this volume). When promoter activity is measured, the percentage of clock-regulated genes in *Arabidopsis* increases to as much as 35% (Michael & McClung, 2003). The increase in the number of genes from those that are rhythmically regulated at the mRNA abundance level to those regulated at the promoter activity level implies a significant component of post-transcriptional regulation.

10.7.1 *Photoperiodic time measurement*

One clock-controlled process deserves special attention, namely the phenomenon of photoperiodism. Photoperiodic time measurement (PTM) is the ability of plants and animals to sense the season of the year by measuring the duration of the day and/or night in the natural environment and respond appropriately so as to adapt to seasonal changes in their environment (Thomas & Vince-Prue 1997). This phenomenon is a fundamental adaptation of organisms to their environment, especially in the cases of reproduction and development. The first demonstration that organisms could sense the changing season by measuring the duration of the day or night was in higher plants, where the seasonal flowering of a variety of plant species, predominantly tobacco and soybeans, was shown by Garner and Allard in 1920 to be controlled by photoperiod (Garner & Allard, 1920).

The model for PTM that has become generally accepted is that a circadian (daily) clock acts as the timer that measures the length of the night and triggers developmental events, as first proposed by Bünning (1936). When Bünning first proposed this idea, the model seemed too fantastic to be realistic, but subsequent studies supported Bünning's idea (Hamner & Takimoto, 1964; Thomas & Vince-Prue, 1997). A wide range of physiological data in plants and animals supports the hypothesis that circadian clocks and PTM are intrinsically linked, as do more recent genetic analyses (Imaizumi *et al.*, 2003; Valverde *et al.*, 2004).

To date, PTM is well-documented in higher plants, but it seems logical that simple plants and even unicellular organisms might also benefit from being able to respond to seasonal changes in their environment. Recent studies support that logic. Among unicellular algae, photoperiodic responses have been reported for the dinoflagellate *Lingulodinium* (cyst formation; Balzer & Hardeland, 1991) and the green alga *Chlamydomonas* (germination; Suzuki & Johnson, 2002). More recently, photoperiodic control of spore formation has been discovered in the moss *Physcomitrella* (Hohe *et al.*, 2002; Masashi *et al.*, 2004). Therefore, circadian-controlled processes extend beyond the daily regime to encompass seasonal phenomena, and this capability extends from the 'simplest' unicells to the most complex plants.

Do these clock-controlled processes give any clues about the selection pressures acting on clocks, either now or in the past? Critically speaking, the answer is 'not really.' The case of temporal separation was mentioned above (Section 10.3.1). Another example is in the well-studied cyanobacterium *Synechococcus elongatus* PCC 7942, which exhibits robust rhythms of gene expression, including those encoding proteins involved in photosynthesis (e.g. *psbAI*); there is no significant rhythm of photosynthetic capacity in LL in this strain (Yen *et al.*, 2004)! Another prime example is that of the first circadian rhythm to be documented – the leaf movement rhythm observed by de Mairan to persist in DD. We still do not understand the function of the leaf movement rhythm, even though several hypotheses have been advanced (Enright, 1982).

10.8   Why not an hourglass timer?

It is not immediately obvious why natural selection favoured a self-sustained oscillator and not a temperature-compensated hourglass when devising a timekeeping system. The vast majority of organisms experience the daily light/dark cycle in some form (even if only as brief twilight exposures), and are therefore under entrainment conditions. There are a few cases where a self-sustained oscillator that paces events in the absence of any zeitgeber has been documented, particularly in some Arctic animals (Folk, 1964; Semenov *et al.*, 2001). However, in general, those cases are rare. Under entrained conditions, a temperature-compensated hourglass seems adequate to phase biological processes.

10.8.1   Advantages of a self-sustained oscillator

Why is it that a temperature-compensated hourglass did not become the dominant timekeeping mechanism? One suggestion has been that a self-sustained oscillator allows the organism to anticipate periodic events in the environment (e.g. dawn) before they happen. But a temperature-compensated hourglass (or a series of hourglasses if one wanted to define several phases) could also allow an organism to anticipate any particular local time in a consistent environment (e.g., dawn could be anticipated by timing 12 hours from dusk). If an hourglass can allow anticipatory behaviour, why then use an oscillator? Possibly the reason is that an oscillator allows a much wider range of capabilities than does an hourglass. As one example, a simple hourglass cannot always specify a given local time in a fluctuating environment. In particular, the duration of night and day (photoperiod) is a function of the time of year (except at the Equator). Therefore, an hourglass that always measures a particular duration of night will only be a precise estimator of the time of dawn on *two* days out of the year. On the other hand, an oscillator with appropriate characteristics can do an excellent job of estimating a given phase at all seasons of the year (Pittendrigh & Daan, 1976; Johnson *et al.*, 2003).

One approach towards understanding the evolution of circadian rhythmicity has been based on computer modelling. Most notable in this context is the model of Roenneberg and Merrow (2002), who suggested that the coupling and synchronization of multiple noisy metabolic feedback loops that are inherently oscillatory (and probably ultradian) could have led to the highly precise, self-sustained circadian oscillator. Their model showed that very few evolutionary steps are necessary for a circadian organization to be achieved, starting from the autoregulatory ability of all biochemical pathways within the complex metabolic networks of the cell. While speculative, the model can explain both the similarities and differences in designs of evolutionarily diverse circadian systems (Rosato & Kyriacou, 2002).

10.9   The clock as an adaptation: past and present

So, the circadian clockwork is an adaptation favoured by natural selection to orchestrate elaborate temporal programs and has evolved almost ubiquitously among eukaryotes and perhaps commonly among prokaryotes. However, the current evidence suggests that it evolved separately in different taxa. In other words, the genes that encode clock-specific proteins (i.e. not kinases or phosphatases) are not conserved. This conclusion is unexpected if we presume that a clockwork evolved early in cyanobacteria – why wasn't it conserved? Why wasn't the first eukaryotic clockwork conserved? Most perplexing, why don't algae and plants share common cogs? Maybe, the time gap between common ancestors is so great that clock molecules were only conserved between relatively closely related species (e.g. common ancestors separated by about 0.5 billion years or less). Does this mean that the selective pressure(s) that favoured the evolution of circadian clocks was (were) not present early in evolution, but became important only after the major taxa were established? This is a bizarre suggestion – daily cycles of light, temperature, and humidity *must* have been present from the dawn of life on Earth, and should have been expected to exert a consistent selective pressure that would have maintained a functional clock system.

Perhaps the simplest answer is that there were common ancestral clock-specific genes, but they have diverged to the point that these relationships are unrecognizable. Another possibility is that there were ancestral clock genes that were conserved through the initial stages of the evolution of circadian clocks into the major taxa, but as the clock was established in each taxum, genes (such as cryptochrome?) were recruited into the clockwork as new clock-specific genes. Subsequently, the pioneering ancestral clock genes were lost. At the present time, there is not enough information to answer this profound question. Consequently, for the remainder of this chapter, we will focus on the circadian system as an example of an adaptation and discuss the evidence for its value and whether natural selection continues to act upon it.

10.10   The circadian clock as an adaptation

The circadian clock is an adaptation to our cyclic environment. The term, 'adaptation' is used in two different ways by evolutionary biologists. An 'adaptation' refers to an aspect of the phenotype that is the product of evolution by natural selection in a particular environmental context, and represents a solution to some challenge presented by the environment. In this sense, an adaptation is a feature of an organism that enhances its reproductive success relative to other possible features (Futuyma, 1998). On the other hand, the 'process of adaptation' refers to ongoing phenotypic/genetic evolutionary change that is driven by natural selection in a given environmental context. So, an adaptation is the result of the process of adaptation.

Strictly speaking, an adaptation can only be assumed to be adaptive when it first appears. As time goes on, the feature may persist for any of three reasons. First, the feature might still be adaptive for the original reason (selective pressure remains). Second, the selective pressure has relaxed, but in the absence of selection against the feature, it may persist passively (no longer adaptive). Third, since its original appearance, other features may have become linked to the original feature so that even if the original selective pressure is relaxed, the feature persists because so many other processes depend upon it (no longer adaptive for the original reason). Many rigorous evolutionary biologists do not accept the use of the term 'adaptation' for a feature that falls into either of the latter two categories. Probably most evolutionary biologists would evaluate the adaptive significance of a trait in both the context of its phylogenetic history and in the context of the environment in which the organism naturally lives. Is the clock still an adaptation in the strictest sense? It is easy to imagine that the clock in some organisms may have fallen into the third category – for example, that reproduction and other processes may have become linked into the clock system to such an extent that they cannot be disentangled (as may be the case for the $clock^{-/-}$ mutation that disrupts mammalian oestrous and pregnancy reported by Miller $et\ al.$, 2004).

10.11   Experimental tests of adaptive significance before 1980

The adaptive significance of circadian clocks was a topic of great interest to the previous generation of circadian biologists. We will describe a few experimental approaches in insects and plants where the researchers attempted to go beyond 'adaptive storytelling.' The most famous example is probably that of Pittendrigh and Minis (1972). These workers tested the longevity of $Drosophila$ adults maintained in either constant light or in light/dark cycles of 21 h, 24 h and 27 h. They reported that flies lived significantly longer on the light/dark cycle of 24 h, implying an optimal 'resonance' of the internal clock's period with the period of the environmental cycle. This result supported the hypothesis that having an endogenous clock whose period is similar to that of the environment is adaptive. However, the results were not reproducible; Pittendrigh's lab was unable to repeat the result after Pittendrigh moved his laboratory from Princeton University to Stanford University (Dr. Terry Page, personal communication). At about the same time, Jurgen Aschoff's lab reported a similar approach to test adaptive significance in blowflies, which died sooner on non-24 h light/dark cycles as compared with 24 h cycles (von Saint Paul & Aschoff, 1978) or when exposed to repeated shifting of the zeitgeber (Aschoff *et al.*, 1971).

In plants, several studies from the 1950s addressed the adaptive significance of circadian oscillators. For example, tomatoes were found to grow optimally when maintained on light/dark cycles that were similar to those encountered in nature; in other words, tomatoes on LD 12:12 outgrew those on LD 6:6 h or LD 24:24 (Withrow & Withrow, 1949; Highkin & Hanson, 1954; Hillman, 1956). Remarkably, tomato plants on an LD 12:12 cycle grew even faster than those in continuous light, even though the plants in constant light were receiving twice as many photons (Hillman, 1956). In addition, there was an interdependence between the temperature and the optimal light cycle; at colder temperatures (when the clock might be expected to run slower), the optimal light/dark cycle was longer than at higher temperatures (Went, 1960). The Q_{10} for the effect was about 1.2, which is within the range that would be expected for the temperature dependence of a circadian oscillator's period. Those data indicated that tomato plants were optimally adapted to growth in light/dark cycles that were similar to those found in nature and implied that it was a circadian timekeeper that was responsible for the adaptation.

A defect of these pre-1980 experiments was that the measures of reproductive fitness were indirect, either longevity, growth rate, developmental rate, and so on. These parameters may influence fitness, but these ancillary factors are *not* direct measures of fitness. For example, if a slowly growing plant produces more seeds that successfully germinate than a rapidly growing plant, the slow-grower may be 'fitter.' Moreover, in the case of the plant studies from the 1950s, many species of plants other than tomato do not exhibit reduced growth in non-24 h light/dark cycles. Therefore, the results with tomatoes mentioned in the preceding paragraph were not generalizable to many other plant species. One such species is *Arabidopsis* – *Arabidopsis* plants grow perfectly well in constant light, in fact they often grow faster in constant light than in LD 12:12. A more direct test of fitness in plants would measure fecundity, e.g. by measuring the number and germinating ability of the seeds. A recent report has used that type of assay in *Arabidopsis thaliana* and found that clock-disrupted mutant plants produce fewer viable seeds than wild-type plants on LD 4:20 (Green *et al.*, 2002). However, LD 4:20 is not a photoperiod that most ecotypes of *Arabidopsis thaliana* are likely to encounter in nature, and more reasonable LD 8:16 and LD 16:8 cycles did not show a fecundity difference between wild-type and clock-disrupted mutants. Therefore, it is difficult to assess the significance of these data. (Also see reference added in proof: Dodd et al., 2005.)

10.12   Laboratory studies of circadian clocks and reproductive fitness since 1980

The next generation of research into the adaptive value of circadian systems must apply more direct measures of fitness than was true of the pre-1980 studies. From 1975 to 1995, the isolation of circadian clock mutants in *Drosophila* and *Neurospora* that exhibited dramatically different free-running periods (or even arhythmicity) but which appeared to grow and reproduce as well as wild-type in the laboratory might have discouraged such studies. More recent studies have found some mutations that impact longevity in the laboratory (Klarsfeld & Rouyer, 1998; Hurd & Ralph, 1998), but it is not clear that these are clock-specific effects ('clock genes' may affect some processes that are not clock related). Since 1995, however, several research groups have returned to this topic, using measures that more closely address reproductive fitness *per se*.

10.12.1   Advantages of cycles that resonate with the environment

The adaptive significance of circadian programs in the cyanobacterium *S. elongatus* was tested by competing different strains against each other in different laboratory environments (Ouyang *et al.*, 1998; Woelfle *et al.*, 2004). For asexual microbes such as *S. elongatus*, differential growth of one strain under competition with other strains is a good measure of reproductive fitness. In pure culture, the strains grew at about the same rate in constant light and in light/dark cycles, so there did not appear to be a significant advantage or disadvantage in having different circadian periods when the strains were grown individually. The fitness test was to mix different strains together and to grow them in competition to determine if the composition of the population changes as a function of time. The laboratory environments were different kinds of light/dark (LD) cycles or constant light. The cultures were diluted at intervals to allow growth to continue.

In a test of whether the clock was useful in cyanobacteria, wild-type cells were competed against an apparently arhythmic strain (CLAb). As shown in Fig. 10.2, the arhythmic strain was rapidly defeated by wild-type in LD 12:12, but under competition in constant light, the arhythmic strain grew slightly better than wild-type (Woelfle *et al.*, 2004). Therefore, the clock system does not appear to confer an intrinsic value for cyanobacteria in constant conditions. Period mutants were used to answer the question, 'does having a period that is similar to the period of the environmental cycle enhance fitness?' The circadian phenotypes of the strains shown in Fig. 10.2 had free-running periods of about 22 h (B22a) and 30 h (A30a). These strains had point mutations in *kaiA* (A30a) and *kaiB* (B22a). Wild-type has a period of about 25 h under these conditions.

When each of the strains was mixed with another strain and grown together in competition, a pattern emerged that depended upon the frequency of the light/dark

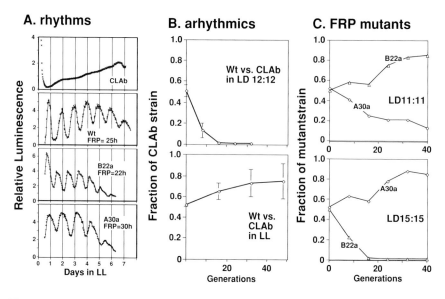

Figure 10.2 Competition of clock-modified strains with wild-type in mixed cultures of the cyanobacterium *S. elongatus*. (A) The circadian phenotypes of luminescence emission from wild-type, the clock-disrupted strain CLAb, and period mutants (mutation in *kaiB* [B22a, FRP ~22 h] and *kaiA* [A30a, FRP ~30 h]). (B) Competition between wild-type and CLAb in LD12:12 (upper) and LL (lower) plotted as the fraction of mutant in the mixed culture versus the estimated number of generations (mean +/− S.D.). (C) Kinetics of competition between wild-type and the period mutant strains in mixed cultures exposed to LD 11:11 (upper) or LD 15:15 (lower). Data are plotted as the fraction of the mutant strain in the mixed culture versus the estimated number of generations (modified from Woelfle *et al.*, 2004).

cycle and the circadian period. When grown on a 22 h cycle (LD 11:11), the 22 h-period mutant could overtake wild-type in the mixed cultures (Fig. 10.2C). On the other hand, in a 30 h cycle (LD 15:15), the 30 h-period mutant could defeat wild-type (Fig. 10.2C). On a 'normal' 24 h cycle (LD 12:12), the wild-type strain could outcompete either mutant (Ouyang *et al.*, 1998; Woelfle *et al.*, 2004). Note that over many cycles, each of these light-dark conditions have equal amounts of light and dark (which is important since photosynthetic cyanobacteria derive their energy from light); it is only the frequency of light versus dark that differs among the LD cycles.

Figure 10.2 clearly shows that the strain whose period most closely matched that of the light/dark cycle eliminated the competitor. Because the mutant strains could defeat the wild-type strain in LD cycles in which the periods are similar to their endogenous periods, the differential effects that were observed are likely to result from the differences in the circadian clock. Since the growth rate of the various cyanobacterial strains in pure culture are not detectably different, these results are most likely an example of 'soft selection' where the reduced fitness of one genotype is seen only under competition (Futuyma, 1998). Taken together, our results show that an intact clock system whose free-running period is consonant with the environment significantly enhances the reproductive fitness of cyanobacteria in rhythmic environments; however this same clock system provides no adaptive advantage in constant environments and may even be slightly detrimental to this organism. It would be interesting to attempt a competition-type experiment in a plant system, perhaps where plants of different genotypes compete for a common light source and where differential growth might lead to shading of the more slowly growing plants.

10.12.2   Testing clock conferred advantages in other species

Attempts to apply measures of reproductive fitness to clock function in other organisms in the laboratory have been partially successful, especially in the fruit fly *Drosophila*. One investigation studied male fecundity as sperm production in singly mated flies and found that clock mutations that disrupt circadian rhythmicity also decrease sperm production (Beaver *et al.*, 2002). Those authors speculate that the significant declines in fertility observed with singly mated flies are not readily detected in the usual laboratory populations where males and females are housed together for several days and therefore have the opportunity to mate multiple times (Beaver *et al.*, 2002). On the other hand, while clock mutations also affect oogenesis in *Drosophila* females, this effect appears to be pleiotropic and does not involve the circadian clock (Beaver *et al.*, 2003). Another group that studied the 'trade-off' between fecundity and longevity in *Drosophila* reported the unexpected result that lifetime egg production was higher in LL (when the flies are arhythmic) than in LD (Sheeba *et al.*, 2000).

A different tack towards the issue of the adaptive significance of circadian systems is the question of whether these clocks could be involved in a reproductive isolation that could drive speciation and/or prevent interspecific hybridization. This 'temporal mating isolation' was addressed in a recent study in *Drosophila* (Tauber *et al.*, 2003). The authors found that species-specific timing of mating behaviour significantly affects mate choice. This phenomenon could help to prevent interspecies mating, thereby increasing the fitness of any individual within a particular species; moreover, it could play (or have played in the past) a role in providing the permissive conditions for speciation (Tauber *et al.*, 2003).

A reasonable criticism of laboratory studies in general is that these settings are artificial and cannot mimic the selective pressures found in nature. This criticism is likely to be more potent for organisms that live in complex environments and/or exhibit complex behaviours. This may partially explain why the competition experiment appears to have been successful for the relatively 'simple' cyanobacteria (Ouyang *et al.*, 1998; Woelfle *et al.*, 2004), but not for *Drosophila* (Klarsfeld & Rouyer, 1998). Clearly, studies in natural settings are most appropriate to answer questions of adaptive significance. Such studies are discussed in the next section.

10.13   Evidence that the clock is still adaptive from studies of organisms in natural environments

One way to assess the adaptive value of circadian clocks would be to search for evidence of natural selection acting upon circadian parameters in nature. If clocks have adaptive value, then natural selection should be acting upon their properties. One type of evidence could be variation in circadian rhythm properties where selective strength varies. For example, where natural selection may be presumed to have relaxed, we would predict that the robustness of circadian expression would decline. In support of this prediction, the expression of circadian rhythms has often been found to be absent or reduced in animals that have evolved for a long time in cave environments in which the conditions are no longer cyclic.

10.13.1   Environmental gradients

Similarly, over a gradient of a relevant environmental condition – what we might call a selective gradient – we might expect to observe gradation of a responding clock property. An excellent example of this type of environmental condition is the latitudinal gradation in annual daylength and temperature. Daylength and temperature are both highly relevant to daily clocks and its associated property of photoperiodism. In support of the prediction that these gradients influence clock properties, there is a positive correlation between the circadian period and the latitude from which samples of the plant *Arabidopsis* has been isolated from nature (Michael *et al.*, 2003). Another latitudinal cline of interest is that of polymorphism in the *Drosophila period* gene. There are differing lengths of the threonine-glycine encoding repeat region of the *period* gene that vary over the latitudes of Europe (Costa *et al.*, 1992). Statistical tests involving the patterns of nucleotide variation around this region in *D. melanogaster* revealed evidence for weak balancing selection (Rosato *et al.*, 1997). Interestingly, the length variants show different circadian temperature compensation properties that appear to be adaptations to the geographical regions in which each particular Thr-Gly allele predominates. Thus, balancing selection at the nucleotide level appears to be mirrored at the behavioural level (Sawyer *et al.*, 1997). This also extends to the same *per* region in *D. simulans* that also shows length variation in Europe, but which is not clinally distributed in this case (Rosato *et al.*, 1994). Nevertheless, the nucleotide variation around this region again shows the footprint of balancing selection (Rosato *et al.*, 1994), which is partially reflected in the temperature compensation profiles of the different variants (Rogers *et al.*, 2004). Natural selection is also shaping the evolution of this region via the phenomenon of 'intramolecular coevolution', both statistically (Peixoto *et al.*, 1993; Nielsen *et al.*, 1994) and experimentally (Peixoto *et al.*, 1998). Here, different repeat lengths found in any one species appear to be compensated by amino acid changes in the immediate flanking regions. The reason for this appears to be related to temperature compensation (Peixoto *et al.*, 1998).

10.13.2   Quantitative trait locus analysis

Another type of data that suggests the continuing action of natural selection comes from analyses of quantitative trait loci (QTL). In the plant *Arabidopsis*, two studies have used recombinant inbred lines (RILs) of different isolates from nature (from northern Europe vs. the Cape Verde Islands). These studies found a much larger variation in circadian period, phase angle and amplitude among crosses between different isolates (RILs) than was found in the parental lines (Swarup *et al.*, 1999; Michael *et al.*, 2003). Therefore, the hybridizations unmasked a large amount of genetic variation that was not obvious in the parental lines. This result implies that natural selection had favoured combinations of alleles that counterbalance each other's effect on clock properties so as to attain similar emergent phenotypes. In other words, this case provides an example of 'inter-molecular coevolution' or compensation (see above). It was only when disparate lines were crossed so as to create RILs that contained allele combinations that had not co-evolved was the underlying genetic variation uncovered. Again, these QTL results indicate that natural selection is still acting upon genetic variation in *Arabidopsis* to optimize allelic combinations, implying that the clock system itself remains of adaptive value (Michael *et al.*, 2003).

10.14   Clocks: where did they come from? what are they doing now?

We are far from understanding how circadian clocks evolved. Based upon comparisons of essential clock genes from different model systems, however, it appears that circadian clocks may be an excellent example of convergent evolution. There is persuasive genetic evidence for common clock origins among animals, but bacteria, fungi, plants, and animals all appear to have evolved clock systems independent from each other. The new evidence from the *Chlamydomonas* genome suggests that the reasonable assumption that eukaryotic algae might have a clock that is intimately related to those of cyanobacteria and/or plants is yet another assumption overturned.

We have not identified conclusively the selective pressure(s) that led to the evolution of these fascinating timekeepers, and in most cases, the adaptive significance of the clocked rhythms is not known. However, recent evidence from both laboratory experiments and studies of organisms in nature has indicated that these clock systems do confer a selective advantage and are still responding to natural selection. This is an exciting time in research on biological clocks; the slow progress in understanding clock mechanisms that impeded our field for decades is over. As if a dam has burst, there is an overwhelming onrush of new information about clock components and their interactions. As we peer into the inner workings of these clocks, let us also illumine how they came to be and how they serve their hosts.

Acknowledgements

We thank Dr. Setsuyuki Aoki for suggestions and comments on the manuscript, and Dr. David McCauley for discussions. We thank the following for research support: in the USA, the N.I.H. and the N.S.F. for support of the laboratory of CHJ, and in the U.K., the BBSRC, the NERC, and the Royal Society Wolfson Research Merit awards for support of the laboratory of CPK.

References

Aschoff, J., von Saint Paul, U. & Wever, R. (1971) Die Lebensdauer von Fliegen unter dem Einfluss von Zeit-Verschiebungen. *Naturwissenschaften*, **58**, 574.

Balzer, I. & Hardeland, R. (1991) Photoperiodism and effects of indoleamines in a unicellular alga, *Gonyaulax polyedra*. *Science*, **253**, 795–797.

Beaver, L.M., Gvakharia, B.O., Vollintine, T.S., Hege, D.M., Stanewsky, R. & Giebultowicz, J.M. (2002) Loss of circadian clock function decreases reproductive fitness in males of *Drosophila melanogaster*. *Proc. Natl. Acad. Sci. USA*, **99**, 2134–2139.

Beaver, L.M., Rush, B.L., Gvakharia, B.O. & Giebultowicz, J.M. (2003) Noncircadian regulation and function of clock genes *period* and *timeless* in oogenesis of Drosophila melanogaster. *J. Biol. Rhythm.*, **18**, 463–472.

Brosche P., Seiler U., Sundermann J. & Wunsch J. (1989) Periodic changes in earths rotation due to oceanic tides. *Astron. Astrophys.* **220**:318–320.

Bünning, E. (1936) Die Endogene Tagesrhythmik als Grundlage der Photoperiodischen Reaktion. *Ber. Detsch. Bot. Ges.*, **54**, 590–607.

Byrne, T.E., Wells, M.R. & Johnson, C.H. (1992) Circadian rhythms of chemotaxis to ammonium and of methylammonium uptake in *Chlamydomonas*. *Plant Physiol.*, **98**, 879–886.

Chan, R.C., Chan, A., Jeon, M., Wu, T.F., Pasqualone, D., Rougvie, A.E. & Meyer, B.J. (2003) Chromosome cohesion is regulated by a clock gene paralogue TIM-1. *Nature*, **423**, 1002–9.

Christensen, S. & Silverthorne, J. (2001) Origins of phytochrome-modulated *Lhcb* mRNA expression in seed plants. *Plant Physiol.*, **126**, 1609–1618.

Costa, R., Peixoto, A.A., Barbujani, G. & Kyriacou, C.P. (1992) A latitudinal cline in a Drosophila clock gene. *Proc. Roy. Soc. Lond. B*, **250**, 43–49.

Daniel, X., Sugano, S. & Tobin, E.M. (2004) CK2 phosphorylation of CCA1 is necessary for its circadian oscillator function in Arabidopsis. *Proc. Natl. Acad. Sci. USA*, **101**, 3292–3297.

Darwin, C. (1880) *The Power of Movement in Plants*, John Murray, London.

DasSarma, S., Kennedy, S.P., Berquist, B., Ng, W.V., Baliga, N.S., Spudich, J.L., Krebs, M.P., Eisen, J.A., Johnson, C.H. & Hood, L. (2001) Genomic perspective on the photobiology of *Halobacterium* species NRC-1, a phototrophic, phototactic, and UV-tolerant haloarchaeon. *Photosynthesis Res.*, **70**, 3–17.

de Mairan, J.J. (1729) Observation botanique. *Histoire de l'Academie Royale de Sciences (Paris)*, p. 35 (see *Sleep*, 2, 155–160, 1979, for a translation in English).

Ditty, J.L., Williams, S.B. & Golden, S.S. (2003) A cyanobacterial circadian timing mechanism. *Annu. Rev. Genet.*, **37**, 513–43.

Dodd, A.N., Salathia, N., Hall, A., Kérei, E., Tóth, R., Nagy, F., Hibberd, J.M., Millar, A.J. & Webb, A.A.R. (2005) Plant Circadian Clocks Increase Photosynthesis, Growth, Survival, and Competitive Advantage. *Science*, **309**, 630–633.

Dvornyk, V., Vinogradova, O. & Nevo, E. (2003) Origin and evolution of circadian clock genes in prokaryotes. *Proc. Natl. Acad. Sci.*, **100**, 2495–2500.

Edmunds, L.N. (1988) *Cellular and Molecular Bases of Biological Clocks*, Springer-Verlag, New York.

Enright, J.T. (1982) Sleep movements of leaves: in defense of Darwin's interpretation. *Oecologia (Berl)*, **54**, 253–259.

Folk, G.E. Jr. (1964) Daily physiological rhythms of Arctic carbivores exposed to extreme changes in Arctic daylight. *Fed. Proc.*, **23**, 1221–1228.

Futuyma, D.J. (1998) *Evolutionary Biology, Third Edition*, Sinauer, Sunderland, MA.

Garner, W.W. & Allard, H.A. (1920) Effect of the relative length of day and night and other factors of the environment of growth and reproduction in plants. *J. Agric. Res.*, **18**, 553–606.

Gehring, W. & Rosbash, M. (2003) The coevolution of blue-light photoreception and circadian rhythms. *J. Mol. Evol.*, **57**, Suppl 1, S286–9.

Golden, S.S. & Canales, S.R. (2003) Cyanobacterial circadian clocks–timing is everything. *Nat. Rev. Microbiol.*, **1**, 191–9.

Goto, K. & Johnson, C.H. (1995) Is the cell division cycle gated by a circadian clock? The case of *Chlamydomonas reinhardtii. J. Cell. Biol.*, **129**, 1061–1069.

Green, R.M., Tingay, S., Wang, Z.Y. & Tobin, E.M. (2002) Circadian rhythms confer a higher level of fitness to Arabidopsis plants. *Plant Physiol.*, **129**, 576–584.

Hamner, K.C. & Takimoto, A. (1964) Circadian rhythms and plant photoperiodism. *Amer. Nat.*, **98**, 295–322.

Harmer, S.L., Hogenesch, J.B., Straume, M., Chang, H. S., Han, B., Zhu, T., Wang, X., Kreps, J.A. & Kay, S.A. (2000) Orchestrated transcription of key pathways in Arabidopsis by the circadian clock. *Science*, **290**, 2110–2113.

Highkin, H.R. & Hanson, J.B. (1954) Possible interaction between light-dark cycles and endogenous daily rhythms on the growth of tomato plants. *Plant Physiol.*, **29**, 301–302.

Hillman, W.S. (1956) Injury of tomato plants by continuous light and unfavorable photoperiodic cycles. *Amer. J. Bot.*, **43**, 89–96.

Hohe, A., Rensing, S.A., Mildner, M., Lang, D. & Reski, R. (2002) Day length and temperature strongly influence sexual reproduction and expression of a novel MADS-box gene in the moss *Physcomitrella patens. Plant Biol.*, **4**, 595–602.

Hurd, M.W. & Ralph, M.R. (1998) The significance of circadian organization for longevity in the golden hamster. *J. Biol. Rhythm.*, **13**, 430–436.

Imaizumi, T., Tran, H.G., Swartz, T.E., Briggs, W.R. & Kay, S.A. (2003) FKF1 is essential for photoperiodic specific signaling in *Arabidopsis. Nature*, **426**, 302–306.

Ishiura, M., Kutsuna, S., Aoki, S., Iwasaki, H., Andersson, C.R., Tanabe, A., Golden, S.S., Johnson C.H. & Kondo, T. (1998) Expression of a gene cluster *kaiABC* as a circadian feedback process in cyanobacteria. *Science*, **281**, 1519–1523.

Iwasaki, H., Taniguchi, Y., Kondo, T. & Ishiura, M. (1999) Physical interactions among circadian clock proteins, KaiA, KaiB and KaiC, in Cyanobacteria. *EMBO. J.*, **18**, 1137–45.

Jacobsen, S., Lüning, K. & Goulard, F. (2003) Circadian changes in relative abundance of two photosynthetic transcripts in the marine macroalga *Kappaphycus alvarezii* (Rhodophyta). *J. Phycol.*, **39**, 888–896.

Johnson, C.H. (2001) Endogenous timekeepers in photosynthetic organisms. *Annu. Rev. Physiol.*, **63**, 695–728.

Johnson, C.H. (2004) Precise circadian clocks in prokaryotic cyanobacteria. *Curr. Issues Mol. Biol.*, **6**, 103–110.

Johnson, C.H. & Golden, S.S. (1999) Circadian programs in cyanobacteria: adaptiveness and mechanism. *Ann. Rev. Microbiol.*, **53**, 389–409.

Johnson, C.H., Roeber, J. & Hastings, J.W. (1984) Circadian changes in enzyme concentration account for rhythm of enzyme activity in *Gonyaulax. Science*, **223**, 1428–1430.

Johnson, C.H., Golden, S.S., Ishiura, M. & Kondo, T. (1996) Circadian clocks in prokaryotes. *Mol. Microbiol.*, **21**, 5–11.

Johnson, C.H., Elliott, J.A. & Foster, R.G. (2003) Entrainment of circadian programs. *Chronobiol. Int.*, **20**, 741–774.

Kloppstech, K. (1985) Diurnal and circadian rhythmicity in the expression of light-induced plant nuclear messenger RNAs. *Planta*, **165**, 502–506.

Kondo, T. (1982) Persistence of the potassium uptake rhythm in the presence of exogenous sucrose in *Lemna gibba* G3. *Plant Cell Physiol.*, **23**, 467–472.

Klarsfeld, A. & Rouyer, F. (1998) Effects of circadian mutations and LD periodicity on the life span of Drosophila melanogaster. *J. Biol. Rhythm.*, **13**, 471–478.

Krasinsky, G. (2002) Dynamical history of the Earth-Moon system. *Celestial Mech. Dynam. Astron.*, **84**, 27–55.

Larimer, F.W., Chain, P., Hauser, L., Lamerdin, J., Malfatti, S., Do, L., Land, M.L., Pelletier, D.A., Beatty, J.T., Lang, A.S., Tabita, F.R., Gibson, J.L., Hanson, T.E., Bobst, C., Torres, J.L., Peres, C., Harrison, F.H., Gibson, J. & Harwood, C.S. (2004) Complete genome sequence of the metabolically versatile photosynthetic bacterium *Rhodopseudomonas palustris. Nat. Biotechnol.*, **22**, 55–61.

Laskar, J. (1999) The limits of Earth Orbital calculations for geological time use. *Phil. Trans. Roy. Soc. Lond. A*, **357**, 1735–1759.

Lathe, R. (2004) Fast tidal cycling and the origin of life. *Icarus*, **168**,18–22.

Leipe, D.D., Aravind, L., Grishin, N.V. & Koonin, E.V. (2000) The bacterial replicative helicase DnaB evolved from a RecA duplication. *Genome. Res.*, **10**, 5–16.

Liu, Y., Tsinoremas, N.F., Johnson, C.H., Lebedeva, N.V., Golden, S.S., Ishiura, M. & Kondo, T. (1995) Circadian orchestration of gene expression in cyanobacteria. *Genes and Dev.*, **9**, 1469–1478.

Lumsden, P. & A. Millar, A. (eds) (1998) *Biological Rhythms and Photoperiodism in Plants*, BIOS Scientific Publishers, Oxford.

Masashi, S., Kazuhiro, I. & Aoki, S. (2004) Photoperiod-regulated expression of the *PpCOL1* gene encoding a homolog of CO/COL proteins in the moss *Physcomitrella patens. BBRC*, **324**, 1296–1301.

Michael, T.P. & McClung, C.R. (2003) Enhancer trapping reveals widespread circadian clock transcriptional control in Arabidopsis. *Plant Physiol.*, **132**, 629–639.

Michael, T.P., Salomé, P.A., Yu, H.J., Spencer, T.R., Sharp, E.L., McPeek, M.A., Alonso, J.M., Ecker, J.R. & McClung, C.R. (2003) Enhanced fitness conferred by naturally occurring variation in the circadian clock. *Science*, **302**, 1049–1053.

Miller, B.H., Olson, S.L., Turek, F.W., Levine, J.E., Horton, T.H. & Takahashi, J.S. (2004) Circadian *clock* mutation disrupts estrous cyclicity and maintenance of pregnancy. *Curr. Biol.*, **14**, 1367–1373.

Mitsui, A., Kumazawa, S., Takahashi, A., Ikemoto, H. & Arai, T. (1986) Strategy by which nitrogen-fixing unicellular cyanobacteria grow photoautotrophically. *Nature*, **323**, 720–722.

Mittag, M., Kiaulehn, S. & Johnson, C.H. (2005) The circadian clock in *Chlamydomonas reinhardtii*: What is it for? What is it similar to? *Plant Physiol.*, **137**, 399–409.

Mori, T. & Johnson, C.H. (2001) Circadian programming in cyanobacteria. *Sem. Cell Develop. Biol.*, **12**, 271–278.

Nielsen J., Peixoto A.A. Piccin A., Costa R., Kyriacou C.P. & Chalmers D. 1994. Big flies, small repeats: the 'Thr-Gly' region of the *period* gene in *Diptera. Mol. Biol. Evol.*, **11**, 839–853.

Nikaido, S.S. & Johnson, C.H. (2000) Daily and circadian variation in survival from ultraviolet radiation in *Chlamydomonas reinhardtii. Photochem. Photobiol.*, **71**, 758–765.

Nishimura, H., Nakahira, Y., Imai, K., Tsuruhara, A., Kondo, H., Hayashi, H., Hirai, M., Saito, H. & Kondo, T. (2002) Mutations in KaiA, a clock protein, extend the period of circadian rhythm in the cyanobacterium *Synechococcus elongatus* PCC 7942. *Microbiology.*, **148**, 2903–2909.

Ouyang, Y., Andersson, C.R., Kondo, T., Golden, S.S. & Johnson, C.H. (1998) Resonating circadian clocks enhance fitness in cyanobacteria. *Proc. Natl. Acad. Sci. USA*, **95**, 8660–8664.

Pattanayek, R., Wang, J., Mori, T., Xu, Y., Johnson, C.H. & Egli, M. (2004) Visualizing a circadian clock protein: crystal structure of KaiC and functional insights. *Mole. Cell*, **15**, 375–388.

Peixoto, A., Campesan, S., Costa, R., & Kyriacou, C.P. (1993). Molecular evolution of a repetitive region within the *per* gene of Drosophila. *Mol. Biol. Evol*, **10**, 127–139.

Peixoto, A.A., Hennessey, M., Townson, I.,Hasan, G., Rosbash, M., Costa, R., & Kyriacou, C.P. (1998). Molecular coevolution within a clock gene in Drosophila.. *Proc. Nat. Acad. Sci. USA*, **95**, 4475–4480.

Pittendrigh, C.S. (1965) Biological clocks: the functions, ancient and modern, of circadian oscillations.

In: *Science and the Sixties. Proceedings of the Cloudcraft Symposium*, pp. 96–111. Air Force Office of Scientific Research.

Pittendrigh, C.S. (1993) Temporal organization: reflections of a Darwinian clock-watcher. *Annu. Rev. Physiol.*, **55**, 17–54.

Pittendrigh, C.S., & Minis, D.H. (1972) Circadian systems: longevity as a function of circadian resonance in Drosophila melanogaster. *Proc. Natl. Acad. Sci. USA*, **69**, 1537–1539.

Pittendrigh, C.S. & Daan, S. (1976) A functional analysis of circadian pacemakers in nocturnal rodents. IV. Entrainment: pacemaker as clock. *J. Comp. Physiol.*, **106**, 291–331.

Roenneberg, T. & Merrow, M. (2002) Life before the clock: modeling circadian evolution. *J. Biol. Rhythm.*, **17**, 495–505.

Rogers A.S., Rosato E, Costa R. & Kyriacou C.P. (2004). Molecular analysis of circadian clocks in Drosophila*simulans*. *Genetica*, **120**, 223–232.

Rosato, E. & Kyriacou, C.P. (2002) Origins of circadian rhythmicity. *J. Biol. Rhythm.*, **17**, 506–511.

Rosato E., Peixoto A.A., Barbujani G., Costa R. & Kyriacou C.P. (1994). Molecular evolution of the *period* gene in Drosophila*simulans*. *Genetics*, **138**, 693–707.

Rosato, E., Peixoto, A.A., Costa, R., & Kyriacou, C.P. (1997). Mutation rate, linkage disequilibrium, and selection in the repetitive region of the *period* gene in Drosophila melanogaster. *Genet. Res.*, **69**, 89–99.

Sathyanarayanan, S., Zheng, X., Xiao, R. & Sehgal, A. (2004) Posttranslational regulation of Drosophila PERIOD protein by protein phosphatase 2A. *Cell*, **116**, 603–615.

Sawyer, L.A., Hennessy, J.M., Peixoto, A.A., Rosato, E., Parkinson, H., Costa, R. & Kyriacou, C.P. (1997) Natural variation in a Drosophila clock gene and temperature compensation. *Science*, **278**, 2117–2120.

Semenov, Y., Ramousse, R., Le Berre, M,. Vassiliev, V. & Solomonov, N. (2001) Aboveground activity rhythm in Arctic black capped marmot (*Marmota camtschatica bungei Katchenko 1991*) under polar day conditions. *Acta. Oecologica. Int. J. Ecol.*, **22**, 99–107.

Sheeba, V., Sharma, V.K., Shubha, K., Chandrashekaran, M.K. & Joshi, A. (2000) The effect of different light regimes on adult life span in Drosophila melanogaster is partly mediated through reproductive output. *J. Biol. Rhythm.*, **15**, 380–392.

Shimizu, M., Ichikawa, K. & Aoki, S. (2004) Circadian expression of the *PpLhcb2* gene encoding a major light-harvesting chlorophyll *a/b*-binding protein in the moss *Physcomitrella patens*. *Plant. Cell. Physiol.*, **45**, 68–76.

Suzuki, L. & Johnson, C.H. (2001) Algae know the time of day: circadian and photoperiodic programs. *J. Phycol.*, **37**, 1–10.

Suzuki, L. & Johnson, C.H. (2002) Photoperiodic control of germination in the unicell *Chlamydomonas*. *Naturwissenschaften*, **89**, 214–220.

Swarup, K., Alonso-Blanco, C., Lynn, J.R., Michaels, S.D., Amasino, R.M., Koornneef, M. & Millar, A.J. (1999) Natural allelic variation identifies new genes in the Arabidopsis circadian system. *Plant J.*, **20**, 67–77.

Sweeney, B.M. (1987) *Rhythmic Phenomena in Plants, Second Edition*, San Diego, Academic Press, 172 pp. 000–000.

Taniguchi, Y., Yamaguchi, A., Hijikata, A, Iwasaki, H., Kamagata, K., Ishiura, M., Go, M. & Kondo, T. (2001) Two KaiA-binding domains of cyanobacterial circadian clock protein KaiC. *FEBS Lett.*, **496**, 86–90.

Tauber, E., Roe, H., Costa, R., Hennessy, J.M. & Kyriacou, C.P. (2003) Temporal mating isolation driven by a behavioral gene in Drosophila. *Curr. Biol.*, **13**, 140–145.

Tauber, E., Last, K.S., Olive, P.J.W. & Kyriacou, C.P. (2004) Clock gene evolution and functional divergence. *J. Biol. Rhythm.*, **19**, 445–458.

Thomas, B. & Vince-Prue, D. (1997) *Photoperiodism in Plants, Second Edition*, San Diego, Academic Press, 428 pp. 000–000.

Valverde, F., Mouradov, A., Soppe, W., Ravenscroft, D., Samach A. & Coupland, G. (2004) Photoreceptor regulation of CONSTANS protein in photoperiodic flowering. *Science*, **303**, 103–106.

von Saint Paul, U. & Aschoff, J. (1978) Longevity among blowflies *Phormia terraenovae* R.D. kept in non-24-hour light-dark cycles. *J. Comp. Physiol.*, **127**, 191–195.

Wells, J.W. (1963) Coral Growth and Geochronometry. *Nature*, **197**, 948–950.

Went, F.W. (1960) Photo- and thermoperiodic effects in plant growth. In: *Cold Spring Harbor Symposia on Quantitative Biology, Biological Clocks*, Vol. 25, pp. 221–230. Cold Spring Harbor Press, Cold Spring Harbor NY.

Withrow, A.P. & Withrow, R.B. (1949) Photoperiodic chlorosis in tomato. *Plant Physiol.*, **24**, 657–663.

Woelfle, M.A., Ouyang, Y., Phanvijhitsiri, K. & Johnson, C.H. (2004) The adaptive value of circadian clocks: An experimental assessment in cyanobacteria. *Current Biol.*, **14**, 1481–1486.

Yang, Y., He, Q., Cheng, P., Wrage, P., Yarden, O. & Liu, Y. (2004) Distinct roles for PP1 and PP2A in the *Neurospora* circadian clock. *Genes Dev.*, **18**, 255–260.

Yen, U.C., Huang, T.C. & Yen, T.C. (2004) Observation of the circadian photosynthetic rhythm cyanobacteria with a dissolved-oxygen meter. *Plant Sci.*, **166**, 949–952.

Zhong, H.H. & McClung, C.R. (1996) The circadian clock gates expression of two Arabidopsis catalase genes to distinct and opposite circadian phases. *Mol. Gen. Genet.*, **251**, 196–203.

Index